Physiology of Human Female Lactation

Nikolai Petrovitch Alekseev

Physiology of Human Female Lactation

Springer

Nikolai Petrovitch Alekseev
Department of Physiology
Saint Petersburg State University
Saint Petersburg, Russia

ISBN 978-3-030-66363-6 ISBN 978-3-030-66364-3 (eBook)
https://doi.org/10.1007/978-3-030-66364-3

Translation from the Russian language edition: Физиология лактации женщины by Nikolai Petrovitch Alekseev Copyright © 2019 Urait. All Rights Reserved.

This Springer imprint is published by the registered company Springer Nature Switzerland AG.
The registered company address is: Gewerbestrasse 11, 6330 Cham, Switzerland

Preface

The book presents data on peripheral and central mechanisms of regulation of milk secretion and excretion from the alveolar-ductal system of a woman's breast. Worldwide, the number of women wishing to lactate and breastfeed children has increased. A large experimental material indicates that lactation and breastfeeding is a unique state of the female organism, which is an integral part of the reproductive process and has no equal as a way to ensure the ideal nutrition of infants. In addition to exclusive nutritional value, maternal milk has the ability to maintain both active and passive immunity. In this regard, the relevance of the book increases greatly, since these books are the foundation for the development of practical measures to correct the milk productivity of lactating women for effective feeding of infant. The proposed book is the first monograph in the world specifically devoted to the physiology of lactation of women. This book fully collects modern data on the structure and function of the breast of women so that specialists can use this information in the compilation of individual schemes for correcting secretion or excretion of milk in nursing mothers. The book is of interest to researchers, lactation specialists, obstetricians-gynecologists, pediatricians, and students, graduate students, and residents of universities and medical universities.

Contents

1 Introduction

All that is beautiful in human life is from the rays of the sun and from mother's milk.

Russian writer Maxim Gorky

Abstract

Materials on lactation of nursing women in different periods of human society development are presented in a short form. Systematic studies of the physiology of lactation and breastfeeding begun in the 1990s, and since then, the number of experimental works on this topic has been steadily growing. Scientific studies have shown that in addition to ideal nutrition, breastfeeding reduces the number of cases or severity of diseases. Breast milk also contributes to the development of the mental abilities of children. For most women, lactation reduces the risk of breast cancer.

1.1 Brief History of Research on the Physiology of Female Lactation and Breastfeeding

The final stage, the peak of the reproductive cycle in mammals, is the lactation period—feeding the born cubs with mother's milk. This function is performed by the mammary gland that emerged as a result of evolution. Lactation function of the mammary glands consists of two main processes: milk secretion and milk excretion. In the wild, the lactation period appears to be as important for reproduction as pregnancy and childbirth. The lack of milk in the female or the inability to withdraw it due to abnormalities in the structure or function of her mammary glands leads to the death of the offspring. The same harsh situation was evident in the human community at the dawn of its existence. It seems that it is no accident that primitive artists and sculptors, as an ideal, depicted women on rocks or created stone sculptures of women with fairly large mammary glands, thereby emphasizing that such a woman can provide effective breastfeeding of offspring. In the future, as humanity developed and, first of all, economic activity appeared, and relations

N. P. Alekseev, *Physiology of Human Female Lactation*,
https://doi.org/10.1007/978-3-030-66364-3_1

between people in communities improved, ways began to be found to help nursing mothers who had problems with lactation, i.e., with the secretion or excretion of milk from the breast. Thus, with the domestication of wild animals, it became possible to give the child milk from cows, goats, and donkeys as a substitute for breast milk. In addition, the milk of lactating women-nurses who had an excess of milk and could successfully feed their own child and the child of another mother with a lack of milk or the mother who died was used. To date, the scientific literature has a number of reviews (Drake 1935; Caulfield 1952; Wickes 1953a, b, c, d, e; Cowie 1974; Lawrence and Lawrence 1999, 2011; Zakharova and Machneva 2016) on the history of lactation of women—breastfeeding in different countries of the world in different eras. The first printed works on this topic began to appear in Europe (and probably in the world) during the Renaissance. According to available data, the vast majority of women in Europe and America in the sixteenth and nineteenth centuries, despite sometimes great difficulties, tried to maintain lactation and feed children with breast milk because it was noticed that milk mixtures made from milk domestic animals or liquid cereal gruel often had harmful effects on the baby with a fatal outcome. To date, it is known that this is due to the immaturity of the gastrointestinal tract and the immune system of newborns. The use of wet nurse's milk was more favorable, but for material reasons, not all women could hire a wet nurse. At the same time, women from families with high income, despite their ability to feed their children themselves, often hired wet nurses. For centuries, these mothers have put forward a variety of explanations for their reluctance to lactate and breastfeed their child: breastfeeding is not fashionable; it can damage the mother's health, spoil the shape of her breasts and figure, and prevent her from going to balls, going to the theater, etc. (Cowie 1974).

A similar situation with lactation was observed during this period in Russia. The majority of women tried to breastfeed their children. However, some women from families with high income, who have enough milk to feed a child for the abovementioned far-fetched reasons, hired wet nurses. Women from families with low material income in case of lack of breast milk used milk of farm animals for additional feeding of the child. In the second half of the nineteenth century in Europe, surrogates (formula) for feeding infants began to appear which were very simple in composition and received in Russia, for example, a mixture of Bidert, Nestle flour, and Liebig's children's soup (Gundobin 1901; Mother and Child 1901). However, the cost of these mixtures was very high, and their effectiveness did not differ much from the milk of farm animals (Lipsky 1896). Therefore, they were not widely distributed among nursing women. In general, it should be noted that, despite the existing traditional difficulties, the percentage of lactating and breastfeeding women in Russia at the beginning of the twentieth century was quite large. Thus, according to the well-known Russian pediatrician G. N. Speransky (1928) in Moscow in 1926, about 98% of children were breastfed up to a year. Statistical data for other regions of Russia are not available in the medical literature of that time. However, it can be assumed with a high degree of probability that the percentage of breastfeeding for the whole of Russia slightly differed from the metropolitan data. It should also be noted that the religious factor was of great importance for maintaining lactation and breastfeeding in Russian women. According to the provisions of the main confessions of Russia (Christianity, Islam, Buddhism, Judaism), breastfeeding an infant is the highest good for both the child and the mother (Gundobin 1901; Fateeva and Pustograev 2005; Shaikh and Ahmed 2006; Eidelman 2006; Segawa 2008). Given that religion has had and still has a great influence on the way of life of the population, its encouragement of breastfeeding was and is a strong initiating and supporting psychological incentive for lactation of women.

The intensive development of fundamental and applied sciences that began in the twentieth century and, in particular, the improvement of methods of biochemical analysis and synthesis of biologically significant compounds for the human body allowed us to start developing nutritional mixtures for infants whose mothers were not able to breastfeed for various good reasons. As a result of decades of hard work, multicomponent mixtures (formula) have been created that are more or less adapted to infant nutrition. Unfortunately, the advertising activities of companies that produce infant formula were organized in such a way that women who did not have problems with lactation were also involved in using the formula. The reclaim for the formula mentioned two main factors that pushed lactating women to switch to feeding babies with formula from a bottle: the convenience of feeding the baby with formula and the greater nutritional value of these mixtures than the mother's milk. It was really easy and quick to make sure that bottle feeding was simple and comfortable compared to stressful and time-consuming breastfeeding. In addition, the transition to bottle feeding greatly facilitated the work of medical specialists on the nutrition of children in the first months of life, but long-term studies were required to test the second advantage of feeding a child with artificial mixtures. Intensive development of production of relatively cheap multicomponent mixtures has equalized the ability of women from families with high and low incomes not to breastfeed. The spread of formula in the developed world has captured more and more breastfeeding women so that by the early 1970s of the twentieth century, for example, in the United States in the first weeks after birth, only 24% and in Australia 48% of women breastfed their children (Ryan et al. 1991; Hartmann 2007). The attitude of the public in developed countries at that time to breastfeeding and woman lactation is well illustrated by the memoirs of the world's leading lactation specialist, Professor P. Hartman. He wrote:" I have applied to the Australian national health grants Committee for a women's lactation grant. I clearly remember a member of the National Health and Medical Research Council, Grants Committee asking me during an interview why I wanted to do research on these unusual women, i.e., breastfeeding mothers" (Hartmann 2007).

However, at the same time, the scientific literature has accumulated a sufficient amount of experimental data indicating the fallacy of claims that the nutritional value of the formula exceeds or does not differ from the mother's milk. Moreover, research has found that women's milk has active biological properties that are not present in artificial mixtures. Women's milk contains hormones, biologically active substances, immune complexes, and living cells, including stem cells (Hassiotou et al. 2012, 2013; Witkowska-Zimny and Kaminska-El-Hassan 2017), (live cells in 1 ml of milk, 10^4–13×10^6), and has a strong beneficial effect on the child's body, ensuring the normal course of the metabolic process and maintaining resistance to infections and other external adverse factors. All these data pointed to the principal impossibility of artificially producing a formula that completely replaces natural women's milk. This was the reason in the 1970s for a sharp turn toward breastfeeding in all developed countries—the "revolution of the 70s in child feeding" (Wharton 1982), and the practice of artificial feeding of young children was described by the World Health Organization in one of its reports (1981) as "the most widespread uncontrolled experiment in the world." Fortunately, the women of Russia, as well as the women of the Union Republics that were part of the Soviet Union at that time, did not participate in this "experiment." However, the appearance of multicomponent milk formulas in the 1980s in Russia and their corresponding advertising negatively affected lactation and breastfeeding. The percentage of breastfeeding women began to decline, and in the 1990s of the twentieth century in Russia, it was about 40% (Fateeva and Pustograev 2005).

By the beginning of the 1990s, there was an increase in international cooperation aimed at supporting and stimulating women's lactation and breastfeeding children. Here, first of all, it is necessary to note the meeting in 1990 in Italy in the city of Spedale degli Innocenti of representatives of various organizations on the problems of breastfeeding. The famous "Declaration of Innocenti" was adopted, calling on women of all countries to "practice exclusively breastfeeding children from birth to 4–6 months of life, and above this age, breastfeeding with adequate complementary foods up to 2 years and older." This provision was based on numerous scientific data on the exclusive role of breastfeeding for the physical and mental health and development of children, as well as for the health of women themselves, contained in publications, the number of which increased dramatically at the time of the adoption of the Declaration and has steadily increased every year since. Figure 1.1a, b as an illustration shows the distribution of scientific articles starting from the 1950s of the twentieth century by year in the search engine PubMedline for the following keywords: woman's lactation physiology (a) and breastfeeding (b). Despite the fact that not all publications reflect these charts, they largely characterize trends in scientific research on breastfeeding and on various aspects of the physiology of women's lactation.

As a commentary in Fig. 1.1a, b, the statement of one of the world's leading specialists in breastfeeding and physiology of female lactation, Professor R. Lawrence (Lawrence and Lawrence 2009), is well suited. She writes: a thin stream of scientific work on female lactation from the middle of the twentieth century to its end turned into a full-flowing river that went off the banks. It should also be noted that along with the increase in the number of publications in obstetric and pediatric journals, specialized scientific journals devoted to women's lactation and breastfeeding have appeared: The *Journal of Human Lactation* (since 1985), the *International Breastfeeding Journal* (since 2003), and *Breastfeeding Medicine* (since 2006).

Extensive promotion of breastfeeding, actively conducted in the scientific and popular press, as well as on radio and television in the world, has brought positive results. In most cases, women want to breastfeed their newborns, because there is a wealth of scientific evidence that leaves no doubt about the benefits of breast milk. Thus, in addition to ideal nutrition, breastfeeding reduces the incidence or severity of

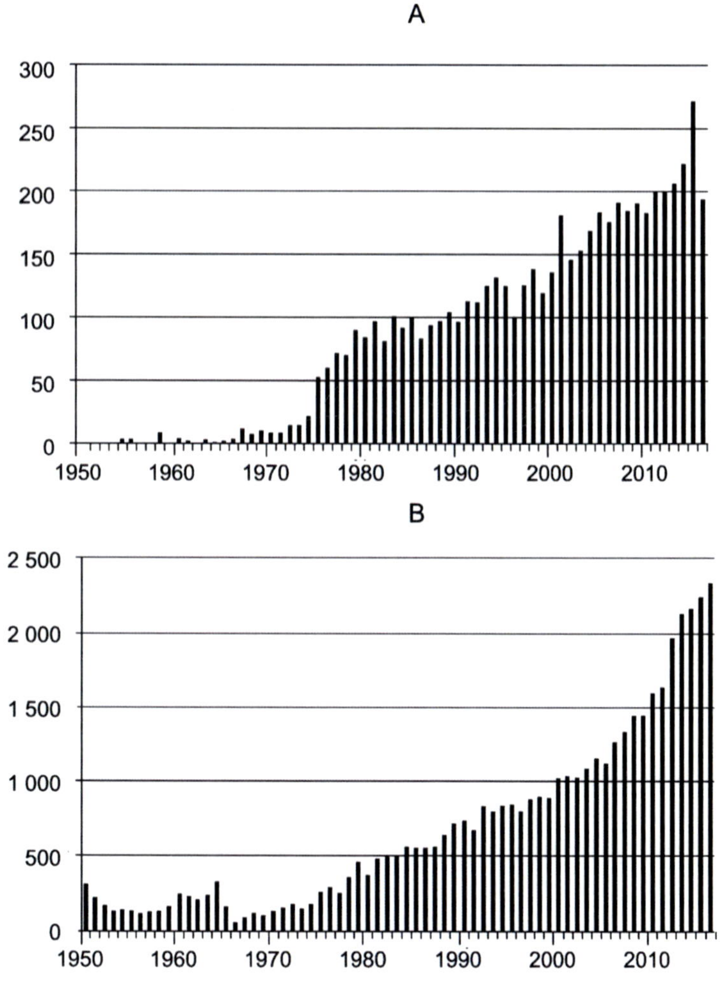

Fig. 1.1 Dynamics of articles output by year on the keyword: woman physiology lactation (**a**) (**b**) in the search engine PubMedline. On the ordinate axis, the number of articles, and on the abscissa axis, the year of publication of articles

a number of diseases, including bacterial meningitis (Istre et al. 1985; Cochi et al. 1986); bacteremia (Istre et al. 1985; Takala et al. 1989); diarrheal diseases (Howie et al. 1990; Popkin et al. 1990; Dewey et al. 1995; Beaudry et al. 1995; Lopez-Alarcon et al. 1997; Bhandary et al. 2003; Kramer et al. 2003); infectious diseases of the respiratory system, including asthma (Lopez-Alarcon et al. 1997; Oddy et al. 1999, 2002, 2003a; Gdalevich et al. 2001; Blaymore Bier et al. 2002; Chulada et al. 2003; Bachrach et al. 2003); infectious diseases of the urinary tract (Pisacane et al. 1992; Marild et al. 2004); infectious diseases of the auditory system (Saarinen 1982; Duncan et al. 1993; Owen et al. 1993; Paradise et al. 1993; Aniansson et al. 1994);

necrotic enterocolitis (Lucas and Cole 1990; Schanler et al. 1999); various forms diabetes (Kostraba et al. 1993; Perez-Bravo et al. 1996; Gerstein 1994; Pettit et al. 1997); lymphoma; leukemia, Hodgkin's disease (Davis 1998; Smulevich et al. 1999; Bener et al. 2001); obesity (Gillman et al. 2001; Singhal et al. 2002; Armstrong and Reilly 2002; Stettler et al. 2002; Toschke et al. 2002; Arenz et al. 2004; Grummer-Strawn and Mei 2004); and sudden infant death syndrome (McVea et al. 2000; Alm et al. 2002). Infants who were breastfed later in adulthood are less prone to alcoholism (Goodwin et al. 1999). Breastfeeding contributes to the development of children's mental abilities (Lucas et al. 1998; Anderson et al. 1999; Jacobson et al. 1999; Reynolds 2001; Horwood et al. 2001; Mortensen et al. 2002; Rao et al. 2002; Bier et al. 2002; Oddy et al. 2003b; Batstra et al. 2003; Feldman and Eidelman 2003; Horta et al. 2018). For most breastfeeding women, the risk of breast cancer is reduced (Newcomb et al. 1994; Enger et al. 1998; Tryggvadottir et al. 2001; Lee et al. 2003; Jernstrom et al. 2004).

However, in our time, during lactation, some women (sometimes the percentage of such women is quite high) have the same difficulties that were observed in nursing women in the previous centuries. The main ones are associated with a lack of milk secretion and its removal from the breast by the child. Moreover, these problems are most often found at the beginning of a woman's lactation. Therefore, as medical practice shows, the degree of effectiveness of eliminating these problems depends on whether a woman will continue to lactate and breastfeed the child or will switch to an easier way of feeding artificial infant formula from a bottle due to the difficulties encountered. Here, it should be noted that the treatment and correction of the work of any organ is carried out more effectively if the structure and mechanisms of its functioning are known. To date, scientists of various specialties have received quite extensive material on the structure and function, as well as mechanisms, for regulating the work of a woman's breast. However, this information is scattered in various and numerous publications, which makes it difficult to use it in the practice of breastfeeding. Therefore, the purpose of this book is to present up-to-date data on the structure and function of a woman's breast, so that medical professionals, as well as participants in breastfeeding support groups, can use this information when drawing up individual schemes for correcting the secretion or excretion of milk in nursing mothers. In addition, this book can be an indispensable tool for medical students and postgraduates in the study of woman lactation or breastfeeding. It should be noted that the book does not include experimentally and statistically unbiased "anecdotal" data on the structure and function of a woman's breast, which, unfortunately, are sometimes contained in textbooks and manuals on women's lactation.

References

Alm B, Wennergren G, Norvenius SG, Skjaerven R, Lagercrantz H, Helweg-Larsen K, Irgens LM (2002) Breast feeding and the sudden infant death syndrome in Scandinavia. Arch Dis Child 86 (6):400–402

Anderson JW, Johnstone BM, Remley DT (1999) Breast-feeding and cognitive development: a meta-analysis. Am J Clin Nutr 70(4):525–535

Aniansson G, Alm B, Andersson B, Hakansson A, Larsson P, Nylen O, Petersson H, Rigner P, Svanborg M, Sabharval H et al (1994) A prospective cohort study on breastfeeding and otitis media in Swedish infants. Pediatr Infect Dis J 13(3):183–188

Arenz S, Ruckerl R, Koletzko B, von Kries R (2004) Breast-feeding and childhood obesity. A systematic review. Int J Obes Relat Metab Disord 28(10):1247–1256

Armstrong J, Reilly JJ (2002) Breastfeeding and lowering the risk of childhood obesity. Lancet 359 (9322):2003–2004

Bachrach VR, Schwarz E, Bachrach LR (2003) Breastfeeding and the risk of hospitalization for respiratory disease in infancy: a meta-analysis. Arch Pediatr Adolesc Med 157(3):237–243

Batstra L, Neeleman J, Hadder-Algra M (2003) Can breast feeding modify the adverse effects of smoking during pregnancy on the child's cognitive development? J Epidemiol Community Health 57(6):403–404

Beaudry M, Dufour R, Marcoux S (1995) Relation between infant feeding and infections during first six months of life. J Pediatr 126:191–197

Bener A, Denic S, Galadari S (2001) Longer breast-feeding and protection against childhood leukaemia and lymphomas. Eur J Cancer 37(2):234–238

Bhandary N, Bahl R, Mazumdar S, Martines J, Black RE, Bhan MK (2003) Effect of community-based promotion of exclusive breastfeeding on diarrhoeal illness and growth: a cluster randomised controlled trial. Lancet 361(9367):1418–1423

Bier JA, Oliver T, Ferguson AE, Vohr BR (2002) Human milk improves cognitive and motor development of premature infants during infancy. J Hum Lact 18(4):361–367

Blaymore Bier JA, Oliver T, Ferguson A, Vohr BR (2002) Human milk reduces outpatient upper respiratory symptoms in premature infants during their first year of life. J Perinatol 22 (5):354–359

Caulfield E (1952) Infant feeding in colonial America. J Pediatr 41(6):673–687

Chulada PC, Arbes SJ Jr, Dunson D, Zeldin DC, Chulada PC (2003) Breast-feeding and the prevalence of asthma and wheeze in children: analyses from the Third National Health and Nutrition Examination Survey, 1988-1994. J Allergy Clin Immunol 111(2):328–336

Cochi S, Fleming DW, Hightower AW et al (1986) Primary invasive *Haemophilus influenzae* type b disease: a population-based assessment of risk factors. J Pediatr 108:887–896

Cowie AT (1974) Overview of the mammary gland. J Invest Dermatol 63:2–9

Davis MK (1998) Review of the evidence for an association between infant feeding and childhood cancer. Int J Cancer Suppl 11:29–33

Dewey KG, Heinig MJ, Nommsen-Rivers LA (1995) Differences in morbidity between breast-fed and formula-fed infants. J Pediatr 126:696–702

Drake TGH (1935) Infant welfare laws in France in the 18th century. Ann Med Hist 7:49–55

Duncan B, Ey J, Holberg CJ, Wright AL, Martinez FD, Taussing LM (1993) Exclusive breast-feeding for at least 4 months protects against otitis media. Pediatrics 91(5):867–872

Eidelman AI (2006) The Talmud and human lactation: the cultural basis for increased frequency and duration of breastfeeding among Orthodox Jewish women. Breastfeed Med 1:36–40

Enger SM, Ross RK, Paganini-Hill A, Bernstein L (1998) Breastfeeding experience and breast cancer risk among postmenopausal women. Cancer Epidemiol Biomarkers Prev 7(5):365–369

Fateeva EM, Pustograev NN (2005) Encyclopedia of breastfeeding in Orthodox Russia. M, Oranta (in Russian)

Feldman R, Eidelman AL (2003) Direct and indirect effects of breast milk on the neurobehavioral and cognitive development of premature infants. Dev Psychobiol 43(2):109–119

Gdalevich M, Mimouni D, Mimouni M (2001) Breast-feeding and the risk of bronchial asthma in childhood: a systematic review with meta-analysis of prospective studies. J Pediatr 139 (2):261–266

Gerstein HC (1994) Cow's milk exposure and type I diabetes mellitus. A critical overview of the clinical literature. Diabetes Care 17(1):13–19

Gillman MW, Rifas-Shiman S, Camargo CA Jr, Berkey CS, Frazier AL, Rockett HR, Field AE, Colditz GA (2001) Risk of overweight among adolescents who were breastfed as infants. JAMA 285(19):2461–2467
Goodwin DW, Gabrielli WF, Penick EC, Nickel EJ, Chhibber S, Knop J, Jensen P, Schulsinger F (1999) Breast-feeding and alcoholism: the Trotter hypothesis. Am J Psychiatry 156:650–652
Grummer-Strawn LM, Mei Z (2004) Does breastfeeding protect against pediatric overweight? Analysis of longitudinal data from the Centers for Disease Control and Prevention Pediatric Nutrition Surveillance System. Pediatrics 113(2):e81–e86
Gundobin NP (1901) The life of a child. Home Doctor 1:2–114. (in Russian)
Hartmann PE (2007) The lactating breast: an overview from down under. Breastfeed Med 1:3–9
Hassiotou F, Beltran A, Chetwynd E, Stuebe AM, Twigger AJ, Metzger P, Trengove N, Tat LC, Filgueira L, Blancafort P, Hartmann PE (2012) Breastmilk is a novel source of stem cells with multi-lineage differentiation potential. Stem Cells 30:2164–2174
Hassiotou F, Geddes DT, Hartmann PE (2013) Cells in human milk: state of the science. J Hum Lact 29:171–182
Horta B, de Sousa BA, de Mola CL (2018) Breastfeeding and neurodevelopmental outcomes. Curr Opin Clin Nutr Metab Care 21:1–4
Horwood LJ, Darlow BA, Mogridge N (2001) Breast milk feeding and cognitive ability at 7-8 years. Arch Dis Child Fetal Neonatal Ed 84:F23–F27
Howie PW, Forsyth JS, Ogston SA, Clarc A, Florey CD (1990) Protective effect of breast feeding against infection. BMJ 300:11–16
Istre GR, Conner JS, Broome CV, Hightower A, Hopkins RS (1985) Risk factors for primary invasive Haemophilus influenzae disease: increased risk from day care attendance and school-aged household members. J Pediatr 106(2):190–195
Jacobson SW, Chiodo LM, Jacobson JL (1999) Breastfeeding effects on intelligence quotient in 4- and 11-year-old children. Pediatrics 103(5):e71–e78
Jernstrom H, Lubinski J, Lynch HT, Ghadirian P, Neuhausen S, Isaacs C, Weber B, Horsman D, Rosen B, Foulkes WD, Friedman E, Gershoni-Baruch R, Ainsworth P, Daly M, Garber J, Olsson H, Sun P, Narod SA (2004) Breast-feeding and the risk of breast cancer in BRCA1 and BRCA2 mutation carriers. J Natl Cancer Inst 96(14):1094–1098
Kostraba JN, Crickshanks KJ, Lawler-Heavner J, Jobim LF, Rewers MJ, Gay EC, Chase HP, Kligensmith G, Hamman RF (1993) Early exposure to cow's milk and solid foods in infancy, genetic predisposition, and risk of IDDM. Diabetes 42(2):288–295
Kramer MS, Guo T, Platt RW, Sevkovskaya Z, Dzikovich I, Collet JP, Shapiro S, Chalmers B, Hodnett E, Vanilovich I, Mezen I, Ducruet T, Shishko G, Bogdanovich N (2003) Infant growth and health outcomes associated with 3 compared with 6 mo of exclusive breastfeeding. Am J Clin Nutr 78(2):291–295
Lawrence RA, Lawrence RM (1999) Breastfeeding: a guide for medical profession, 5th edn. Mosby
Lawrence RA, Lawrence RM (2009) Breastfeeding: a gide for gqmedical profession, 5th edn. Mosby, 968 p
Lawrence RA, Lawrence RM (2011) Breastfeeding: a gide for medical profession, 7th edn. Mosby, 1114 p
Lee SY, Kim MT, Kim SW, Song MS, Yoon SJ (2003) Effect of lifetime lactation on breast cancer risk: a Korean women's cohort study. Int J Cancer 105(3):390–393
Lipsky AF (1896) Encyclopedic dictionary. Brockhaus and Efron 48:104–106
Lopez-Alarcon M, Villalpando S, Fajardo A (1997) Breast-feeding lowers the frequency and duration of acute respiratory infection and diarrhea in infants under six months of age. J Nutr 127:436–443
Lucas A, Cole TJ (1990) Breast milk and neonatal necrotising enterocolitis. Lancet 336 (8730):1519–1523
Lucas A, Morley R, Cole TJ (1998) Randomised trial of early diet in preterm babies and later intelligence quotient. BMJ 317:1481–1487

Marild S, Hansson S, Jodal U, Oden A, Svedberg K (2004) Protective effect of breastfeeding against urinary tract infection. Acta Paediatr 93(2):164–168
McVea KL, Turner PD, Peppler DK (2000) The role of breastfeeding in sudden infant death syndrome. J Hum Lact 16(1):13–20
Mortensen EL, Michaelsen KF, Sanders SA, Reinisch JM (2002) The association between duration of breastfeeding and adult intelligence. JAMA 287:2365–2371
Mother and Child (1901) Part 2 child. Family Practitioner 9:69–103 (in Russian)
Newcomb PA, Storer BE, Longnecker MP, Mittendorf R, Greenberg ER, Clapp RW, Burker KP, Willet WC, MacMahon B (1994) Lactation and a reduced risk of premenopausal breast cancer. N Engl J Med 330(2):81–87
Oddy WH, Holt PG, Sly PD et al (1999) Association between breast feeding and asthma in 6 years old children: findings of prospective birth cohort study. BMJ 319:815–819
Oddy WH, Peat JK, de Klerk NH (2002) Maternal asthma, infant feeding, and risk of asthma in childhood. J Allergy Clin Immunol 110:65–67
Oddy WH, Peat JK, de Klerk NH et al (2003a) Breast feeding and respiratory morbidity in infancy: a birth cohort study. Arch Dis Child 88:224–228
Oddy WH, Kendal GE, Blair E, de Clerk NH, Stanley FJ, Landau LL, Silbum S, Zubrick S (2003b) Breast feeding and cognitive development in childhood: a prospective birth cohort study. Paediatr Perinat Epidemiol 17:81–90
Owen MJ, Baldwin CD, Swank PR, Pannu AK, Johnson DL, Howie VM (1993) Relation of infant feeding practices, cigarette smoke exposure, and group child care to the onset and duration of otitis media with effusion in the first two years of life. J Pediatr 123(5):702–711
Paradise JL, Elster BA, Tan L (1993) Evidence in infants with cleft palate that breast milk protects against otitis media. Pediatrics 91(5):867–872
Perez-Bravo F, Carrasco E, Gutierrez-Lopez MD, Martinez MT, Lopez G, de los Rios MG (1996) Genetic predisposition and environmental factors leading to the development of insulin-dependent diabetes mellitus in Chilean children. J Mol Med 74(2):105–109
Pettit DJ, Forman MR, Hanson RL, Knowler WC, Bennett PH (1997) Breastfeeding and incidence of non-insulin-dependent diabetes mellitus in Pima Indians. Lancet 350(9072):166–168
Pisacane A, Graziano L, Mazzarella G, Scarpellno B, Zona G (1992) Breast-feeding and urinary tract infection. J Pediatr 120(1):87–89
Popkin BM, Adair L, Akin JS, Black R, Briscoe J, Flieger W (1990) Breastfeeding and diarrheal morbidity. Pediatrics 86:874–882
Rao MR, Hediger ML, Levine RJ, Naficy AB, Vik T (2002) Effect of breastfeeding on cognitive development of infants born small for gestational age. Acta Paediatr 91(3):267–274
Reynolds A (2001) Breastfeeding and brain development. Pediatr Clin N Am 48(1):159–171
Ryan AS, Rush D, Krieger FW (1991) Recent declines in breastfeeding in the United States, 1984 through 1989. Pediatrics 88:719–727
Saarinen UM (1982) Prolonged breast feeding as prophylaxis for recurrent otitis media. Acta Paediatr Scand 71(4):567–571
Schanler RJ, Shulman RJ, Lau C (1999) Feeding strategies for premature infants: beneficial outcomes of feeding fortified human milk versus preterm formula. Pediatrics 103(6):1150–1157
Segawa M (2008) Buddhism and breastfeeding. Breastfeed Med 3:124–128
Shaikh U, Ahmed O (2006) Islam and infant feeding. Breastfeed Med 1:164–167
Singhal A, Farooqi IS, O'Rahilly S, Cole TJ, Fewtrell M, Lucas A (2002) Early nutrition and leptin concentrations in later life. Am J Clin Nutr 75(6):993–999
Smulevich VB, Solionova LG, Belyakova SV (1999) Parental occupation and other factors and cancer risk in children: I. Study methodology and non-occupational factors. Int J Cancer 83 (6):712–717
Speransky GN (1928) Method of rational child feeding. Moscow, p. 42 (in Russian)
Stettler N, Zemel BS, Kumanyka S, Stallings VA (2002) Infant weight gain and childhood overweight status in a multicenter, cohort study. Pediatrics 109(2):194–199

Takala AK, Escola J, Palmgren J et al (1989) Risk factor factors of invasive *Haemophilus influenzae* type b disease among children in Finland. J Pediatr 115:694–701

Toschke AM, Vignerova J, Lhotska L, Osancova K, Koletzko B, Von Kries R (2002) Overweight and obesity in 6- to 14-year-old Czech children in 1991: protective effect of breast-feeding. J Pediatr 141(6):764–769

Tryggvadottir L, Tulinus H, Eyfjord JE, Sigurvinsson T (2001) Breastfeeding and reduced risk of breast cancer in an Icelandic cohort study. Am J Epidemiol 154(1):37–41

Wharton BA (1982) Food for the suckling: revolution and development. Acta Paediatr Scand Suppl 299:5–10

Wickes IG (1953a) A history of infant feeding. I. Primitive peoples, ancient works, Renaissance writers. Arch Dis Child 28:151–158

Wickes IG (1953b) A history of infant feeding. I.I. Seventeenth and eighteenth centuries. Arch Dis Child 28:232–240

Wickes IG (1953c) A history of infant feeding. III. Eighteenth and nineteenth century writers. Arch Dis Child 28:332–340

Wickes IG (1953d) A history of infant feeding. IV. Nineteenth century continued. Arch Dis Child 28:416–422

Wickes IG (1953e) A history of infant feeding. V. Nineteenth century concluded and twentieth century. Arch Dis Child 28:495–502

Witkowska-Zimny M, Kaminska-El-Hassan E (2017) Cells of human breast milk. Cell Mol Biol Lett 22(11):1–11

Zakharova IN, Machneva EB (2016) Breast-feeding traditions in multinational Russia. Materials of the II scientific and practical conference with international participation "Breastfeeding in the modern world". Moscow, pp 1–3. http://www.russianbreastfeedingweek.org/parents/the-history-of-breastfeeding (in Russian)

2 Origin and Development of the Mammary Glands

Abstract

The chapter presents data on the development of breast structure and function during various periods of a woman's life: in the prenatal period, in the first 2 years after birth, during puberty, and during pregnancy. The onset of breast development, as well as early morphogenesis in the embryo, is due to mesenchymal-epithelial interaction with the help of autocrine/paracrine growth factors. The mammary glands of human fetuses achieve a rather high degree of development in structural and functional relation, by the time of birth. After birth, infant mammary glands for some time begin to function while continuing their development. The most intensive development of the breast begins with puberty and the appearance of a menstrual cycle in girls. The development and functioning of the mammary glands changes during this cycle under the influence of the steroid hormones, prolactin and oxytocin. Final breast development occurs during pregnancy and is highly dependent on reproductive and metabolic hormones.

2.1 Origin of the Mammary Glands

According to paleontological research, mammals are descended from reptilian synapsids (Synapsida), which first appeared about 300 million years ago in the second half of the Carbon period. Synapsids were small lizard-like animal that multiplied by laying eggs. Synapsids were a branch of early terrestrial vertebrates-amniotes, which also propagated by laying eggs. But their eggs were different in structure from those of crocodiles, turtles, dinosaurs, and birds that lived at that time. These eggs, called amniotic (hence the name of the animals), had additional shells that contribute to better gas exchange, utilization of nutrients and waste products of the embryo, as well as water retention. Synapsids were better adapted to land and tree life than other reptiles. Some of the synapsids have become able to remain active in the cold season due to the development of a four-chamber heart and a constant

N. P. Alekseev, *Physiology of Human Female Lactation*,
https://doi.org/10.1007/978-3-030-66364-3_2

body temperature. One of these groups has formed a hairline. As a result of modification of the structure and function of a certain type of skin glands, mammary glands were formed, due to which this group was named mammals. This happened at the beginning of the Triassic period (about 250 million years ago). The first mammals still laid eggs and hatched their young, but they were already feeding the young with milk. An idea of what their mammary glands looked like can be made on the basis of the structure of the mammary glands of an ancient almost extinct group of oviparous mammals that has survived to date, including two families: platypuses and echidnas that live in Australia and its surrounding islands. Their glandular department consists of separate tubular glands (in the amount of 100–200). Each gland has a single duct. The excretory ducts are grouped into bundles and open directly on the surface of the abdominal wall, forming a so-called glandular (milky) field, from the surface of which the baby licks the milk. Subsequently, as a result of the evolutionary process, viviparous mammals appeared, the cubs of which developed before birth in a special organ formed during pregnancy—the placenta. Significant alterations occurred in placental mammals and in the structure of the mammary gland. The structure of the glandular part changed, and special formations appeared on the surface of the skin—nipples containing milk ducts opening outward (Grachev and Galantsev 1973).

The question of the origin of mammary glands and lactation in mammals began to be discussed in the scientific literature from the mid-nineteenth century. Most authors believe that the mammary glands are derived from the skin glands. However, there is no consensus on the type of skin glands. There are several hypotheses, among which there are two noteworthy ones. According to the first hypothesis, the most common, the mammary glands are derived from the sweat apocrine glands located next to the hair follicles. This position is based on the fact that the glandular apparatus of the apocrine sweat glands has a great similarity in structure and function with the tubular glandular apparatus of the mammary glands that currently exist in ancient groups of lower mammals: platypuses and echidnas. According to the second hypothesis, mammary glands, as well as other skin glands (sweat, sebaceous), were descended from nonspecialized skin glands of animallike reptiles, and in phylogeny, the first among specialized skin glands, apparently, were mammary glands. This is indicated by the fact that in embryogenesis in placental mammals, the laying of mammary glands occurs earlier than the sweat and sebaceous ones (Raynaud 1961). It unfortunately, the lack of sufficient paleontological data, does not allow us to present a corresponding series of evolutionary transformations from ancestral forms of skin glands to the mammary glands of modern placental mammals. More information of hypotheses about the origin of mammary glands can be found in the currently available reviews: Grachev and Galantsev (1973), Galantsev and Gulyaeva (1987), Blackburn (1991), Oftedal (2002a, b, 2012), Lefèvre et al. (2010), and Oftedal and Dhouailly (2013).

2.2 Development of the Breast

The main development of the breast occurs in the postnatal period and reaches the final formation in a woman after childbirth in the first days (weeks) of lactation. Before the final stage, there are three main phases in the development of breast structure and function: intrauterine development, development in the first 2 years of life, and development at puberty (Lawrence and Lawrence 1999). A detailed description of the structure of a woman's lactating breast will be presented in Chap. 3. However, for a better understanding of the material presented in this section and subsequent sections of Chap. 2, it is advisable to first present in schematic form (Fig. 2.1) the structure and location of the main structures of the breast of a woman during lactation. It should be noted that due to the dense packaging and different size of the structures that make up the inner part of the gland, it is fundamentally impossible to visually depict all the elements of the breast on any one scale.

The formation of the milk occurs in epithelial secretory cells, which with another type of epithelial cells—myoepithelial cells that provide milk excretion—form special hollow structures, alveoli or acinuses, connected by a system of ducts

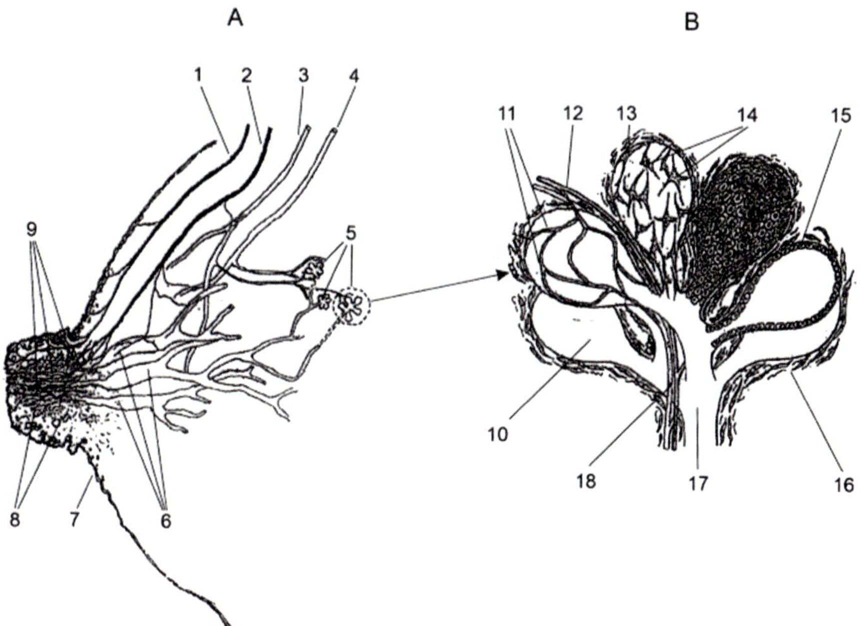

Fig. 2.1 Schematic representation of the general structure of a woman's breast (**a**) and a separate lobule of glandular tissue (**b**). (**a**) 1 nerve trunk containing afferent and efferent nerve fibers of the autonomous nervous system, 2 nerve trunk containing afferent nerve fibers of somatic nervous system, 3 blood vessel, 4 lymph vessel, 5 breast ducts, 6 breast lobes, 7 epidermis, 8 nipple ducts,9 smooth muscles. (**b**) 10 alveolus, 11 blood capillaries, 12 arteriole, 13 interlobular stroma, 14 secretory cells, 15 myoepithelial cells, 16 interlobular stroma, 17 mammary gland duct, 18 venule

(Fig. 2.1b). Here, stem cells are localized, providing the appearance of secretory and myoepithelial cells. A very important structure of the alveolar-ductal system is the basal lamina that covers the alveoli and ducts. The alveolar-ductal system is located under the skin and in addition to the basal lamina is surrounded by a number of connective tissue and cellular structures united by the common name—stroma. Elements of the stroma are extremely important for the normal development and functioning of the breast. Connective tissue structures of the stroma include bundles of various types of collagen and elastin fibers. The stroma cell includes fibroblasts, fat cells, macrophages, lymphocytes, mast cells, Cajal cells, or cells similar to Cajal cells.

Alveoli, united by a single duct, form lobules. Lobule ducts connect to larger ducts to form lobes. The ducts of the lobes of the breast do not just end on the surface of the skin, but pass inside a special skin-muscle formation—a nipple that protrudes above the surface of the skin. The pigmented area of the skin near the base of the nipple is called the areola. When approaching the nipple in the subareolar region, the lobe ducts can join and expand to form the main ducts (Ramsay et al. 2005), the number of which is 4–14. The main ducts have a diameter varying between 1 and 5 mm. Wide, non-branching sections of the main ducts are sometimes called milk sinuses. At the entrance to the nipple, the ducts are largely narrowed and have different diameters along the way in the nipple. The breast nipple and areola do not have hair. The inner part of the gland is abundantly supplied with blood and lymphatic vessels. Blood capillaries pass next to individual alveoli. Somatic afferent nerve fibers in the skin of the nipple and areola of the gland form receptors of various modalities: mechanoreceptors and heat, cold, and pain receptors. Afferent nerve fibers of the autonomic nervous system innervate the walls of broad and medium-sized milk ducts, as well as blood and lymphatic vessels. Efferent nerve fibers of the autonomic nervous system innervate smooth muscle fibers in the breast nipple and in the walls of blood vessels (Eriksson et al. 1996).

2.2.1 Prenatal Development of the Breast

Research on the development of women's breast began in the second half of the nineteenth century (Raynaud 1961; Khizhnyakova 1965; Howard and Gusterson 2000). However, the degree of study of this problem to date is inferior to that of laboratory animals, in particular, in rats and mice. The main reason was and is the limitations associated with the use of methods for the study of a woman's breast. Thus, there are certain ethical and technical difficulties in obtaining histological material in different periods of development of a woman's breast. It was not possible to fully use a number of methods that are successfully used to study the development of mammary glands in laboratory animals. These include primarily methods of immunohistochemistry, molecular genetics, and tissue culture. In addition, great difficulties in studying a woman's breast arise from the fact that the size, shape, and rate of development of the human breast vary significantly between individuals at the macro- and microscopic levels. This is believed to be due to genetic causes, variations in nutrition, and environmental conditions (Howard and Gusterson 2000). However, in laboratory animals, genetic methods allow to breed and maintain lines of animals with certain characteristics while ensuring an appropriate diet and stable environmental conditions. It should be noted here that one of the most important conclusions that has been made on the basis of scientific achievements of recent decades is that the evolutionary process, once stumbled upon an effective version of the structural and functional organization of a system, uses it with varying degrees of modification (sometimes insignificant), both in animals and in humans. Experimental data available to date indicate that this position is also true for the structure and function of the breast (Grachev and Galantsev 1973). Looking ahead, we can note that a comparison of data obtained on the mammary glands of rats and mice and the mammary glands of women reveals similarities in their development, as well as morphofunctional characteristics (Cardiff and Wellings 1999). Therefore, the mammary glands of mice and rats are often used as a model to explain the functioning of women's mammary glands in normal and pathological conditions. However, along with the similarities, as we will see later, there are also noticeable differences (Johnson 2010), which suggests using the mammary glands of mice and rats as a model when studying the breast of a woman with certain adjustments.

In accordance with morphological studies of various authors in the formation of the human breast in the prenatal period, there are several successive stages corresponding to the development of the embryo (human embryo before the ninth week of development) and the human fetus. However, the time of appearance of any breast structure varies depending on the criteria chosen by the authors to determine the age of the embryo or fetus, for example, the time from the beginning of conception, from the last menstrual cycle of a woman in labor, or the length of the embryo and fetus (Howard and Gusterson 2000; Russo and Russo 2004a, b). It is difficult to determine the time with sufficient accuracy based on the first two criteria. More successful was the use of the ratio between the length of the embryo and the development of breast structures (Russo and Russo 2004a, b), proposed by E. Hughes (1950).

While in a pregnant woman, the development of the structure and function of the breast is approaching the final stage in the fetus, who is in the womb, the mammary gland is just beginning to develop. To date, it is an established fact that the final development of the breast during pregnancy is largely dependent on reproductive hormones: estrogen, progesterone, placental lactogen, prolactin, oxytocin, as well as the so-called metabolic hormones, i.e., growth hormone, glucocorticoids, thyroid hormone, and insulin (Neville et al. 2012). This will be discussed in more detail in the following sections of the book. However, the beginning of breast development, as well as early morphogenesis in the embryo, is due to the interaction of ectoderm cells (embryonic epidermis) and mesenchyma (embryonic stromal cells) with the help of known autocrine/paracrine growth factors such as epidermal growth factor (EGF) and transforming growth factor-α (TGF-α), as well as other as yet unknown factors. Moreover, the initial influence is exerted by mesenchyma. This conclusion was made based on the results of experiments on the effect of different combinations of mesenchymal and ectodermal tissue sites on breast development in laboratory animals (rabbits and mice). So in rabbits, before the appearance of the first noticeable signs of mammary gland differentiation, the combination of mesenchyma taken from the area where the mammary gland embryo should occur with the epidermis of the head or neck sections was accompanied by the development of mammary gland structures. At the same time, a similar combination of ectoderm sites, the supposed place of origin of the embryo and mesenchyma of other areas of the embryo, did not lead to the development of mammary gland structures. However, after the first signs of a mammary gland embryo appear, the ectoderm from this site can form the beginnings of the mammary gland when combined with the mesenchyma of other parts of the embryo's body (Robinson et al. 1999; Hovey et al. 2002; Cunha et al. 2004). Similar studies on the mammary glands of human embryos have not been conducted. There are only indirect data indicating that the early development of the human embryo is determined by growth factors. In particular, data from immunohistochemical experiments indicate the presence of growth factors EGF, TGF-α, as well as TGF-β, tenacina-C, and BCL (Osin et al. 1998; Howard and Gusterson 2000).

Analysis of past and current morphological works (Hughes 1950; Raynaud 1961; Grachev and Galantsev 1973; Howard and Gusterson 2000), devoted to the study of prenatal development of the breast of the human embryo and fetus, revealed difficulties in selecting images from microphotographs, which could be sufficiently clearly present simultaneously all the main cellular elements at different stages of breast development. Therefore, for better clarity, when presenting the material, generalized schematic drawings will be provided as illustrations in most cases. The first anatomical structure of the developing breast is the so-called milky strip, which can be seen in a human embryo 4 mm long at the fourth week of the embryo's life (Hughes 1950) (Fig. 2.2). Milky strip are thickening of the ectoderm, consisting of two to four layers of cells and located on both sides of the body from the rudiment of the forelimb in the caudal direction to the inguinal folds. It should be noted here that the milky strip seem to be important not only for the development of the mammary gland but also for other structures of the body, since the milky strip are found in the embryos of reptiles and birds (Howard and Gusterson 2000). In a 7 mm embryo, ectodermal cells appear in the middle of the milky strip in the longitudinal direction, which are slightly larger than neighboring cells and are based on compacted mesenchymal cells. As a result, a narrow ridge is formed in the middle of the milky strip in the longitudinal direction, called the milky line (Fig. 2.2). When the size of the embryo increases to 9 mm, the milky line grows along the entire length of the milk strip (Fig. 2.2). Then the milky line from the caudal side begins to regress so that at the size of the embryo of 10–12 mm, it is observed only in the thoracic region of the embryo. The mesenchyma under the ectodermal cells is differentiated into four layers. On the cross section, the reduced milky line has the form of a biconvex lens or disk (Fig. 2.2) in an embryo of 13–15 mm, and the reduced length of the milky line thickens in the cranial part where the ectodermal cells begin to invaginate into the mesenchyma. As a result, an epithelial thoracic (milky) nodule is formed, which gradually increases in volume. In an embryo at the age of 2 months, having a length of 20–30 mm takes a globular or spherical shape.

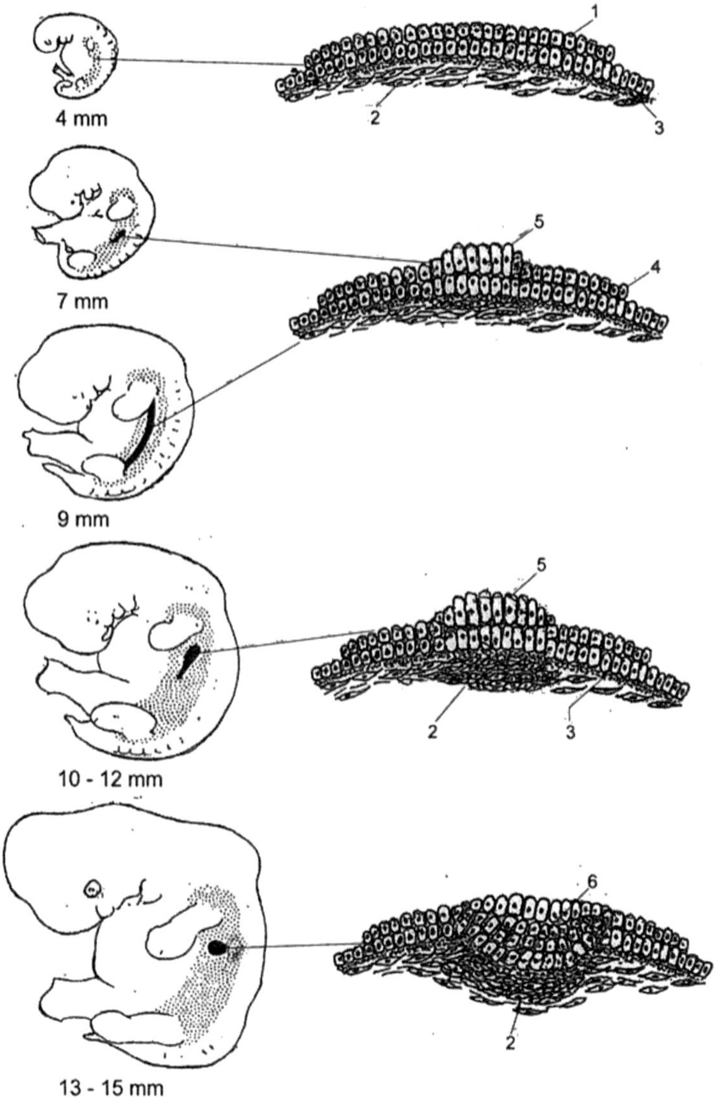

Fig. 2.2 Development of the mammary bud in a human embryo (from Porter J. C., 1974). At the left, the human embryo at different stages of development. Numbers denote embryo length in mm. On the right, mammary glands at different stages of embryo development. 1 milky strip, 2 mesenchyma, 3 basal membrane, 4 ectoderm, 5 milky line, 6 milky nodule

The epithelial cells forming it are located mainly in the radial direction. In parallel with cellular transformations, the milky nodule shifts in the ventral direction.

It should be noted that sometimes the milky line begins to segment, just as it does, for example, in mice or rats. As a result, along the length of the milky line, in

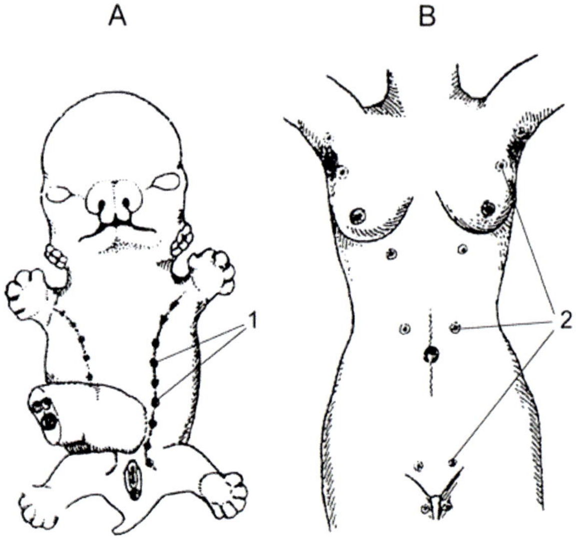

Fig. 2.3 The phenomenon of polythelia in the embryonic period (**a**) and its manifestation in a woman in adulthood (**b**) (from Grachev and Galantsev 1973)

addition to the thoracic pair of milky nodules, several pairs of additional milky nodules are formed, which develop further. The degree of development of additional milky nodules can be different. According to the classification given by Y. Kajava (1915), additional milky nodules can reach the following development: (1) normal mammary glands (polymastia); (2) additional mammary glands that do not have areola region, but the nipple and the glandular tissue are developed; (3) the mammary gland without nipple but with the presence of the areola and glandular tissue; (4) only sections marked glandular tissue (ectopic glandular tissue); (5) the nipple and areola present, but the glandular tissue replaced by fat tissue; (6) there are nipples only (polythelia); and (7) the presence of areola only (polythelia areola). For Fig. 2.3 as an example, the phenomenon of polythelia in the embryonic period (A) and how it manifests in a woman in adulthood (B) are presented. It should be noted that the above abnormalities are most clearly detected during lactation.

Currently available data indicate that abnormalities in the development of the human breast occur as a result of violations at the genetic level (mutations of the TBX-3 gene), which in turn leads to violations in ectodermal-mesenchymal interactions. Moreover, the abovementioned anomalies can occur as a purely phenotypic, involving only the mammary gland, or as a pleiotropic syndrome, in which there is a violation in the development of the breast, limbs, teeth, and genitals (Bamshad et al. 1997).

The development of the milky nodule from 3 to 5 months is relatively slow (Raynaud 1961). However, at 5 months of life with a fetal length of 120–150 mm, noticeable changes in the structure of the milky nodule begin, leading to the formation of the milky bud, from which the development of the ducto-alveolar system of the future mammary gland is directly carried out (Fig. 2.4). The bud is covered with a basement membrane (shell), which is adjacent to dense layers of

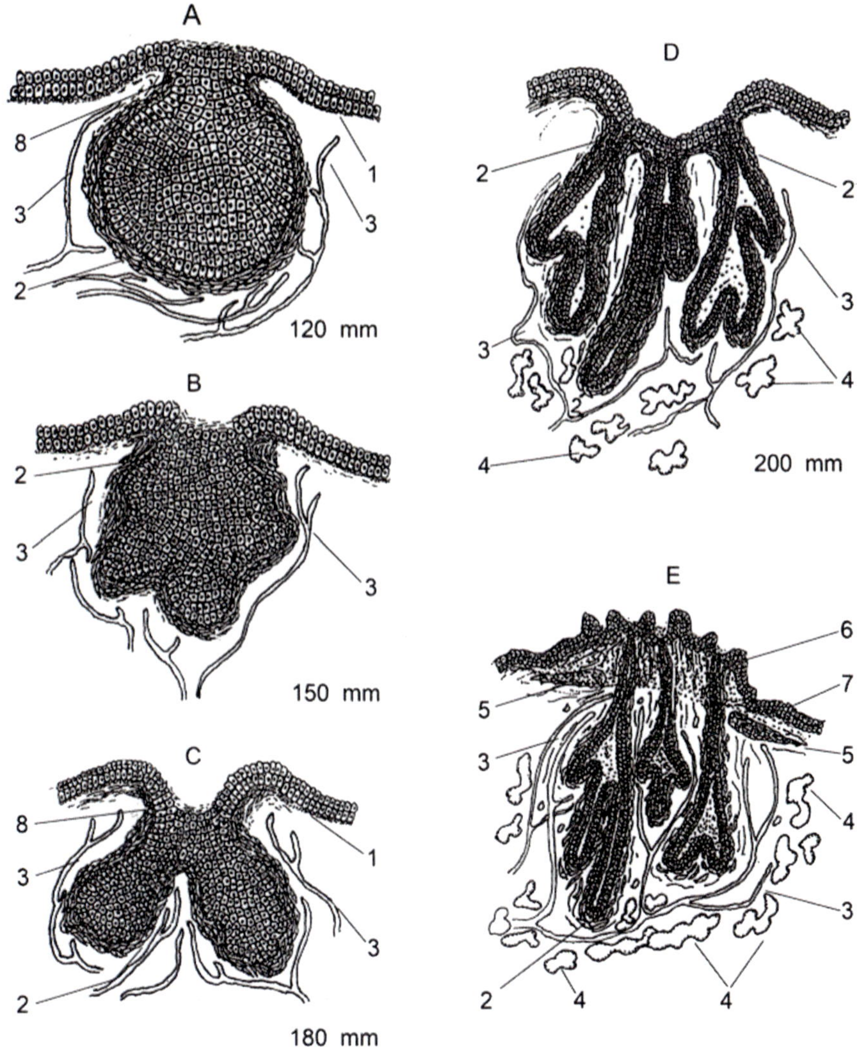

Fig. 2.4 Human fetal breast development. (**a**) Milk bud, (**b**, **c**) the appearance of secondary buds, (**d**) the formation of cell filaments with cavities, (**e**) the formation of the nipple. 1 Basal membrane, 2 layer of fibrocytes and collagen fibers, 3 blood and lymphatic capillaries, 4 adipose tissue, 5 subareolar muscles, 6 breast nipple, 7 epidermis, 8 neck of the milk bud. All the numbers below show the length of the fetus in mm

primary mesenchyma. The bud is attached to the outer epidermis by a short neck formed by epithelial cells and slightly protrudes above the chest surface of the fetus (Fig. 2.4a).

Electron microscopic studies have shown that the cells that make up the bud in the center have a polygonal shape and are connected to each other by thin microvilli and desmosomes. Their cytoplasm is rich in glycogen and has large nuclei. The cells on the periphery of the mammary bud do not differ in structure from the central cells, but they are smaller.

During development, the mesenchymal cells surrounding the bud begin to differentiate into fibroblasts. Collagen fibers are located in the mesenchyma surrounding the milky bud (Tobon and Salazar 1974; Kellokumpu-Lehtinen et al. 1987). The milky bud is bordered by thin blood and lymphatic vessels (Kukhtinova 1963; Osin et al. 1998; Howard and Gusterson 2000). Nerve fibers are found near the milky bud (Kellokumpu-Lehtinen et al. 1987). Near the bud, there are no rudiments of hair follicles (Howard and Gusterson 2000). During this period, the basal cells of the milky bud, as well as mesenchymal cells surrounding the bud, show a high level of expression of the antiapoptotic factor BCL-2 (Nathan et al. 1994; Howard and Gusterson 2000; Naccarato et al. 2000). It is assumed that BCL-2 protects cells from apoptosis and promotes further differentiation of mesenchymal cells into fibroblasts. It should be noted here that at the stage of differentiation of the milky bud, another important difference in the prenatal development of the mammary gland in humans and rodents is revealed. So, in a human fetus, the development of the milky bud to the breast at the time of birth occurs regardless of the sex of the child. Although at the time of formation of the milky bud, male fetuses begin to synthesize and secrete androgens in the testicles (Kellokumpu-Lehtinen et al. 1979, 1980), for example, in the embryos of male mice, the destruction of the milky bud is observed, and by the time of birth, males do not have mammary glands.

Studies using the method of tissue recombination in mice, first conducted by K. Kratochwil and co-worker (Kratochwil 1971, 1977; Kratochwil and Schwartz 1976; Durnberger and Kratochwil 1980; Robinson et al. 1999), showed that the destruction of the milky bud in male mice is caused by epithelial-mesenchymal interactions that are triggered by androgens. It turned out that initially the epithelium of the milky bud induces androgen receptors on mesenchymal cells located between the milky buds so that the milky bud becomes surrounded by mesenchymal cells with androgen receptors. These cells in response to the action of androgens (testosterone) on the 14th day of embryo development condense around the neck of the milky bud and through paracrine factors by the 16th day of development cause regression of the epithelial cells of the bud.

As the milky bud develops in the human fetus, a small depression forms in the center of the protruding part of the bud (Fig. 2.4b), which increases with the growth of the fetus and enters the bud (Fig. 2.4c). At the same time (the length of the fetus is 150–180 mm) on the side surface of the primary bud, secondary buds begin to appear and grow in length, taking the form of cell strands and being the precursors of ducts (Howard and Gusterson 2000). In a fetus about 200 mm long (the end of the second trimester of pregnancy), these cell strands begin to branch at the ends, and cavities form in some of them as a result of destruction (apoptosis) of the inner layer of cells (Fig. 2.4d). The cell strands are surrounded by dense layers of mesenchyma, consisting mainly of fibroblasts and collagen fibers. It should be noted that the cell strands grow into the dermis layer, but they do not reach the adipose tissue (Tobon and Salazar 1974). In contrast to the human fetus in rodents, the mesenchyma surrounding the branching duct consists of fat cells with small inclusions of fibrous tissue (Parmar and Cunha 2004). It turned out that the environment of fibroblasts is essential for the development of the ductal system of the human breast. Studies of epithelial-stromal interactions using tissue culture techniques have shown that the placement of epithelial cells of a woman's breast on a substrate of collagen gel containing fibroblasts led to the formation and growth of ducts. Moreover, the development of the ducts occurred regardless of whether fibroblasts were taken from the mammary gland of a woman or a mouse. At the same time, the formation of ducts from epithelial cells of the woman's breast located on a substrate of the adipose tissue of the mouse mammary gland did not occur. A detailed analysis of these extremely interesting works is given in the review of E. Parmar and D. Kuna (Parmar and Cunha 2004).

Not enough is known about the mechanisms of morphogenesis of the ductal system of the human fetal breast. Available data (Howard and Gusterson 2000) indicate that the morphogenesis of the ducts is due to epithelial-mesenchymal interactions. During this period, in addition to the expression of antiapoptotic factor BCL-2, intense expression of receptors for epidermal growth factor and transforming growth factor-α (TGF-α) was detected in the cells of the strands. To date, due to the methodological advantages, the morphogenesis of the ductal system in mouse embryos has been studied in detail. In particular, using immunochemical methods (Hens et al. 2007), it was found that from the epithelial cells of the milky bud in a 14.5-day-old mouse embryo, PTHrP (parathyroid hormon-related protein) is secreted, which increases the expression of cellular receptors to BMR (bone morphogenetic protein) in mesenchymal cells surrounding the milky bud. This accordingly increases the sensitivity of mesenchymal cells to BMR. The effect of one of the types of BMR-BMR4 on the mesenchyma, in turn, through the paracrine factor causes, in contrast to the milky bud of the human fetus, the growth of only one external cell strand from the mouse milky bud, which later begins to branch. BMR4 also causes an increase in the expression of MSX2 factor, which is involved in epithelial-mesenchymal interaction during breast development, providing inhibition of hair follicle growth in the area of the skin surface from which the nipple and areola are formed. It is interesting to note that the algorithm of epithelial-mesenchymal interactions in the morphogenesis of the ductal system is similar to that in inhibiting the development of the milky bud in males. That is, first, epithelial cells induce receptors in the mesenchymal cells to the corresponding protein, and then the protein in interaction with the receptor causes a change in the development of the milky bud through paracrine factor. It is possible that the same mechanism of interaction is characteristic of the development of the ductal system of the milky bud and the human fetus.

Ultrastructural studies (Tobon and Salazar 1974; Howard and Gusterson 2000) have shown that when the fetal length is about 200 mm (second trimester of pregnancy), the cells forming the ductal system begin to differentiate into epithelial and myoepithelial. In turn, two types of epithelial cells were distinguished by their level of development (Tobon and Salazar 1974). The cells showing the highest level of development are adjacent to the duct cavity (luminal epithelial cells). The sections of cells that exit into the duct cavity are conical in shape and are provided with short and thin microvilli. Dense contacts are formed between neighboring cells. The cells contain a developed rough endoplasmic reticulum with cisterns containing granular material. There are large mitochondria with well-formed crystals. However, the Golgi apparatus is not yet well developed. In some luminal cells in the apical region at the end of the second trimester of pregnancy, vesicles filled with granular electron-dense material are detected, and the contents of the vesicles exit into the duct cavity, which indicates the beginning of secretory activity of the fetal breast. The second group of epithelial cells is adjacent to the basal membrane on one side and the other to the basal or sometimes to the lateral surface of the luminal epithelial cells. Very rarely, the apical surfaces of the second group of cells go into the duct cavity. This type of cell has a light cytoplasm, which has a rounded core in the center, an insufficiently developed rough endoplasmic reticulum, and a Golgi apparatus. Rare mitochondria are found in cells. Myoepithelial cells are adjacent to one surface of the basement membrane, and the other is in contact with epithelial cells, as well as with each other, using thin cytoplasmic outgrowths and desmosomes. Numerous bundles of myofilaments are found inside myoepithelial cells. The nuclei of myoepithelial cells have a characteristic folded shape.

In the final trimester of pregnancy, significant changes are observed in the development of the structure and function of the fetal mammary glands, reaching a peak at the time of birth (Raynaud 1961; Tobon and Salazar 1974; Howard and Gusterson 2000). When the length of the fetus is 300–350 mm (eighth month of pregnancy), the ducts of the breast increase in size, repeatedly branching at the ends with the formation of cavities. The ducts are surrounded by a dense layer of fibrocytes and collagen fibers and separated from each other by layers of the adipose tissue. The blood and lymphatic network of vessels becomes denser, while the vessels penetrate between the ducts, entwining them with a dense network along the course of the ducts (Kukhtinova 1963; Naccarato et al. 2000). As we will see later, the functioning of the breast is largely regulated by the nervous system. However, the experimental data available to date on the development of innervation of external and internal structures of the human fetal breast, as well as the embryos of mammary glands of other mammals, are not available.

Ultrastructural studies carried out on the mammary glands of fetuses of the final trimester (Tobon and Salazar 1974; Howard and Gusterson 2000) showed the presence of well-formed cavities in the branched ducts, with some sections of the ducts resembling alveoli in structure and lined with epithelial cells with numerous microvilli extending into the cavity. There are already well-formed tight contacts between the cells. The cytoplasm of epithelial cells contains a large number of large mitochondria with clearly defined crists and a light matrix. The cisterns of the rough endoplasmic reticulum are filled with granular electron-dense material. The cytoplasm contains glycogen particles. During this period, the secretory activity of epithelial cells is intensified. In particular, the secretion of two types of products is noted (Tobon and Salazar 1974). One of them consists of dark granules enclosed in bubbles located in the Golgi apparatus. Another product is represented by lighter granules grouped into ovoid-shaped vesicles located mainly in the apical regions of cells. Very often, the histological images show how the contents of these vesicles pass into the duct cavity. In addition, in the cytoplasm of cells, there are a large number of fat drops that also exit into the duct cavity. In addition to epithelial cells with intense secretory activity, cells with less pronounced signs of secretion were observed. These cells were classified as cells in the middle stage of differentiation (Tobon and Salazar 1974). This type of cell has a light cytoplasm, which has a rounded core in the center, an insufficiently developed rough endoplasmic reticulum, and a Golgi apparatus. Their surfaces rarely went out into the duct cavity, and in most cases, these cells were localized in the basal area of the duct wall.

Myoepithelial cells were located on the periphery of the ducts, and their main axis ran parallel to the basal shell. They had a multi-lobed nucleus. The cytoplasm contained rare mitochondria and a rough endoplasmic reticulum. Numerous intracellular myofilaments, collected in bundles, were located along the main axis of the cells.

Along with the development of the alveolar-ductal system of the fetal breast, which is the internal part of the breast, in the final trimester, the formation of a very important external structure—the breast nipple—can be seen. The first sign of the appearance of the nipple (elevation above the skin surface) is observed in a fetus, which is 150 mm long (Brouha 1905) (Fig. 2.4b) At this stage, as already noted, a small depression is formed in the center of the milky bud, and the surrounding epidermal area is slightly raised. When the length of the fetus is about 180 mm, the nipple protrusion increases (Fig. 2.4c) and is visible to the naked eye. At 7–8 months of pregnancy with a fetal length of 200 mm, the recess in the center of the milky bud increases (Fig. 2.4d). Cells located closer to the surface desquamate, and cells of deeper layers form the epidermal part of the milk ducts. The milky bud as a separate formation disappears at this stage of development. In place of the site of the future nipple, a depression is formed. Then, at the ninth month of development, as a result of thickening of the underlying part of the dermis, the space drawn in begins to come out so that the outlets of the milk ducts are approximately at the level of the skin surface (Fig. 2.4e). In some cases, the final process of leaving the central part of the nipple outside does not occur, which later leads to various types of endothelia (retraction of the nipples). The phenomenon of nipple retraction (as well as other abnormalities in human breast development) is caused by disorders at the genetic level as a result of mutation of the TBX-3 gene (Linden et al. 2009).

A large experimental material shows that the intensification of development and the beginning of secretory activity of the fetal breast in the final trimester of pregnancy occurs not only as a result of increased epithelial-mesenchymal interactions but also due to the inclusion of a hormonal factor. For a better understanding of the development of the fetal breast, we will run a short period ahead, and figuratively speaking, from the mammary glands of the fetus located in the womb, we will "climb up" to the mammary glands of the mother of the final trimester of pregnancy. It was previously noted that in a pregnant woman, the development of the breast during pregnancy largely depends on the impact of the main reproductive hormones, which include estrogen, progesterone, placental lactogen, prolactin, and oxytocin. The level of these hormones increases significantly in the blood of pregnant women in the third trimester (Figs. 2.11 and 2.12). According to numerous data obtained from the study of mammary glands of various mammals, including humans (Neville et al. 2012), these hormones are necessary for the morphogenesis and functioning of various breast structures. In particular, in women, estrogen contributes to the development of the ductal system. Progesterone is necessary for alveolar morphogenesis and prolactin for the development of alveolar structures, the beginning of secretory activity in the final trimester of pregnancy, and the continuation and maintenance of lactation in the postpartum period. Placental lactogen contributes to the initiation of secretion in the third trimester of pregnancy, and finally, oxytocin is necessary for the excretion of milk in the postpartum period.

The study of changes in the concentration of reproductive hormones in the human fetus showed that starting from the fourth month of pregnancy, there is an increase in the content of estrogens in the fetal blood and by the time of birth in the blood plasma of the umbilical vein reached 108.9 ng/ml and in the blood of the umbilical artery, 23.3 ng/ml. In addition, the concentration of estrogens from this period increased in the mother's blood (Shutt et al. 1974). However, the presence of cell receptors on the epithelial and stromal cells of the fetal breast was detected only at the beginning of the seventh month of pregnancy (Keeling et al. 1998; Naccarato et al. 2000). Progesterone receptors were identified in the fetal breast at birth and were observed on the epithelial cells of the mammary glands of newborns during the next 2–3 months (Keeling et al. 1998). A detailed study of the content of the most important reproductive hormone prolactin in the pituitary and bloodstream of the fetus also revealed an increase in its concentration starting from 4 months of pregnancy (Aubert et al. 1975). A particularly rapid increase in prolactin content was observed starting from the third trimester of pregnancy. By the time of birth, its concentration in the fetal blood reached 150 ng/ml (Aubert et al. 1975). There was also a significant increase in the concentration of placental lactogen, which could reach 500 ng/ml at 6 months of pregnancy (Hovey et al. 2002). There is currently no information about the presence of cell receptors for prolactin, which also have a high affinity for placental lactogen, in the epithelial cells of the fetal breast.

In connection with the circulation of reproductive hormones in the bloodstream of the fetus, the question arose about the source or sources of these hormones. The first suggestion made at the beginning of the last century by J. Halban (1905) was that maternal reproductive hormones in various mammals pass through the placental barrier of the fetus at the end of pregnancy and increase their content in the fetal body. This assumption was accepted by a number of researchers, and, despite the absence of any experimental verification, it is still sometimes included as an established fact in reviews, monographs, and manuals on lactation topics (Khizhnyakova 1965; Alipov et al. 1988; Grachev and Galantsev 1973; Lawrence and Lawrence 1999; Howard and Gusterson 2000, etc.). Experimental material available to date indicates that the source of progesterone and estrogens, as well as placental lactogen for both the fetus and the mother in the final trimester of pregnancy, is the placenta (Hovey et al. 2002). As for prolactin, the study of its passage through the placenta in a sheep found that its placenta is practically not permeable to maternal prolactin (Alexander et al. 1973). Experiments with labeled

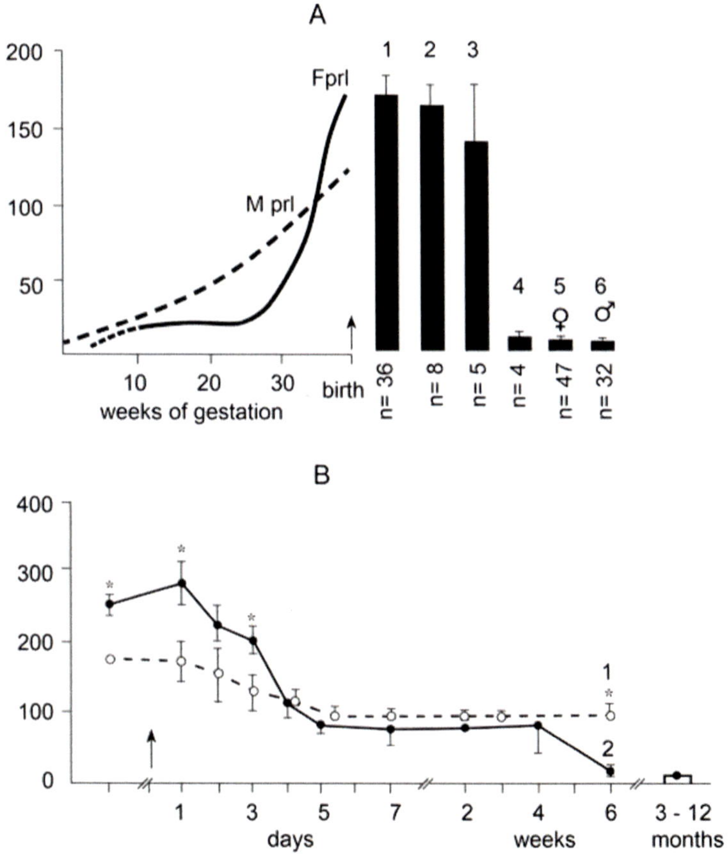

Fig. 2.5 Changes in the content of prolactin in the blood of human fetuses and newborns. (**a**) (Graph) Changes in the content of prolactin in the blood of the fetus (Fprl) and mother (Mprl). On the ordinate axis, the concentration of prolactin in ng/ml, on the abscissa axis of the week of pregnancy. (Charts) The level of prolactin in the blood in the umbilical vein (1), on the first day after birth in normal (2) and acephalic children (3), as well as in the first 1–5 months (4) and 5–15 years in girls (5) and boys (6). *n* the number of subjects (from Aubert et al. 1975). (**b**) The level of prolactin in the blood of full-term (2) and premature (1) children at different times after birth. Ordinate axis, the concentration of prolactin in ng/ml, and the abscissa axis, the time after delivery (from Guyda and Friesen 1973)

prolactin in rhesus monkeys revealed that only 1% of labeled maternal prolactin enters the fetal bloodstream (Josimovich et al. 1974) There is no direct data on the passage of maternal prolactin through the human placenta. As indirect data, we can consider the data obtained in very detailed experiments on the ontogenesis of prolactin in human fetuses (Aubert et al. 1975). Figure 2.5 A taken from this paper illustrates the comparative change in the content of prolactin in the blood of the fetus and mother. In both cases, there is an increase in the content of prolactin, but the dynamics of changes in the concentration of prolactin differs markedly between the mother and the fetus, especially in the final trimester of pregnancy. Based on these data, it was concluded that the secretion of prolactin in the pituitary gland in the mother and fetus is independent, and the passage through the placenta of maternal prolactin does not occur or is significantly limited (Aubert et al. 1975).

Thus, the materials presented in this section allow us to conclude that the mammary glands of fetuses of both sexes at the time of birth in structural and functional terms reach a fairly high degree of development. The outer part of the gland contains the nipple and the area of the skin around the nipple, devoid of hair follicles—the areola. At the top of the nipple, there are outlets of ducts that branch out in the subcutaneous part and form extensions with cavities, resembling in some cases the alveoli of the gland of a lactating woman. The extensions of the ducts are filled with secret, which indicates the beginning of the functioning of epithelial cells. It should be noted that this is a significant difference from the development of mammary glands of fetal mice, whose branching ducts are not filled with secret before birth (Sternlicht et al. 2006).

2.2.2 Development and Functioning of the Breast in the First Two Years of a Child's Life

Note that there is no error in the title of this section. Indeed, as observations show after birth, the mammary glands of infants of both sexes begin to function while continuing their development. Data on this phenomenon began to appear at the end of the eighteenth century (Khizhnyakova 1965). In most babies born at term, the mammary glands are felt as nodular seals-buds (McKiernan and Hull 1981a; McKiernan et al. 1988). Ultrasound measurements of bud size (Kaplan et al. 2016) found that bud size differs in newborn boys and girls. In boys, the length of the long axis (sagittal direction) is 1.11 ± 0.41 cm and the volume of the buds is $0.6\ cm^3$, and in girls, it is 1.2 ± 0.4 cm and $0.7\ cm^3$, respectively. The nipple protrudes above the skin surface. Around the nipple is a pigmented area, devoid of hair—the areola. Externally, the secretory activity of the newborn gland begins to manifest itself most noticeably on day 4–6 after birth. At this time, there is a swelling and slight redness of the skin of the mammary glands, a kind of "neonatal engorgement." When the gland was lightly squeezed, and sometimes spontaneously, a thick muddy liquid of a slightly yellowish color came out of the gland in small drops, which after 7–9 days became similar to milk (Khizhnyakova 1965). Due to the unusual nature of this phenomenon in the people since time immemorial, the secret released from the mammary glands of newborns was called "witch's milk," and the roughening and redness of the skin in the breast area in the scientific world of the nineteenth and early twentieth century was sometimes defined as "mastitis of newborns" (Khizhnyakova 1965). In the future, morphophysiological studies conducted mainly since the second half of the twentieth century have shown that all these phenomena represent a normal physiological state characteristic of postnatal functioning and development of the mammary glands of newborns. Let's look at the data from these studies in more detail.

The study of the internal structures of the breast of newborns found that, on the one hand, there is no difference between their structures in girls and boys. On the other hand, there are wide variations in the degree of their development in newborns of both sexes, both in the first weeks after birth and in subsequent weeks and months (Ognev 1915; McKiernan 1984; McKiernan et al. 1988; Howard and Gusterson 2000). At birth, the milk ducts open at the bottom of small invaginations of the skin of the tip of the nipple. The ducts pass through the tissue of the nipple and in the subareolar space in some cases remain not branched blindly ending, and in others they are divided twice or repeatedly into branches, resembling the ductal system of the fetus before childbirth (Fig. 2.4e). In the first months after birth, the branched sections of the ducts in most cases have cavities or on some distal ends of the ducts structures similar in structure to the alveoli of the mammary glands of lactating women. During this period, part of the main and additional ducts are lined with a single layer of flat "active" epithelium (McKiernan et al. 1988). Very often, on the preparations of the ducts, there is an apocrine type of secretion, i.e., vesicles of secretion on the apical surface of epithelial cells, enclosed in a membrane that is detached from the membrane of the epithelial cell. Duct extensions contain a secretory product. In addition, desquamated cells were usually present in the duct cavity (McKiernan et al. 1988; Howard and Gusterson 2000). Myoepithelial cells were attached to the epithelial cells of the ducts. It is noted (Howard and Gusterson 2000) that in cases when a branched ductal system appears in the gland, which is grouped into separate lobules, one can see the development of a specialized inter- and intralobular fibroblast stroma. These two fibroblast populations differ in their collagen IV synthesis. In newborns, there is a common fairly dense blood and lymphatic network around the glands. In addition, blood and lymphatic vessels penetrate inside, located between the lobes, and also pass next to the milk ducts. There is no information about innervation of external and internal structures of the breast of newborns. Thus, it is clear from the histological studies that the mammary gland of newborns does not regress immediately after birth, but remains active for several months, and secretion can be observed in some cases at 18 months (McKiernan et al. 1988). According to the data presented in the previous section, for the development of the gland and the formation of milk, a number of hormones are necessary, primarily estrogens, progesterone, and prolactin. It is interesting to find out the change in the concentration of these hormones during the development and "lactation" of the breast in newborns, as well as to determine the source or sources of these hormones. As already noted, for human fetuses, the main source of steroid hormones is the placenta (Hovey et al. 2002). On the first day after birth, the concentrations of the steroid hormones estrogen (17-β estradiol) and progesterone in the blood of infants of both sexes were quite high and amounted to 9.2 ± 1.3 ng/ml and 247 ± 8.2 ng/ml, respectively (Hiba et al. 1977). However, over the next 3 days, the hormone content decreased significantly (Fig. 2.6b). It is interesting to note that a similar dynamics in the concentrations of these hormones was observed at this time in the blood of mothers of newborns (Fig. 2.6a).

The content of prolactin before and immediately after birth in the blood of fetuses and term-born children of both sexes is also very high, 200–270 ng/ml (Aubert et al.

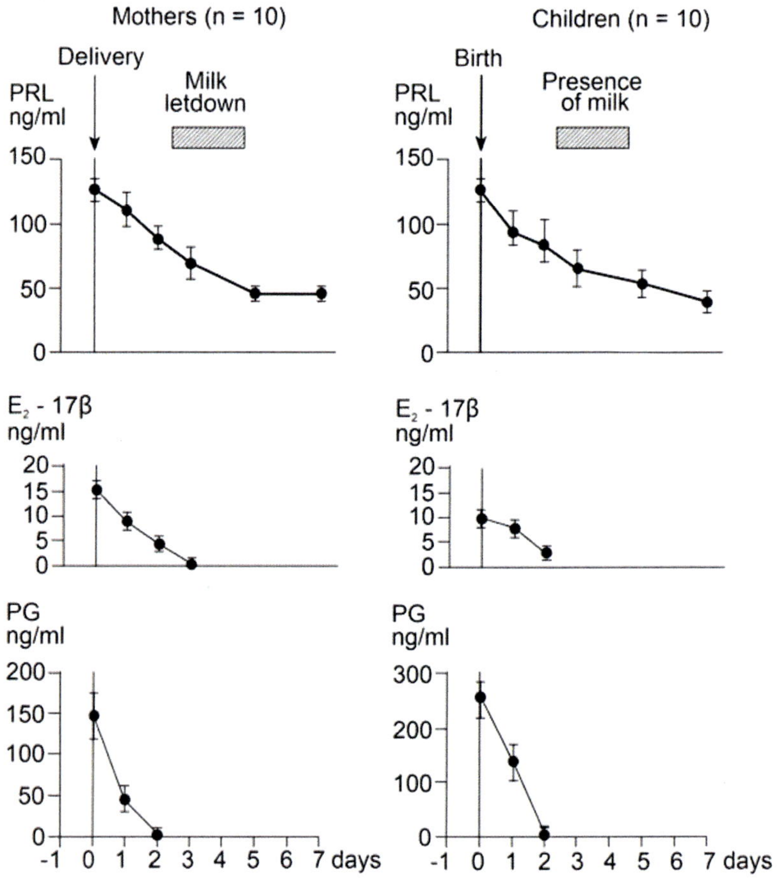

Fig. 2.6 Dynamics of steroid hormones and prolactin in the blood of mothers and newborns. *PRL* prolactin, *GH* growth hormone, *E_2-17β* estradiol (17β), *PG* progesterone. The ordinate, the concentration of the hormone in ng/ml, abscissa, time in days. Shaded horizontal rectangles show, respectively, the engorgement breast in mothers and the appearance of milk when the gland is squeezed in newborns (from Hiba et al. 1977)

1975; Guyda and Friesen 1973; Perlman et al. 1978; McKiernan and Hull 1981b; Lucas and Cole 1990) (Fig. 2.5b). However, it was lower in premature infants and was dependent on the timing of birth (Perlman et al. 1978; McKiernan and Hull 1981b; Lucas and Cole 1990). For example, in babies born at term (39–42 weeks) on the first day, the concentration of prolactin was 267 ± 20 ng/ml, and in premature babies (30–32 weeks)—156 ± 8 ng/ml (Perlman et al. 1978). During the next 5–6 days, the concentration of prolactin in the blood of full-term and preterm newborns decreased to a level that exceeds about eight to ten times its concentration in the blood of children of 3 or more months of age (10 ng/ml), remained at this level for the first 4–5 weeks, and then decreased to values of about 10 ng/ml (Aubert et al. 1975; Guyda and Friesen 1973; Perlman et al. 1978; McKiernan and Hull 1981a, b;

Lucas and Cole 1990) (Fig. 2.6b). At the same time, the increased concentration of prolactin was kept longer in the blood plasma of premature babies.

Determining the dynamics of prolactin concentration during this period in the blood of mothers nursing full-term children (Fig. 2.6a) (Hiba et al. 1977) also found a decline in the concentration of prolactin from the maximum value to a level similar to its concentration in the blood plasma of newborns. Another similarity in the development of lactation function of the breast of newborns and their mothers was the coincidence of the time of the beginning of the appearance of milk in the gland of newborns and the "tide" of milk in the nursing mother (Fig. 2.6). In turn, these two processes coincided with the achievement of a minimum concentration of estrogen and progesterone in the blood of mothers and newborns. Apparently, as in lactating women, high concentrations of steroid hormones inhibit the action of prolactin in the formation of a secret ("witch's milk") in the mammary glands of infants. Based on the close relationship between the dynamics of the concentration of prolactin in the blood of nursing mothers and newborns and the intensification of secretion in their mammary glands, it was suggested that the high level of prolactin in newborns is due to an increased content of this hormone in the mother's blood, entering the child's body through milk. It should be noted here that, just as in the case of the dynamics of prolactin in blood plasma in fetuses, this assumption was accepted by a number of researchers without experimental evidence as an established fact and to this day is sometimes included in reviews, monographs, and manuals on lactation topics. (Khizhnyakova 1965; Alipov et al. 1988; Grachev and Galantsev 1973; Lawrence and Lawrence 1999; Howard and Gusterson 2000, etc.). The study of this issue showed that the concentration of prolactin in the milk of nursing women in the initial 3 days of lactation was quite high and amounted to 157 ± 18 ng/ml, but over the next 13 days, it decreased to 24 ng/ml (Healy et al. 1980) and was at this level in the following weeks and months of lactation (Cox et al. 1996; Cregan et al. 2002a, b). However, the content of prolactin in the blood of newborns who were fed mother's milk from the first days of lactation with an increased content of prolactin and mature donor mother's milk with a reduced concentration of prolactin did not show significant differences (Lucas and Cole 1990). Thus, the high level of prolactin in the blood of newborns and the associated development of their mammary glands, as well as the appearance of a secret in them, is mainly due to the active secretion of prolactin by the central nervous system of the child, and not to the passive effect of maternal prolactin (McKiernan and Hull 1981b).

The amount of secret that could be extracted from the mammary glands of newborns varied significantly and depended mainly on the size of the nodular seal, but did not exceed 1.8 g (Davis and Moncrieff 1938). Information about the content of various components of neonatal milk is not as detailed as data on the composition of milk of lactating women. The morphological composition of the secret in most of the studied newborns on 7–8 days after birth had similar characteristics. In particular, the milk contained various shapes and sizes of fat globules, individual epithelial cells with a diameter of 15–20 mcm, neutrophils, monocytes, lymphocytes, and macrophages. In addition, the remains of destroyed cells and their nuclei in the free state were observed (Khizhnyakova 1965; Pittard et al. 1988). The ion composition, concentration of lactose, and total protein were similar to the corresponding components of maternal milk. A study of the content of immune system proteins-immunoglobulins showed that neonatal milk contains the highest concentration of IgG (80–100 mg/l). IgA is contained in a concentration of 0.43–17.6 mg/l, which is about 1000 times less than in the mother's milk, and IgM was not detected. Concentrations of lactoferrin and lysozyme did not differ from those in mother's milk (Yap et al. 1980, 1981; Pittard et al. 1988). The total fat content in newborn milk was less than in mother's milk, but the concentration of short-chain fatty acids was higher compared to mother's milk (Davis and Moncrieff 1938; McKiernan and Hull 1982).

A survey of different genders and ages of newborns found that neonatal milk can be squeezed out of the glands mainly in the first month of life of children (Khizhnyakova 1965; McKiernan and Hull 1981a). However, there is evidence in the literature that if the mammary glands of newborns are regularly massaged, i.e., the external mechanoreceptors of the breast are stimulated (which, however, is unnatural for the mammary glands of newborns), the secret from the glands is removed for a longer time (Dossett 1960; Bluestein and Wall 1963; Madlon-Kay 1986). This information can be considered the first experimental data indicating a significant role of the nervous system in the external regulation of milk secretion. Here, it should be noted that in comparison with the secretory products of "evolutionarily related" sweat and sebaceous skin glands, which are necessary from the first days of life for the normal functioning of the skin of children, there is no "demand" for the secret of the mammary glands of newborns. Therefore, with a decrease in the blood of newborns prolactin (Fig. 2.6) and the lack of natural stimulation and regular removal of milk from their mammary glands, the secretory activity of the mammary glands gradually fades. There is a "neonatal" involution of the breast. Histological studies show that the apocrine single-layer type of the epithelium in the ducts is gradually replaced by multilayer epithelium. The number of branches of the ducts decreases so that by the age of 2 years, there are small areas of blindly ending bulavoid ducts surrounded by a fibroblast stroma (Naccarato et al. 2000; Howard and Gusterson 2000). In this state, the mammary gland grows isometrically with the growth of the rest of the body until puberty begins.

2.2.3 Breast Development in Puberty

With the approach of puberty, the pathways in the development of mammary glands in boys and girls diverge. The first sign of the beginning of puberty in girls is the acceleration of the growth of their mammary glands; isometric growth is replaced by allometric (i.e., faster growth of other parts of the body). This is due to an increase in the blood content of mainly ovarian steroid hormones, in particular, estrogens (Lee et al. 1976; Thorner et al. 1977), which, as already mentioned, induce the growth of ducts, stroma, and blood vessels in the mammary gland of a woman. Moreover, the beginning of allometric growth of the mammary glands coincides with the beginning of an increase in the concentration of estrogens (Fig. 2.7). In boys, the testes, along

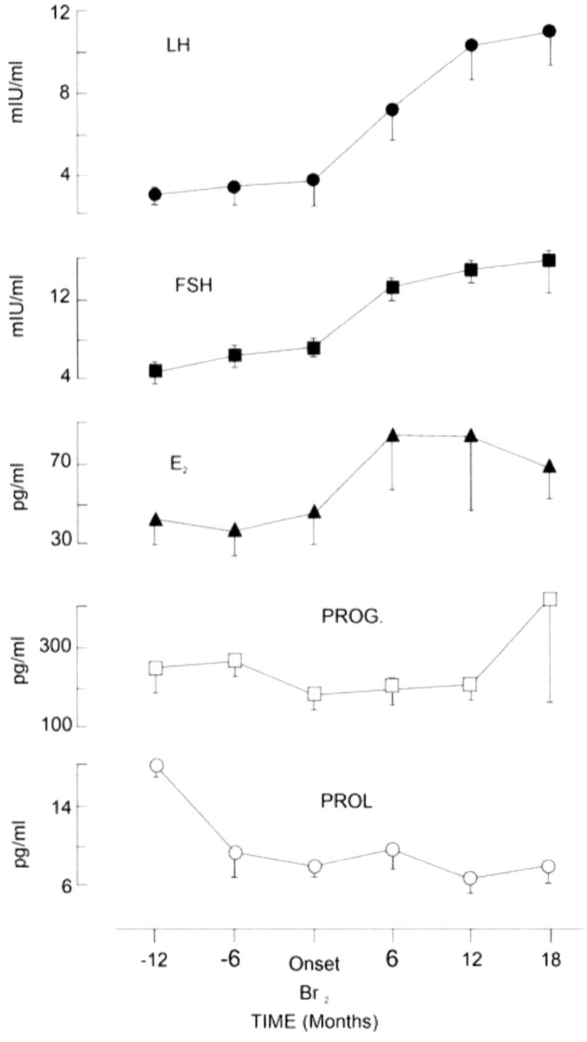

Fig. 2.7 Changes in steroid and gonadotropin hormones in the blood of girls during the prepubertal period. *LH* luteinizing hormone, *FSH* follicle-stimulating hormone, *Es* estrogen, *PR* progesterone, *Prol* prolactin (from Lee et al. 1976). The ordinate, the concentration of the LH hormone in mIU/ml, FSH hormone in mIU/ml, Es hormone in pg/ml, Pr hormone in pg/ml, Prol hormone in pg/ml, abscissa, time in months, onset Br z—onset of puberty

with male sex steroid hormones, also produce estrogens. However, the allometric development of the mammary glands does not occur, because the relationship between blood concentration and functionality of male sex hormones, mostly testosterone and estrogen, is such that the proliferative action of estrogen on breast tissue of boys inhibited testosterone (Zhou et al. 2000; Dimitrakakis et al. 2003; von Schoultz 2007; Johnson and Murad 2009). However, clinical observations show that an imbalance in the action of testosterone and estrogen can lead to breast growth. Moreover, this imbalance can occur at different ages in men. This phenomenon is called gynecomastia. The reasons for the imbalance can be different: from genetic defects in breast tissue to side effects of pharmacological drugs used to treat various diseases (Johnson and Murad 2009).

It is interesting to note that the development of glandular tissue in the prepubescent period in girls resembles the development of glandular structures in newborns in the first month after birth (see Sect. 2.2.2). Thus, as puberty approaches in girls, the number of estrogen receptors increases in the nuclei of cells forming the walls of glandular tissue ducts, stroma cells (fibrous and adipose tissue), and endothelial cells of blood vessels (Li et al. 2010). The site of active proliferation of glandular tissue is the end sections of the milk ducts that have a bulbous shape (bud), where the involution of glandular tissue ended in the first 2 years after the birth of the child (Fig. 2.8). The ducts lengthen, and a gap appears in the distal part, which increases so that, eventually, two buds are formed. In addition, lateral bud-shaped outgrowths are additionally formed on the lateral sections of the ducts. In some cases, terminal buds can divide into three to five daughter alveolar buds, forming lobular structures (Rudland 1991) (Fig. 2.8). Bud-shaped outgrowths in all cases have cavities, and their walls are formed by a heterogeneous layer of cells. The cells lining the cavity are closely adjacent to each other; the outer cells are looser (Rudland 1991). Cytochemical studies have found that cells facing into the cavities of the ducts, lateral and alveolar buds, are more intensely colored when treated with monoclonal antibodies to the membranes of fat globules, while the color of antibodies to actin of smooth muscles is more clearly expressed on cells lying closer to the surface of these structures. In addition, intermediate groups of cells with less pronounced color or no color were observed. These data allowed us to make an assumption (Rudland 1991) that in developing glandular tissue, there are undifferentiated forms of cells (progenitor cells), which, as they develop, turn into epithelial (secretory) or myoepithelial cells. Later, using a large number of histochemical markers of cells that are part of the ducts and alveolar buds of women's breast, this assumption was confirmed and supplemented. In particular, in the epithelial tissue of glandular lobes, stem cells were found, which as a result of differentiation could give progenitor cells, differentiating in turn into epithelial (secretory), or myoepithelial, cells (Clarke 2005; Wagner and Smith 2005; Stingl et al. 2005; Villadsen et al. 2007; Matulka et al. 2007; Cregan et al. 2007; Hassiotou et al. 2012).

In parallel with the development of glandular tissue, there is an increase in the volume of the stroma, as well as the number of blood and lymph vessels, due to which the mammary glands begin to prominently protrude above the surface of the chest. Here, it should be noted that the mammary glands before puberty and immediately after puberty consist of 90% stroma (Russo and Russo 2004a, b). However, in some cases (fortunately, quite rare) when approaching puberty, there is a significant, giant increase (up to 25 cm in diameter and several kilograms in weight) of both breasts or one breast in girls—the phenomenon of juvenile gigantomastia (Dancey et al. 2008). In this case, the growth of the mammary glands occurs intensively for the first 6 months and then changes to a slow but constant increase in their volume for many years. The study of the structure of mammary glands in gigantomastia shows the growth of both glandular tissue ducts and stroma. The growth of ducts is accompanied in most cases by the formation of cystic

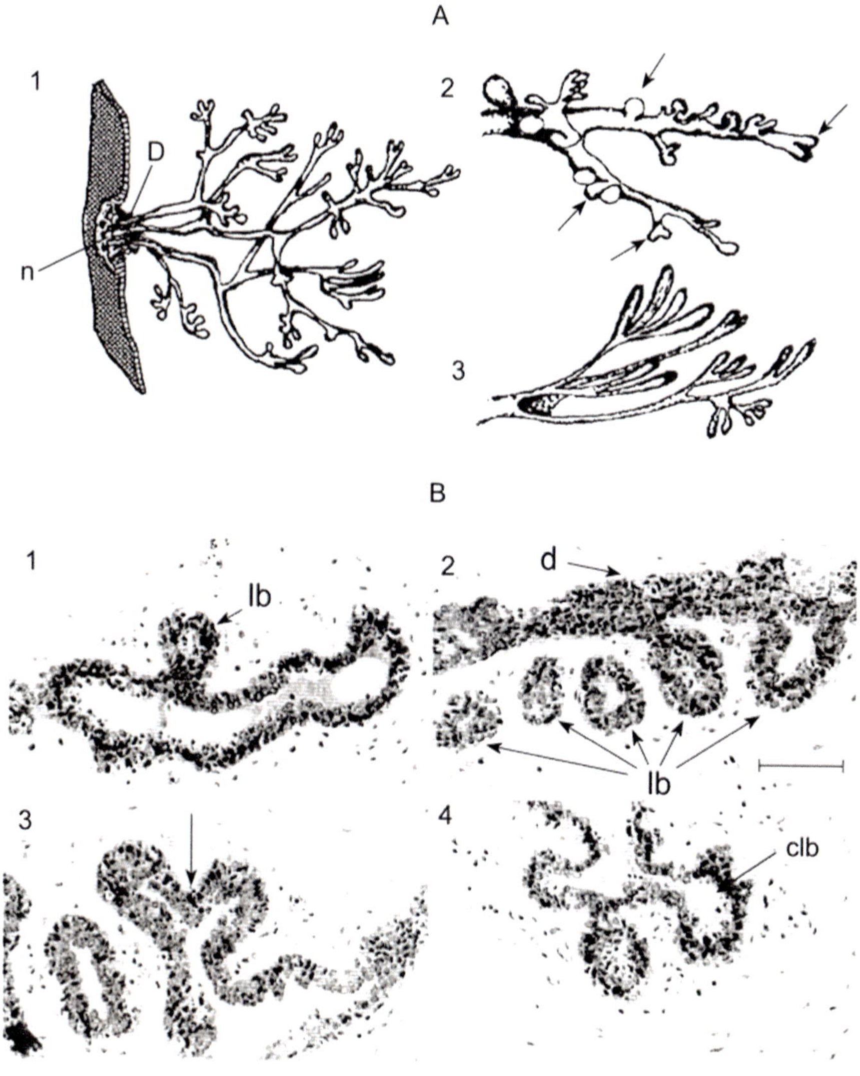

Fig. 2.8 Development of the ductal-alveolar system of the breast in the prepubertal period in girls. (**a**) Schematic drawing of the ductal-alveolar system. 1 General view of the external area and the ductal-alveolar system of the breast; 2, 3 sections of the ductal system. *n* breast nipple, D-milk duct, arrows show the lateral buds (from Russo and Russo 2004a, b). (**b**) Photomicroscopic data on the development of the ductal-alveolar system (from Rudland 1991). 1 A lateral bud (**lb**) formed on the lateral surface of the duct. 2 A series of lateral buds (**lb**) that appeared on the lateral surface of the duct (**d**). 3 Dichotomous division of the bud (shown by the arrow). 4 Formation of a lobular structure. (**clb**) Lateral bud cavity. For all images, the calibration is 75 micron (2)

structures. Edematous areas are observed in the surrounding stroma. According to current data, the causes of juvenile gigantomastia may be different and not fully understood (Dancey et al. 2008). In particular, it is believed that juvenile gigantomastia may occur due to hypersensitivity of breast tissues to steroid hormones, side effects of pharmacological drugs used to treat various diseases, hereditary factors, as well as other unknown reasons. Endocrine therapy proved to be ineffective in eliminating juvenile gigantomastia (Arscott et al. 2001; Dancey et al. 2008). Most often, the removal of sections of the breast is used surgically, which excludes further feeding of the child with this gland in most cases.

In addition to hormones, other regulatory factors are undoubtedly involved in the development of a woman's breast during puberty. However, there is no experimental data on this issue in the literature.

2.2.4 Development and Functioning of the Breast of a Mature Woman

The most intensive (normal) development of the breast begins with the achievement of puberty and the appearance of the menstrual cycle in girls. In the prepubertal period, along with an increase in the concentration of estrogens, there is an increase in the content of gonadotropins in the blood-follicle-stimulating hormone (FSH) and luteinizing hormone (LH) (Fig. 2.7) (Lee et al. 1976). FSH provides growth and differentiation of follicles in the ovary-folliculogenesis. Folliculogenesis is a continuous process until the supply of follicles, which is in the amount of about 400,000, is exhausted. Under the influence of FSH in the right or left ovary, girls begin to develop several follicles. However, after a certain time, with the exception of one follicle, the growth of other follicles stops and their degradation—atresia. The remaining follicle, called the dominant one, secretes increasing amounts of estrogen (17β-estradiol). Increasing the level of estrogen in the blood further stimulates the development of the endometrium in the uterus, and eventually the endometrium becomes hypertrophied. After a certain time, the dominant follicle reaches its maximum size, but ovulation, i.e., the rupture of the follicle wall and the exit of the ovum, does not occur. Then the follicle undergoes atresia, the level of estrogens falls, while the endometrium is detached and the first bleeding or menarche occurs. Anovulatory cycles can be observed in girls for 1–1.5 years after menarche. The lack of ovulation is explained by the fact that the pituitary gland is not able to release LH in response to estrogens until the gonadotropin-releasing factor is maximally activated. By the time of menarche, this activation has not yet reached the threshold that is necessary for the induction of massive release in an increased concentration of LH (Short 1984).

After the onset of menarche in the mammary gland, the number of alveolar buds formed by the terminal duct can increase to 11. Each terminal duct, along with the alveolar buds, is surrounded by a connective tissue stroma. This structure is called the type 1 milk lobule (Russo and Russo 2004a, b).

Numerous studies have shown that the appearance of an ovulatory menstrual cycle is accompanied by noticeable changes in a woman's blood concentrations of steroid and gonadotropin hormones, which significantly affect the development of the breast (Fig. 2.9). It should be noted that in textbooks and reviews, when graphically representing the dynamics of hormone concentrations during the menstrual cycle, only information about the content of estrogen, progesterone, FSH, and LH in a woman's blood is given. However, since the 1980s of the last century, there are quite convincing data on changes in the concentration of prolactin (Guyda and Friesen 1973; Cole et al. 1977; Vekemans et al. 1977; Khamadyanov et al. 1980; Dada et al. 1981; Backstrom et al. 1982; Braund et al. 1984; Tanner et al. 2011, etc.) and oxytocin (Mitchell et al. 1981; Amico et al. 1981; Shukovski et al. 1989; Stock et al. 1991; Salonia et al. 2005, etc.) in the blood of women during the menstrual cycle. Moreover, the level of concentrations of these hormones changes impulsively during the cycle by three to four times, i.e., by the same amount as the concentration of other hormones (Fig. 2.9). The peculiarity of oxytocin was its extremely low concentration in the blood in comparison with other hormones (Fig. 2.9). However, even after childbirth during full lactation, when oxytocin is most in demand, its base concentration and change during milk excretion are of the same order as in non-lactating women. This indicates a very high sensitivity of glandular tissue to oxytocin. It is believed that in women, the selection of the dominant follicle is carried out during the first days after the beginning of menstruation. Just as in the anovulatory cycle, the dominant follicle secretes estrogen (17β-estradiol), which causes its level in the blood to increase. When the dominant follicle reaches maturity, estradiol secretion (Fig. 2.9) is sufficient to induce a positive feedback effect, resulting in a massive release of pituitary LH with a simultaneous increase in FSH and prolactin. The period of time from the beginning of menstruation to the final maturation of the dominant follicle (about 14 days) is called the follicular phase (Fig. 2.9). The next phase is called ovulatory. During the ovulatory phase, which falls in the middle of the menstrual cycle and lasts about 3 days, there is a massive release of LH, which completes the development of the follicle. LH stimulates the production of prostaglandins and proteolytic enzymes necessary for breaking the follicle wall and releasing the mature ovum (ovulation proper). By the middle of the cycle, the concentration of oxytocin increases (Fig. 2.9). Ovulation usually occurs in the next 24 h after the largest wave of LH release (Fig. 2.9). The time between ovulation and the beginning of menstrual bleeding is called the luteal phase of the cycle, which lasts 13–14 days. After the rupture of the follicle, its walls fall, the cells of the follicle accumulate lipids, and lutein pigment gives it a yellowish color. The transformed follicle is now called the yellow body. The yellow body begins to secrete progesterone and estrogen. Accordingly, their concentration in the blood increases. At the same time, the concentration of prolactin increases. The level of these hormones reaches the highest value in the middle of the luteal phase, while the concentration of FSH and LH decreases (Fig. 2.9). By the middle of the luteal phase, oxytocin concentrations usually decrease (Amico et al. 1981; Shukovski et al. 1989). However, there is evidence that in some women with a normal cycle (Stock et al. 1991), the content of oxytocin in the blood may conversely increase by the middle of the

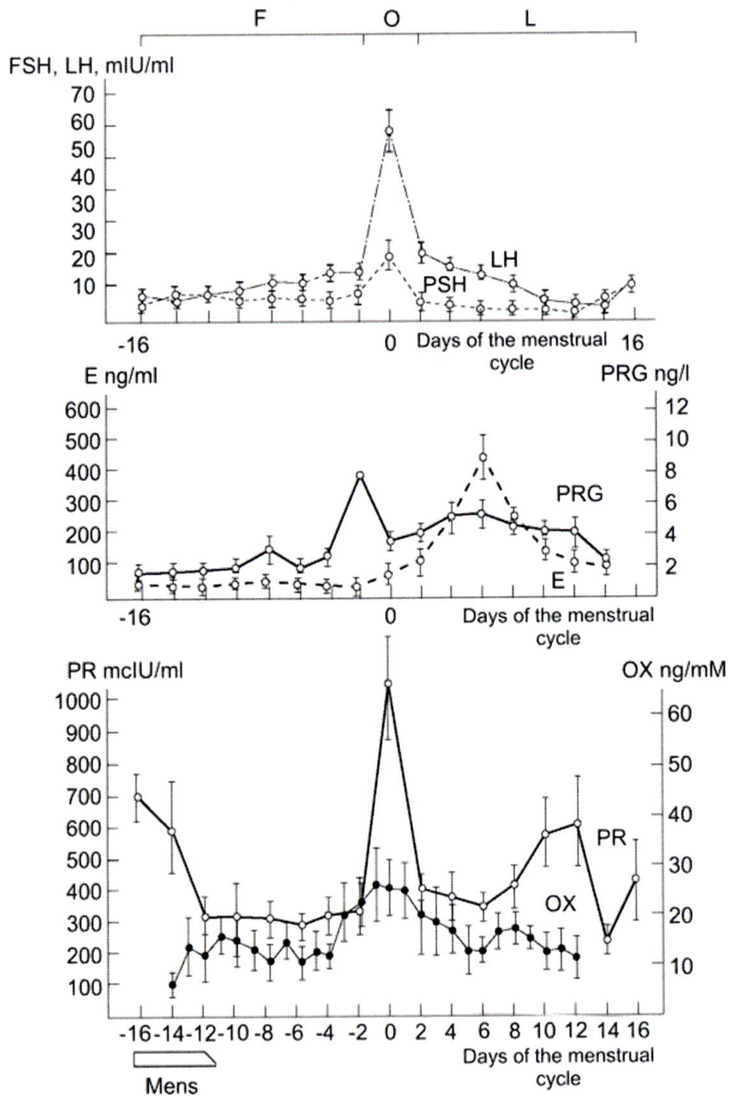

Fig. 2.9 Changes in the concentration of follicle-stimulating hormone (FSH) and luteinizing hormone (LH), upper graph; estradiol (E) and progesterone (PRG), middle graph; prolactin (PR) (from Khamadyanov et al. 1980) and oxytocin (OX) lower graph (from Shukovski et al. 1989) in the blood serum of women during the menstrual cycle. *F* follicular phase, *O* ovulatory phase, *L* luteal phase. Ordinate axis, the concentration of hormones. Abscissa axis, zero mark (0)—the middle of the menstrual cycle. On the lower graph, "menses" is the time of menstruation

luteal phase. A feature of the luteal phase was an increase in the concentration of oxytocin in the blood during mechanical stimulation of the nipple and areola of the breast. Moreover, the average peak level of oxytocin increase in the blood in the luteal phase (Leake et al. 1984b; Amico and Finley 1986) and in the postpartum period (Leake et al. 1983; Weitzman et al. 1980) had the same value.

An increased concentration of estrogen and progesterone changes the properties of the two outer layers of the endometrium. Glandular cells of the endometrium proliferate and begin to secrete, and the uterus prepares for implantation of a fertilized ovum. If pregnancy does not occur, the yellow body stops functioning, and the level of estrogens and progesterone decreases, which leads to edema and necrotic changes in the endometrium. The uterine arteries narrow, limiting the nutrition of the overgrown endometrium. There is a rejection of the outer layers of the endometrium, which is accompanied by bleeding. Thus, with the onset of ovulatory cycles, the reproductive system of a woman every month comes to a state of readiness for reproduction of offspring until about 51 years old, i.e., until the time when about 99% of the follicles undergo atresia.

Since the mammary gland is an integral part of the reproductive system, its development and functioning must also change during the cycle under the influence of steroid hormones, prolactin and oxytocin, which affect the tissues of the gland through the corresponding cellular receptors found in the cells of the alveolar buds, ducts and stroma. To date, two types of estrogen receptors are known, Erα and Erß, and progesterone, PRA and PRB, one type for oxytocin, and seven types of prolactin receptors. Receptors for steroid hormones are localized in the cell nuclei (Li et al. 2010). Receptors for prolactin and oxytocin are located in the outer membrane of cells and consist of three sites—domains: extracellular domain, intramembrane domain, and intracellular domain (Freeman et al. 2000; Gill et al. 2001; Gimpl et al. 2008). The dynamics of the concentration of steroid hormones, prolactin and oxytocin (Fig. 2.9), indicates that structural and functional changes are more likely to occur in the luteal phase of the cycle.

Detailed morphological and immunohistochemical studies of the internal structure of the breast at the light level (Vogel et al. 1981; Going et al. 1988; Ramakrishnan et al. 2002; Navarrete et al. 2005; Welsch et al. 2007) showed that at the beginning of the cycle in the follicular phase, the cells forming the ducts and alveolar buds are located in two to three layers and have a polygonal structure. At the same time, it is difficult to distinguish the difference between cells facing inside the cavity (luminal cells) and those located in the basal region (myoepithelial cells). In alveolar buds, which have a small luminal space, sometimes there were products of secretion; in particular, peptides with antimicrobial properties were found in the alveolar cavity: beta defensins, cathelicidin LL37, lactoferrin, and adrenomedullin. Therefore, it can be assumed that the presence of antimicrobial peptides in the breast's capacitive system will effectively block the entry of infection into the ducts of the non-lactating breast. In addition to antimicrobial peptides, fat droplets and lactalbumin molecules were observed (Welsch et al. 2007). However, there was no active secretion. The lobes are clearly bounded from the slightly edematous intralobular stroma by the cellular layer of fibroblasts. As we approach the ovulatory phase, the difference between luminal and myoepithelial cells begins to show. The luminal cells become more elongated and are arranged radially. Their cytoplasm is light and the nuclei are located in the basal part of the cells; on the apical part, there are rare microvilli. Myoepithelial cells are often vacuolated. In the alveolar buds and ducts appear cavities containing a small amount of secretion, with signs of active secretion of luminal cells. The lobules are clearly distinguished from the interlobular stroma, in which there is no edema. With the beginning of the luteal phase, cell differentiation increases. Lobules increase in volume due to the appearance of additional primary ducts with alveolar buds. The cavities in the alveolar buds and ducts expand. There are cells in which mitosis and apoptosis are observed. By the end of the luteal phase (day 25–27), the number of microvilli increases in the areas of the apical parts of the luminal cells facing the cavity of the alveolar buds or ducts. Inside the cells, the volume of the rough endoplasmic reticulum and the number of secretory vacuoles and glycogen granules increase. In the alveolar buds, there is an intense apocrine secretion. As a result, the cavities in the alveolar buds and ducts noticeably expand, and their walls stretch accordingly. The number of cells with mitotic activity increases. The number of alveolar buds formed by the terminal duct increases, and in addition to the lobes containing 11 alveolar buds, there are lobules with an average of 47 and 81 alveolar buds. These lobules are called type 2 and 3 milk lobes (Russo and Russo 2004a, b) (Fig. 2.13a). Edema and infiltration increase in the intralobular stroma.

The results of determining the proliferative activity of breast cells in various phases of the menstrual cycle by autoradiographic method using a thymidine label (3H-thymidine) (Going et al. 1988; Potten et al. 1988), as well as the proliferative index, defined as the number Ki-67 (marker of cell proliferation) of positively colored nuclei in 1000 cells (Navarrete et al. 2005), additionally indicate the highest differentiation of cells in the luteal phase of the cycle. Since the differentiation and proliferation of breast cells is largely associated with a hormonal factor, data on the localization of cell receptors for hormones in breast tissues, as well as the dynamics of their density distribution depending on the phase of the menstrual cycle, are of great interest. To date, there are several immunocytochemical studies in which the localization of cellular receptors to steroid hormones in normal mammary glands has been studied in sufficient detail. Samples for research were taken in connection with reduction mammoplasty or in areas of the mammary glands located next to fibroadenomas (Speirs et al. 2002; Li et al. 2010). However, as the authors themselves note, the menstrual status of women whose breast tissue samples were analyzed was not known. According to the data obtained (Li et al. 2010), the color on Erα receptors is observed in 10% in the nuclei of epithelial cells of the inner layer of alveolar buds and the outer layer of interlobular ducts. Color on Erß receptors is observed in 70–85% in the nuclei of epithelial cells of both layers, in alveolar buds and ducts, as well as stromal cells (Speirs et al. 2002; Li et al. 2010). Color on PRA is found in the nuclei of epithelial cells of the inner layer of alveolar buds and the outer layer of cells of interlobular ducts and is also noted in the cytoplasm of cells. The localization of PRB is similar to PRA, but it is concentrated only in the nuclei of epithelial cells. Progesterone receptor staining was not observed in stroma cells (Li et al. 2010). Studies of the dynamics of steroid hormone receptors in different phases of the menstrual cycle are less detailed and were conducted only in the follicular and luteal phases (Soderqvist 1998; Shaw et al. 2002). It turned out that the change in the number of colored nuclei of breast tissue cells is most pronounced only for Erα. In particular, the number of colored nuclei on Erα receptors is significantly higher in lobule and duct cells in the early follicular phase. With an increase in the concentration of estradiol in the blood, the level of colored nuclei on the Erα decreases and reaches the lowest value in the luteal phase. For cell nuclei stained with Erß receptors and progesterone receptors, there was no noticeable difference in their distribution in breast tissue between the follicular and luteal phases. It should be noted here that there is sufficient detailed data on the distribution and change in the density of steroid hormone receptors in various phases of the menstrual cycle, obtained in monkeys (macaques) (Cheng et al. 2005). The structure of the mammary gland (Macpherson and Montagna 1974), the timing and course of the menstrual cycle, as well as the dynamics and values of steroid hormone concentrations during the menstrual cycle (Bosu et al. 1973; Atkinsson et al. 1975; Cline 2007) in these primates largely coincide with the corresponding indicators in women. During the menstrual cycle in monkeys, the level of expression of Erα in the nuclei of epithelial cells of the lobes and ducts as well as in the mammary glands of women is the highest in the initial period of the follicular phase and is 12.5%. Then it decreases in the late follicular phase to 3.3% and the luteal phase to 1.2%. In the nuclei of epithelial cells lining the ducts, the values were 7.5%, 1.4%, and 0.4%, respectively. The change in the expression of Erß occurs in the opposite direction relative to Erα. In the early follicular phase, the expression of Erß is observed in a small number of epithelial cells. However, in the late follicular phase, expression increases to 62%, and in the luteal phase to 82%. In addition to epithelial cells in the late follicular phase and luteal phase, Erß is expressed in the nuclei of myoepithelial cells and fibroblasts. PRA and PRB are present mainly in the nuclei of epithelial cells of lobes and ducts. A small number of stromal cell nuclei express PRA or PRB. PRA and PRB are expressed in the same way at different stages of the menstrual cycle. Thus, the expression on PRA and PRB increases from 4% in the early follicular phase to 13% in the late follicular phase and reaches 17% in the luteal phase of the menstrual cycle. Thus, in contrast to Erα, there is a noticeable difference between the dynamics of expression of Erß, PRA, and PRB in the mammary tissues of monkeys and women. It is not clear whether this discrepancy is due to methodological reasons or whether the difference is related to species characteristics. In this regard, it is necessary to note the extremely important data obtained by X. Russo and colleagues (1999) who studied the distribution of Erα and progesterone receptors in different types of breast lobes. It turned out that the content of Erα and PR is directly proportional to the level of proliferation of cell structures included in the lobule (Fig. 2.13). Thus, the number of receptors is maximum in the cells of type 1 lobules that have the highest proliferative activity, which progressively decreases when type 1 lobules differentiate into type 2 and type 3 lobules (Fig. 2.13a) with a simultaneous decrease in the number of receptors. Since the determination of receptors in cells that form milk lobules in the mammary glands of women to steroid hormones was determined without relation to the types of lobules, this could be the reason for the discrepancy in the results obtained by different authors.

Data on the localization of prolactin receptors in the normally functioning mammary gland of non-pregnant women are fragmentary (Gill et al. 2001; Ueda et al. 2011). Color on prolactin receptors (PrlR) was observed in areas of epithelial cell membranes that exit into the cavity of alveolar buds and ducts. The menstrual status of women from whom glandular tissue samples were taken was unknown.

Information about oxytocin receptors in the normally functioning mammary gland of non-pregnant women is also scarce (Kimura et al. 1998; Bussolati et al. 1996; Reversi et al. 2005). Oxytocin receptor (OR) staining was found in restricted areas of myoepithelial cells of the alveoli and ducts, as well as epithelial cells. OR density was very low. There was no quantitative assessment of OR. The menstrual status of women from whom glandular tissue samples were taken was unknown.

Since the light-microscopic method has certain limitations, information about changes in the ultrastructure of breast tissue in women during the menstrual cycle is of significant interest. There are quite a large number of works in the literature, which appeared mainly in the 1970s–1980s of the last century, devoted to electron microscopic examination of the breast of women. However, the main part of the experimental material in these surveys was taken from areas of the breast with any pathology or located near them (Stirling and Chandler 1976a). And most importantly, in ultrastructural studies, there was no such clear "binding" of the material taken to the time of the menstrual cycle as was done in some works on the light-microscopic level. To date, information about breast ultrastructure at various stages of the menstrual cycle is available only in one work (Fanger and Ree 1974). Material for examination was taken from sections cut from biopsy samples located in glandular tissue next to fibroadenomas or cysts. The main attention in this work was focused on the ultrastructure of epithelial cells that form the walls of the ducts and alveolar buds. It turned out that in contrast to the light-microscopic data in the early follicular phase (2–4 days), the cells forming the walls of the ducts remain completely differentiated. Part of the cells' apical surfaces are largely narrowed duct cavities. The other part, smaller, is adjacent to the basal or lateral parts of these cells. The nuclei of cells communicating with the cavity are ovoid or pear-shaped. The cytoplasm contains a small amount of ribosomes. The rough endoplasmic reticulum contains narrow cisterns that are not filled or contain a small amount of granular material. The Golgi apparatus consists of insufficiently developed empty tubular plates. In the cytoplasm, rare mitochondria, secretory vesicles, glycogen granules, and fat droplets are observed. Thus, the state of intracellular structures indicates that the functional activity of cells is reduced. Epithelial cells are quite close to each other. Microvilli are practically absent in the areas of cells that exit into the cavity. According to the data presented in Fig. 2.9, the concentration of progesterone and estrogen during this period is at the basal level. The content of prolactin is increased, but by the fourth day of the follicular phase, it decreases to the basal concentration.

Closer to the ovulatory phase (10–15 days), epithelial cells become more elongated, and their volume increases. Microvilli appear on the apical areas that exit into the cavity. However, the number and development of intracellular structures are similar to the cells of the early follicular phase. At this time, the beginning of an increase in the concentration of estrogen, prolactin, and oxytocin in the blood of women is noted (Fig. 2.9).

At the beginning of the luteal phase (day 17–19), there is an expansion of cavities in the ducts, as well as an increase in the number and size of microvilli on the apical sections of epithelial cells. However, with the exception of an increase in the content of glycogen granules, there are no noticeable differences in the number and development of intracellular structures compared to cells of the early follicular phase.

As the end of the luteal phase approaches (day 20–25), when the concentration of estrogen and progesterone in the blood of women begins to decrease to the basal level, and the level of prolactin increases (Fig. 2.9), the epithelial cells of the breast reach their highest development. In particular, the size of cell nuclei increases, and their surface becomes uneven. The number of ribosomes increases, as well as the number of polysomes. The cisterns of the rough reticulum are expanded and filled with electron-dense material. In many cells, the Golgi complex grows and consists of well-developed tubular plates and contains aggregates of vesicles. The number of microvilli on the apical part increases significantly, and they reach their maximum size. However, the images do not show examples of apocrine secretion, which is referred to in light-microscopic studies. This may be due to the fact that the secretion reaches its maximum in the last days of the luteal phase (day 25–27). However, the authors (Fanger and Ree 1974) did not have ultrastructural data for this period of the cycle.

More detailed ultrastructural studies of various types of breast cells were performed on normal mammary glands only during the luteal phase. The walls of the alveolar buds, thin terminal ducts, interlobular ducts, and the widest ducts that fit to the nipple have the same structure during this period and are formed by epithelial and myoepithelial cells (Ozzello 1974; Stirling and Chandler 1976a, 1977; Watson et al. 1988; Tsuchiya and Li 2005). Three groups were identified among epithelial cells. The first group included cells whose shape varies from cuboid to cylindrical. Long axes of these cells are located radially to the duct cavity. The apical surfaces of these cells contain numerous microvilli of various shapes and lengths and border on the duct cavity. The density of microvilli tends to increase in cells of the terminal ducts and alveolar buds. Epithelial cells of the first group, depending on their electronic density, are divided into dark and light. It is believed that the difference in electron density is associated with a different functional state. Their intracellular structures in development and structure differ little from the intracellular structures of epithelial cells in the luteal phase, discovered by previous authors (Fanger and Ree 1974). Epithelial cells of the first group are connected to each other by a system of dense contacts that provide isolation of the duct cavities from the surrounding intra- and interlobular environment. The second and third groups include cells with light cytoplasm, which are localized between the basal areas of the first group of cells and myoepithelial cells. Cells of the second type are located more loosely and do not form dense contacts with neighboring cells. The second group of cells has an internal structure similar to that of dark epithelial cells. It is assumed that the second type of cells is progenitors (precursors) for epithelial cells of the first group and myoepithelial cells. The third type of epithelial cells was identified as lymphocytes and macrophages (Ferguson 1985). In the walls of the alveolar buds and ducts of the non-lactating ("resting") mammary gland, the percentage of lymphocytes and macrophages relative to the entire cell population varies from 5% to 15%, and most of these cells are lymphocytes (Ferguson 1985).

Myoepithelial cells form a mesh structure, entwining the alveolar buds and ducts with their cytoplasmic processes, between which desmosomes are observed. On the walls of the ducts, the processes of myoepithelial cells are located mainly along the longitudinal axis of the duct. Myoepithelial cells are characterized by the presence of a multi-lobed nucleus. The cytoplasm of myoepithelial cells in the luteal phase contains the same organelles as epithelial cells. A characteristic feature of the internal structure of myoepithelial cells is the presence of a large number of filaments, which are similar in structure to the myofilaments of smooth muscle cells. A feature of myoepithelial cells was the presence of special cilia (Stirling and Chandler 1976b), similar in structure to the kinocilia of hair cells of the auditory or vestibular system in some animals (Barber 1974). Inside the cilia, nine pairs of fibrils were located along the periphery; the central pair of fibrils was absent. The cilia penetrated the cell and were associated with the basal body. Cilia were localized on the surface of the myoepithelial cell facing the basal sides of the epithelial cells. The diameter of the cilia was about 250 nm, and their maximum length was 830 nm. Cilia were observed in myoepithelial cells located in all parts of the ductal system of the breast. It should be noted that cilia were found on myoepithelial cells and in other mammals (McDermott et al. 2010). The most detailed study of their function was conducted in the development of mammary glands in mice. It turned out that during the morphogenesis of the ductal system of the mammary glands, cilia are found not only in myoepithelial cells but also in epithelial and stromal cells. Defects in intraciliary transport in mutant mice were manifested in impaired branching of mammary ducts during morphogenesis (McDermott et al. 2010). In normal mice, after the completion of morphogenesis of the ducts, the cilia remain only in myoepithelial cells. It is believed that cilia in the process of duct morphogenesis perform a coordinating function between the cells of the milk ducts (McDermott et al. 2010). In the mature mammary gland, they apparently perform a mechanoreceptor function and serve to coordinate the mechanical displacements of the walls of the milk ducts in the process of milk excretion (Stirling and Chandler 1976b). Myoepithelial cells had a noticeable difference in shape depending on their location (Stirling and Chandler 1977; Tsuchiya and Li 2005). Thus, myoepithelial cells located on thick ducts located under the areolar region often formed quite long processes immersed in the stroma. While on the alveolar buds and thin ducts, the contact of myoepithelial cells with the stroma had a flat shape.

The cell layer that forms the walls of the ducts and alveolar buds is completely covered by the basal membrane, or rather the basal shell. The basal shell consists of two layers, called light (lamina lucida) with a thickness of 15–30 nm and dark (lamina densa) with a thickness of 40–80 nm (Tsuchiya and Li 2005). Myoepithelial cells in contact with basal membrane form hemidesmosomes. Areas of epithelial cells are also close to the basal shell, but the formation of hemidesmosomes is not noted. A layer of fibroblasts that separates the intralobular ducts and alveolar buds from the intralobular stroma adjoins the basal membrane through a gap filled with collagen fibers. Capillaries approach the fibroblast layer from the stroma side. The intralobular stroma contains a large number of collagen fibrils. In addition to fibroblasts and capillaries, the intralobular stroma contains macrophages, lymphocytes, mast cells (Ferguson 1985), and relatively recently identified Cajal cells or more precisely Cajal-like cells (ICLC) (Gherghiceanu and Popescu 2005). It should be noted here that earlier studies (Eyden et al. 1986) considered ICLC as a type of fibroblast. However, detailed studies at the ultrastructural level have found significant differences between these cell types. So, ICLC have two to three fairly long, several tens of microns of cytoplasmic processes. Their thickness is 0.1–0.5 microns and they have extensions along the way. Most of the cell is occupied by the nucleus and the endoplasmic reticulum, is mostly smooth, has a small volume, and reaches about 7% of the cell volume. The volume of mitochondria is about 10% of the cell volume. ICLCs are localized throughout the volume of the intralobular stroma and send their processes to various stroma cells, forming contacts with them, as well as surrounding capillaries and milk ducts. Fibroblasts have numerous short cytoplasmic processes. Inside the cell, the mitochondria occupy about 6% of the volume, but the volume of the endoplasmic reticulum (rough) is almost three times larger (17%) than in the ICLC. Fibroblasts, in contrast to ICLC, are located mainly in the border area between the milk ducts and milk buds and the rest of the intralobular stroma.

The interlobular stroma also contains a large number of collagen fibers, but in comparison with the intralobular stroma, it is “poor” in cellular structures. It contains blood and lymphatic vessels and groups of fat cells. According to light-microscopic data, the stroma undergoes noticeable changes in volume and density during the cycle. Studies of the ultrastructure of the intra- and interlobular stroma in different phases of the cycle (Stoeckelhuber et al. 2002) have shown that the collagen fibrils of the intralobular stroma lie loosely and pass in different directions. The thickness of the fibrils is approximately 45 nm. The collagen fibrils of the interlobular stroma are thicker (52 nm in diameter) and are arranged more orderly, parallel to each other. The collagen fibers are bonded by transverse protein strands 4–8 nm in diameter, which have been identified as dermatan sulfate proteoglycans. The location of attachment of dermatan sulfate proteoglycans proteins are localized in the **d** sections of the collagen fibers. The distance between the attachment sites of dermatan sulfate proteoglycan proteins in the collagen fibers of the intralobular and interlobular stroma is equal and constant in the follicular phase of the cycle and is 46 nm. In the luteal phase, this distance increases by 9 nm. Moreover, the increase in distance is more typical for a loose intralobular stroma. It is believed that the change in distance is due to an increase in the water content in the intercollagen space under the action of steroid hormones (Stoeckelhuber et al. 2002). It should be noted here that prolactin, in addition to its lactogenic action, also participates in osmoregulation in the mammary gland (Shennan 1994). Since the content of prolactin increases in the second half of the luteal phase of the cycle, it can be assumed that this hormone can also affect the water content in the stroma. It is assumed that an increase in the water content in the intralobular stroma is essential for the development and functioning of the gland in the luteal phase. Thus, according to the ultrastructural data given above, intracellular secretory processes are intensified at the end of the luteal phase. In this regard, an additional supply of various organic and mineral substances, as well as oxygen, from the bloodstream to the cells is necessary. Since this process is carried out by diffusion through the intercellular medium, an increase in the fluid around the glandular tissue will facilitate the transport of necessary substances to the cells of the acinuses and ducts (Stoeckelhuber et al. 2002).

In addition to hormones, other factors affect the development of women's breast in the period before menopause. In particular, receptors for epidermal growth factor are maximally expressed in the luteal phase of the cycle mainly on stromal and myoepithelial cells, Stat5 (signal transductor and transcription activator). It regulates the expression of cell receptors for prolactin and can prevent apoptosis during epithelial cell differentiation (Nevalainen et al. 2002).

In addition to increasing proliferative processes in the breast tissues of women, vascularization and blood flow rate increase toward the end of the luteal phase (Madjar et al. 1992; Weinstein et al. 2005).

Thus, the growth of glandular and stromal breast tissue, the filling of secret cavities of the alveolar buds and ducts, the appearance of edema in the interlobular stroma, and an increase in vascularization and blood flow intensity cause an increase in the volume and density of the breast before menstruation (Milligan et al. 1975; Fowler et al. 1990). By analogy with "neonatal engorgement," this condition of the gland before menstruation can be called "premenstrual engorgement," which is a normal physiological process, and the degree of "premenstrual engorgement" varies among women. Clinical examinations show that in some cases, there is a slight increase in the density of the glands. In other cases, some women experience very noticeable discomfort at this time. A woman feels heaviness in the gland, when palpating the glands, the seals are felt, and often there are painful sensations. In the clinic, this phenomenon is called cyclic mastalgia (Smith et al. 2004). It should be noted that feelings of pain can only occur when a woman's cortex receives nervous impulses from afferent receptors located in the inner part of the breast. Therefore, the cause of sensations is mainly stimulation of mechanoreceptors and pain receptors located in close proximity to the walls of the milk ducts (Eriksson et al. 1996). Moreover, the intensity of sensations will depend on the degree of filling the ducts with secret and, accordingly, stretching the walls of the milk ducts. With a different volume of secretions formed in the luteal phase of the cycle in women who secrete more secretions, the stretching of the duct walls and, accordingly, the stimulation of the mechanoreceptors will be more intense, and with a sufficiently strong stretch, pain receptors will be activated. In the case of low intensity of secretion, the increase in the volume of the ducts will be insignificant and the sensations will be hardly noticeable or absent at all, respectively. These assumptions are consistent with the results of studies on the dependence of pain in the breast on the diameter of the milk ducts in women in the second half of the menstrual cycle (Peters et al. 2003). In particular, it turned out that pain in women with cyclic mastalgia is directly correlated with the diameter of the milk ducts. Undoubtedly, a certain contribution to the formation of the feeling of "fullness" of the gland in the luteal phase will be made by stimulation of the mechanoreceptors of blood and lymphatic vessels. Here it should be added that there are experimental data (Robinson et al. 1999) indicating a change in sensitivity to tactile and pain stimuli of the breast skin in the luteal phase as a result of an increase in the concentration of progesterone and estrogen. However, the mechanisms of this phenomenon are still not known.

If the child is not conceived, the yellow body ceases to function and undergoes atresia, and the level of estrogens and progesterone decreases. As a result, there is a regression of glandular and stromal breast tissue. Vascularization and blood flow intensity decrease, and puffiness disappears. Stroma is compacted, and the volume of the breast decreases. Then, in the next cycle, the development of the breast is repeated. It is important to note that according to observations after the end of the cycle, the breast tissue of a woman never returns to its original state, but remains somewhat enlarged (Russo and Russo 2004a, b). Moreover, for each cycle, the number of alveolar buds increases, which leads to an increase in the size of the lobes or the formation of new ones. This process lasts in the mammary glands of women until about 35 years (Russo and Russo 2004a, b). As a result of cyclical development of the breast in unborn women, lobules of type 1, 2, and 3 are noted in the glandular tissue (Russo and Russo 2004a, b). Morphometric studies have found that with an increase in the number of alveolar buds in the lobules, the number of cells forming the bud walls decreases. So, on the basal alveolar buds in the lobules of type 1, there are about 34 cells and 2 and 3 types of slices, respectively, 13 and 11 (Russo and Russo 2004a, b). Research of glandular tissue in women with ovulatory cycles shows a wide variety in the structure and content of lobule structures. In particular, in the glandular tissue, separate undifferentiated terminal buds and alveolar buds are found, which do not undergo further development if pregnancy does not occur, milk lobules of type 1, 2, and 3. This position is well illustrated by the diagram in Fig. 2.10 (Russo and Russo 2004a, b).

With the establishment of regular cycles in women, additional mammary glands begin to appear (Fig. 2.3). Just like the main mammary glands, the additional glands increase in volume during the luteal phase of the cycle. The subcutaneous areas where the glands are located become dense. When you click on them, in some cases, there are painful sensations

2.2.5 Development and Functioning of the Breast During Pregnancy

If a woman is pregnant, the yellow body does not undergo atresia, but continues to function, secreting the hormones progesterone and estrogen, which are necessary for maintaining pregnancy, as well as the development of the mother's breast. "Helps" him in this trophoblast (the outer layer of cells of the embryo), whose cells secrete chorionic gonadotropin, which stimulates the formation of progesterone and estrogens and prolongs the functioning of the corpus luteum until the time (about 10 weeks of pregnancy) when its functions take over the placenta. The concentration

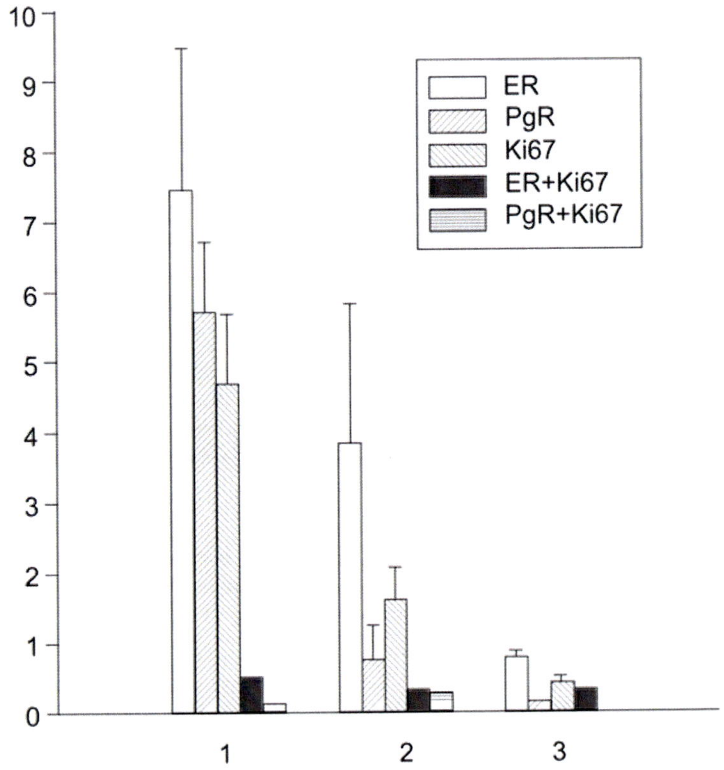

Fig. 2.10 Percentage of alveolar bud cells positive for color for estrogen receptors (ER), progesterone receptors (PR), and Ki67 and simultaneously positive for ER and Ki67 (ER + Ki67) and PR and Ki67 (PR + Ki67) in fractions of different types. Ordinate axis, the percentage of cells positive for color. Abscissa axis 1, 2, 3 type of lobes (from Russo et al. 1999)

of chorionic gonadotropin increases in the blood of women, reaching the highest value (5000–6000 ng/ml) by 8–10 weeks. Then it decreases in the next 20 weeks by about five times and remains at this level until delivery (Fig. 2.11b) (Lin et al. 1995; Neville 1990; Johnson 2010). The concentration of progesterone and estrogen increases throughout pregnancy and by the end of pregnancy exceeds their level in the luteal phase, respectively, by more than 10 and 30 times (Fig. 2.11a). In addition, the anterior pituitary of a pregnant woman increases the production of prolactin, which is synthesized and removed from special cells—lactotrophs. Ultrastructural immunohistochemical studies (Lloyd et al. 1988) have shown that lactotrophs make up approximately 32–55% of the cellular composition of the woman adenohypophysis. During pregnancy, the volume of the adenohypophysis is believed to almost double due to an increase in the number and size of lactotrophs. This is probably one of the reasons that by birth the concentration of prolactin is 15–20 times higher than the level of the hormone in the luteal phase of the cycle (Fig. 2.12a). In addition, an increase in prolactin content may occur due to the removal of inhibition of the

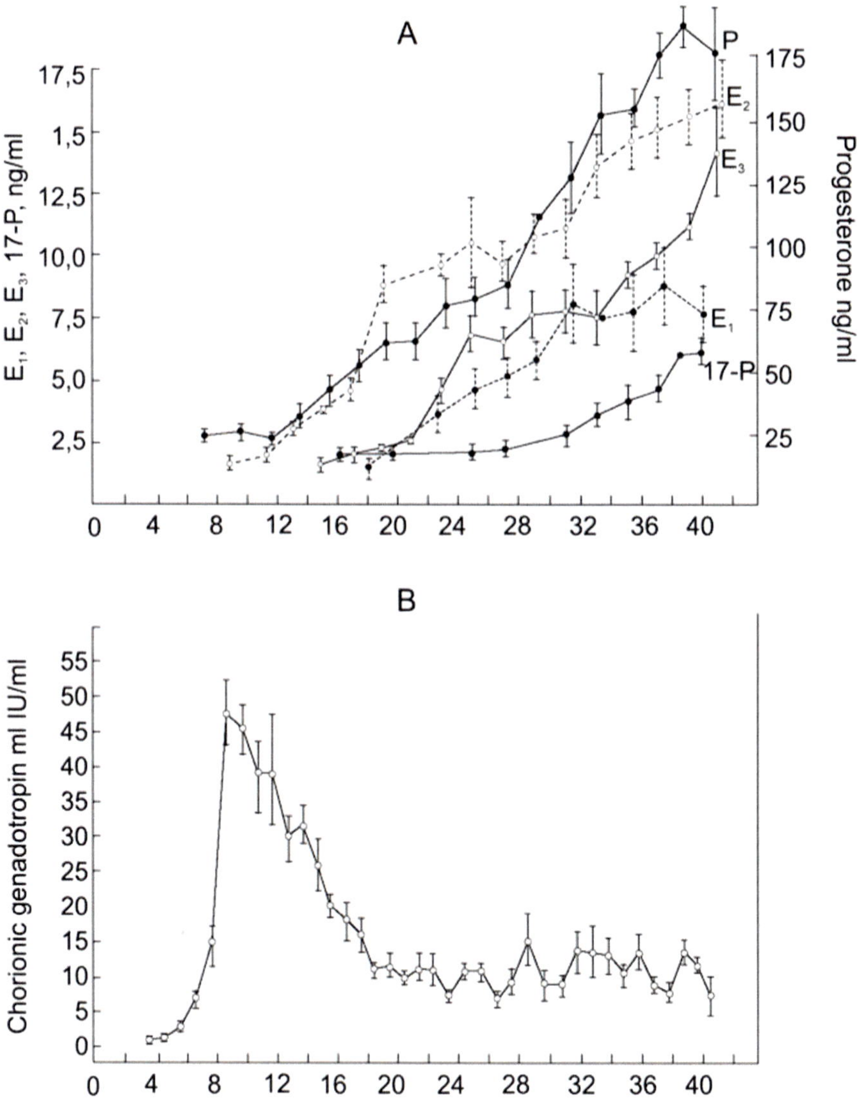

Fig. 2.11 Changes in the concentration of progesterone and estrogens (from Tulchinsky et al. 1972) (upper graph) and chorionic gonadotropin (from Braunstein et al. 1976) (lower graph) in the blood of pregnant women. For the top graph: E1-estrone, E2-estradiol, E3-estriol, 17-P-hydroxyprogesterone, P-progesterone. The ordinate in all graphs, the concentration of hormones; the abscissa, the weeks of pregnancy

inhibiting factor of prolactin secretion-dopamine (Romanò et al. 2013). This will be discussed in more detail in Chap. 4. In addition to chorionic gonadotropin, trophoblast cells begin to secrete a hormone called placental lactogen by the fifth week of

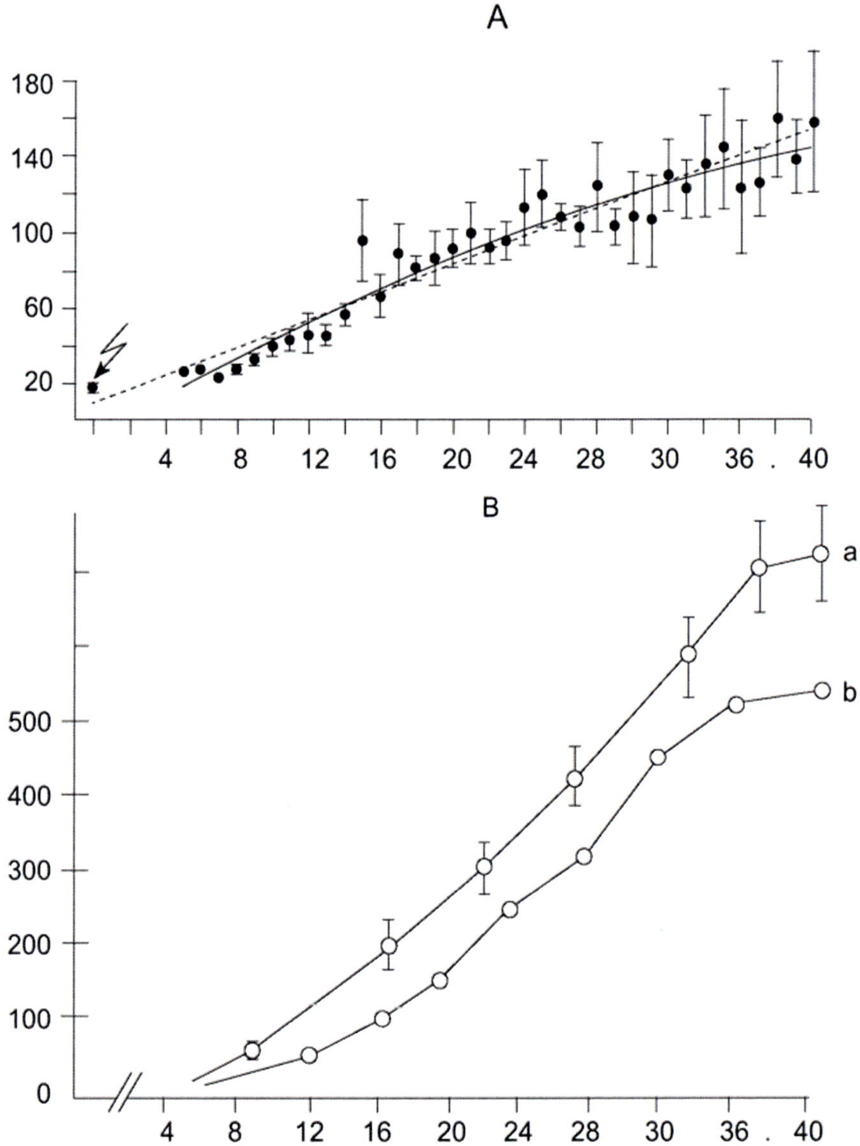

Fig. 2.12 Changes in the concentration of prolactin (from Riggs et al. 1977) (upper graph), placental lactogen, and placental weight (from Selenkow et al. 1969) (lower graph) in the blood of pregnant women. The sinuous arrow on the top graph shows the level of prolactin in non-pregnant women. *PrL* prolactin, *VP* the weight of the placenta. Ordinate axis, concentration of hormones; abscissa axis, the weeks of pregnancy

pregnancy, which directly affects the development and formation of the lactation function of a woman's breast. In its structure, placental lactogen has a similar structure to prolactin. The concentration of placental lactogen correlates with the mass of the placenta and fetus and grows and reaches a plateau at 36 weeks of pregnancy, and its content in the blood is ten times higher than the level of prolactin (Fig. 2.12b). During pregnancy, there is also a change in the concentration of oxytocin in the blood. However, data on changes in the content of oxytocin in the blood of pregnant women are not as unambiguous as data on the level of steroid hormones and prolactin. Most authors (Dawood et al. 1978; Otsuki et al. 1983; De Geest et al. 1985; Stock et al. 1991; Levine et al. 2007) note an increase in the concentration of oxytocin in the blood of women from the beginning of pregnancy to the moment of delivery. The concentration of oxytocin increases with pregnancy and in monkeys (Morris et al. 1980). It should be noted here that when determining the concentration of oxytocin in the blood of pregnant women, various researchers have found significant variations in it. The maximum level of oxytocin at the end of pregnancy also varied by three to ten times (Dawood et al. 1978; Otsuki et al. 1983; De Geest et al. 1985; Stock et al. 1991; Levine et al. 2007). Perhaps the lack of uniformity among different authors is due to the difficulty of determining the level of oxytocin in the blood, since the concentration of oxytocin as already mentioned is two to three orders of magnitude lower than the other hormones.

Due to the fact that the regression of breast tissue after conception does not occur, the "starting" state for breast development during pregnancy will be the structure of the gland in the second half of the luteal phase. An increase in the concentration of steroid hormones, prolactin, placental lactogen, and oxytocin is accompanied by an acceleration of proliferation and differentiation of breast tissue (Neville et al. 2012). As already mentioned, these hormones are necessary for the morphogenesis and functioning of various structures of the woman's breast. In particular, estrogen contributes to the development of the ductal system. Progesterone is necessary for alveolar morphogenesis and prolactin for the development of alveolar structures, the beginning of secretory activity in the final trimester of pregnancy, and the continuation and maintenance of lactation in the postpartum period. Placental lactogen promotes the initiation of secretion in the third trimester of pregnancy. Experiments on the mammary glands of mice during pregnancy showed that oxytocin increases the proliferation and differentiation of myoepithelial cells, while the corresponding effects on epithelial (secretory) cells were insignificant (Sapino et al. 1993). There is no data on the effect of oxytocin on the proliferation and differentiation of breast cells in women. We can only assume that oxytocin has a similar effect on the development of glandular tissue in women. Indirect evidence is an increase in the proportion of myoepithelial breast cells staining in women on Ki-67 (a marker of cell proliferation) to 2% during pregnancy compared to non-pregnant breast (0.2–0.3%) (Suzuki et al. 2000).

During pregnancy, there are two stages of development of glandular breast tissue (Russo and Russo 2004a, b). According to light-microscopic studies, the beginning of the first stage is characterized by an increase in the length of the ducts and their intensive branching. At the same time, there is an active cell proliferation of distal sections of the ducts. The number of mammary lobes is increased. The number of alveolar buds that make up the lobules also increases. Type 2 lobules turn into type 3 lobules, which begin to predominate in glandular tissue by the end of the third month of pregnancy. The intensity of alveolar bud formation and lobule formation is much higher than the corresponding processes in a non-pregnant woman. At the same time, less differentiated bud-like outgrowths are also found in glandular tissue (Fig. 2.10). The volume of the gland increases, and this is mainly due to an increase in the number and size of epithelial cells. Luminal epithelial cells that are part of the alveolar buds differentiate into cells with a typical secretory cell morphology. Accordingly, the alveolar buds turn into full-fledged secretory structures, called acinuses or alveoli. In the future, the term alveola will be used to refer to these structures. The beginning of pregnancy is accompanied by increased vascularization of glandular tissue (Madjar et al. 1992). By midpregnancy, the alveoli continue to increase in size and number so that the central duct from which the alveolar ducts exit is difficult to distinguish under a light microscope (Russo and Russo 2004a, b). In the alveoli, there is an increase in secretory activity.

The beginning of the second stage, which roughly coincides with the middle of pregnancy, is characterized by a slowdown in the formation of new alveoli and lobules. The secret accumulates in the already formed alveoli. The cavity of the alveoli increases, and the walls stretch. In secretory cells, the number of vacuoles filled with lipids increases. At this time, the progressive expression of most of the genes involved in the synthesis of milk components begins (Neville et al. 2012). By the end of the second stage (end of pregnancy), a new wave of mitotic activity is again observed in breast cells with the formation of new alveoli and lobules (Russo and Russo 2004a, b).

The above-stated light-microscopic data on the development of breast glandular tissue are consistent with the results of determining the proliferative activity of breast cells during pregnancy by autoradiographic method using a label to thymidine (^{3}H-thymidine) (Battersby and Anderson 1988). The inclusion of the label in the nuclei of secretory cells of the alveoli and their ducts was quite high in the first half of pregnancy (6%). In the second half of pregnancy, there was a decrease in the inclusion of the label by half. In addition to secretory cells, the inclusion of the label was observed in the nuclei of stroma cells and cells forming the walls of blood vessels. After birth, with the beginning of lactation, the inclusion of the label decreased to 0.2%.

As already noted, one of the main reasons for the intensive development of the breast during pregnancy is the increase in the women's body of steroid hormones, prolactin, placental lactogen, and oxytocin. Their effect on the gland tissue is carried out through the corresponding cellular receptors (Li et al. 2010). In this regard, it is of interest to change their content in the cells of glandular tissue and stroma during pregnancy. Unfortunately, there is no such information for the breast of women during pregnancy. To date, there is one paper (Taylor et al. 2009) that has studied the distribution of PRA, PRB, and Erα receptors in breast tissue samples taken from non-pregnant and pregnant women. However, in both cases, there was no clear "binding" of the distribution and density of receptors to the phases of the menstrual cycle or to the duration of pregnancy. The results of this work indicate a decrease in the density of steroid hormone receptors in pregnant women compared to non-pregnant women with a normal menstrual cycle.

However, there is data on the distribution and dynamics of steroid hormone receptors in breast cells as pregnancy progresses, obtained in monkeys (macaques) (Cheng et al. 2005). As already mentioned, the structure of the breast (Macpherson and Montagna 1974), the time of the menstrual cycle, and the dynamics and values of the concentrations of steroid hormones in the menstrual cycle phases have a coincidence in these primates with the corresponding indicators in women. In addition, there are also similarities with women in terms of pregnancy duration and changes in the concentration of steroid hormones during pregnancy (Bosu et al. 1973; Macpherson and Montagna 1974; Atkinsson et al. 1975; Cline 2007). Therefore, it is more likely to extrapolate data obtained on the mammary glands of macaques to the mammary glands of women during pregnancy. It was previously noted (Cheng et al. 2005) that the expression of Erα in macaque gland structures decreased significantly toward the end of the luteal phase of the cycle. During all phases of pregnancy, Erα expression continued to decrease and could be detected in less than 1% of glandular tissue cells. Expression for Erß in epithelial cells of the alveoli and ducts as well as at the end of the luteal phase was quite high and amounted to about 81% and 78%, respectively. However, by midpregnancy, approximately 10% of alveolar epithelial cells and 8% of ductal epithelial cells had Erß expression. In the second half of pregnancy, Erß expression continued to decrease, and at the end of pregnancy, it was less than 1% in epithelial cells of the alveoli and ducts. Expression for both types of PR increased by midpregnancy and then decreased by the end of pregnancy. At the beginning, middle, and end of pregnancy, the percentage of PRA and PRB expression in alveolar epithelial cell nuclei was about 5%, 9%, and 0.3%, respectively, and in ductal epithelial cell nuclei, about 3%, 7%, and 1% (Cheng et al. 2005).

Due to the great similarity of the structure of placental lactogen and prolactin, it is believed that the effect of these hormones on breast tissue cells is carried out through the same receptors. However, there is no data on the dynamics of cell receptors for prolactin during pregnancy in the mammary glands of women and other primates.

It is important to note that the dynamics of expression of receptors to steroid hormones, as well as in the case of the menstrual cycle, depends on the degree of glandular tissue proliferation. At the beginning of pregnancy, there are quite a lot of type 1 and type 2 lobes in glandular tissue, in which there is a high proliferative activity (Russo et al. 1999). By the middle of pregnancy, the number of type 1 and type 2 lobes decreases significantly, and the number of type 3 lobes increases, in which the proliferative activity is significantly reduced (Russo et al. 1999) (Fig. 2.13). The second half of pregnancy is accompanied by a decrease in the proliferative activity of glandular tissue in the woman and, apparently, in other primates. Accordingly, the expression of steroid receptors in glandular tissue cells is reduced. Data on the dynamics of OR in breast tissue of women during pregnancy are not available in the literature.

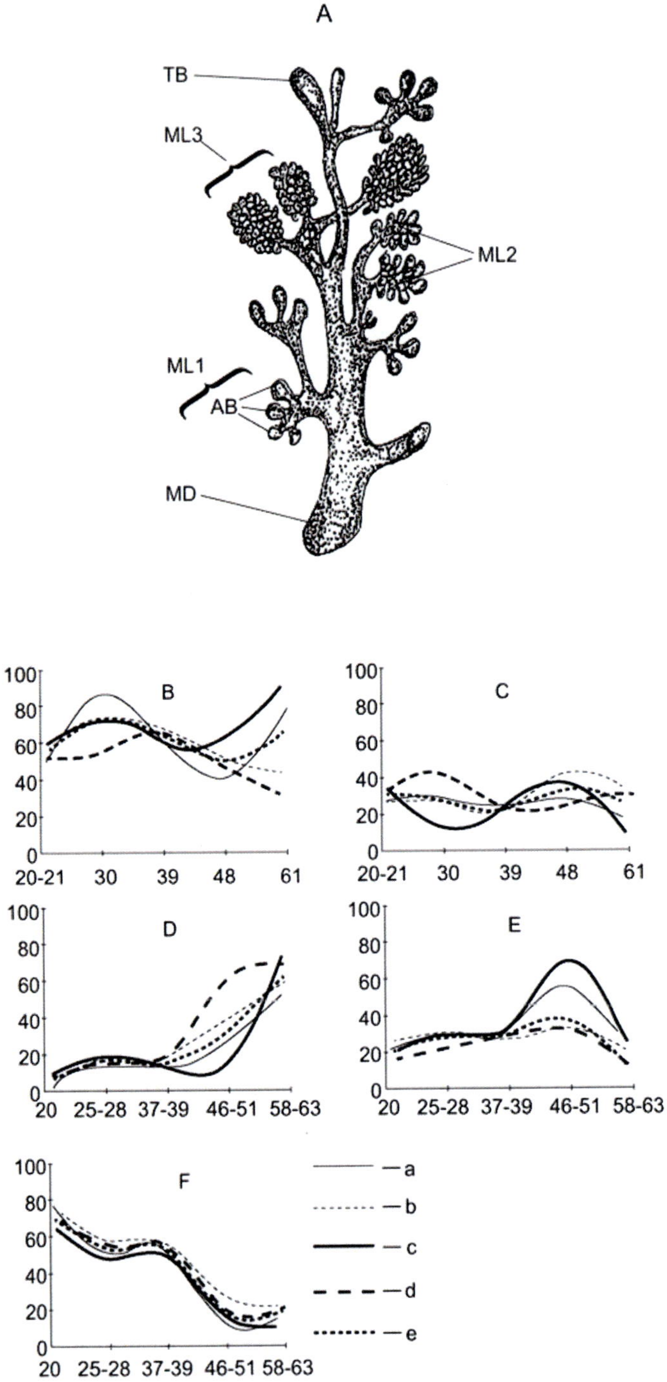

Fig. 2.13 Schematic representation of lobular structures in the breast of a woman (**a**) and changes in their number in nulliparous breast (**b**, **c**) and parous breast (**d**, **e**, **f**) from Russo and Russo (2004a, b). (**a**) TB-terminal bud, ML1, ML2, and ML3—respectively milk lobes of types 1, 2, and

In addition to steroid hormones, prolactin, placental lactogen, and oxytocin (reproductive hormones), which are directly involved in the development of the breast, there are a number of hormones called metabolic hormones that indirectly affect the development and functioning of a woman's breast. This group of hormones includes growth hormone, glucocorticoids, thyroid hormone, and insulin. These hormones primarily coordinate the response of a woman's body to metabolic changes and stressors that occur during pregnancy, as well as after childbirth during lactation. They can alter the response of the breast to reproductive hormones and indirectly regulate the synthesis and secretion of milk by altering the flow of milk precursors to glandular tissue (Neville et al. 2012). In addition to hormones, growth factors undoubtedly influence the development of a woman's breast during pregnancy. However, in the literature, there is data concerning the development of only mammary glands of animals (Johnson 2010).

A study of the development of glandular tissue during pregnancy at the ultrastructural level (Ferguson and Anderson 1983) found a great similarity of its structure in the first trimester of pregnancy with the ultrastructure of the gland in a non-pregnant woman during the luteal phase of the menstrual cycle (Stirling and Chandler 1976a, 1977; Watson et al. 1988) (see more detail in Sect. 2.2.5).

In the second trimester of pregnancy, the number of mitochondria increases in the epithelial cells of the alveoli, and the rough endoplasmic reticulum noticeably grows. The activity of the Golgi complex increases with the formation of large and small vesicles. Most cells contain lipid droplets of various sizes. Moreover, the largest fat droplets are located closer to the apical surface of epithelial cells. In addition to lipid droplets, epithelial cells contain vacuoles of different diameters filled with a dense homogeneous material. With the beginning of the third trimester, the number of epithelial cells containing lipid drops increases. The average size of fat droplets increases, and they begin to occupy a large part of the apical region of epithelial cells. In addition, the number of vacuoles in the apical parts also increases. In some cases, there is a fusion of the vacuole membrane with the apical membrane of the cell and the removal of the contents of the vacuoles into the alveolar cavity. The epithelial cells that form the alveolar cavity are flattened. The cavity of the alveoli increases, and the walls of the alveoli stretch. At the same time, it is noted that the apical intercellular complex, which includes dense contacts and desmosomes, provides contact between cells, and isolates the alveolar cavity from the intralobular

Fig. 2.13 (continued) 3. *AB* alveolar buds, *MD* milk duct. (**b**) Percentage of ML1 in the four quadrants of the mammary gland in nulliparous women. (**c**) Percentage of ML2 in the four quadrants of the mammary gland in nulliparous women. (**d**) Percentage of ML1 in the four quadrants of the breast in parous women. (**e**) Percentage of ML2 in the four quadrants of the breast in parous women. (**f**) Percentage ML3 in the four quadrants of the breast in parous women. For all charts on the ordinate axis: the percentage of corresponding ML to the total number of all ML types. On the abscissa axis: women's age in years. For all graphs, the thin solid lines are the upper outer quadrant; thick, lower outer quadrant; thick dashed, lower inner quadrant; thin dashed, upper inner quadrant; dotted lines, average values space throughout pregnancy, remains unchanged (Ferguson and Anderson 1983). Myoepithelial cells flatten, and their processes form a kind of mesh structure on the surface of the alveoli and ducts. The number of myofilaments inside the cells increases noticeably. However, during pregnancy, proliferation of myoepithelial cells occurs to a lesser extent than epithelial cells. In addition to secretory and myoepithelial cells, the walls of the alveoli and ducts contain lymphocytes and macrophages, which make up about 10% of the total cell population (Ferguson 1985). The volume occupied by the stroma decreases due to the growth and increase in the number of alveoli and ducts. At the same time, the density of the capillary network increases significantly in the intra- and interlobular space. Since the blood supply provides the supply of hormones and other biologically active substances to the glandular tissue that stimulate the development and functioning of the breast, data on the structure and dynamics of glandular tissue vascularization during pregnancy is of great interest. However, data on changes in this parameter during pregnancy were not found in the scientific literature in women. There is detailed information about the dynamics of microvessel density during pregnancy and the beginning of lactation in the mammary glands of rats (Yasugi et al. 1989). Capillaries are mainly concentrated in the intralobular space. Thicker vessels are found in the interlobular space. Vascular density, which was defined as the ratio of the area of the capillary cavity to the total surface of the lobule, was approximately 3%, increased to 7% by midpregnancy, and then began to decrease and reached 4% by the end of pregnancy and the first week of lactation. Illustrative data on the dynamics of the architecture of capillaries located in the glandular tissue of rats (Yasugi et al. 1989) and mice (Djonov et al. 2001) during pregnancy obtained using scanning electron microscopy are impressive. Glandular lobules and alveoli are surrounded by a capillary network as pregnancy progresses, so that they are enclosed in spherical cavities formed by capillaries. Since the capillaries are located at a distance of several tens of micrometers from the cells that form the alveoli and ducts, the diffusion of the necessary mineral and organic substances, including hormones and oxygen, from the capillary bed will be very effective.

It is possible that a similar dynamics of vascular density during pregnancy exists in the woman's breast. This is supported by data on the values of the angiogenic index in fractions of various types (Russo et al. 1999). The angiogenic index was determined as the area of vascular coloration using factor VIII, which is a marker of endothelial cells. It turned out that the angiogenic index had the highest value in the lobes of type 1 (1.04) and significantly decreased in the lobes of type 2 (0.84) and type 3 (0.55). Since the first half of pregnancy is dominated by type 1 and type 2 fractions in glandular tissue (Russo et al. 2004a, b), there will be an increase in blood vessel density during this period. By midpregnancy, the number of type 1 and type 2 lobes decreases significantly, and the number of type 3 lobes increases. Accordingly, the number of blood vessels will reach the highest value. Slowing down the formation of milk lobes in the second half of pregnancy (Russo et al. 1999) is likely to be accompanied by a decrease in further vascularization of glandular tissue. Simultaneously, with the increase in breast vascularization during pregnancy, there is an increase in the velocity of blood flow in the blood vessels that supply the

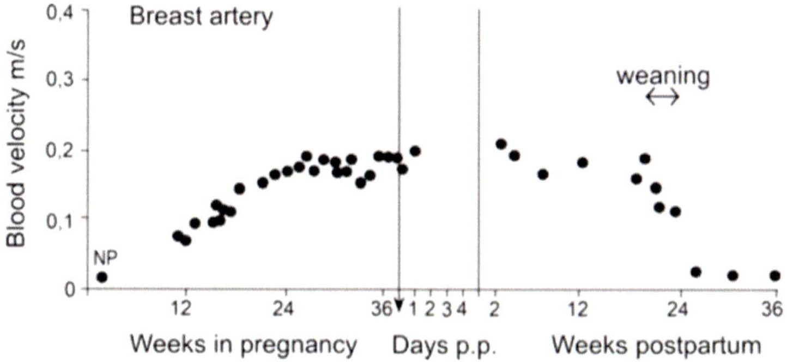

Fig. 2.14 Changes in the average blood velocity in the main branch of the right thoracic lateral artery of the breast of a woman starting from the 12th week of pregnancy, the first days after birth, and during subsequent lactation (from Thoresen and Wesche 1988). NP marked the velocity of blood flow in non-pregnant women; the two-pointed arrow marks the beginning of the introduction of complementary food to the child. The ordinate axis is the blood velocity (m/s), the abscissa axis: the left graph is the weeks of pregnancy and the right graph is the weeks after birth. Between graphs, days after delivery

gland. Figure 2.14 shows the results of direct measurement of the average blood flow rate weekly in the main branch of the thoracic lateral artery of the breast of a woman from the 12th week of pregnancy (Thoresen and Wesche 1988). The blood flow rate started to increase from the second trimester of pregnancy and then slowed down from the middle of pregnancy and reached a plateau by the beginning of the third trimester of pregnancy. After delivery, the rate increased rapidly on the first day and within 4–5 days was about 0.5 times higher than before delivery. Then the blood flow rate decreased to the values of the third trimester of pregnancy and was at this level before the introduction of complementary food to the child, followed by a slow decline.

In addition to the development of blood supply during pregnancy, the mammary gland of a woman has an intensive development of the lymphatic system. However, there is no information on this issue in the literature. In addition, the development of peripheral and central structures of the female nervous system during pregnancy, which regulate the secretion and excretion of milk in the postpartum period, remains unexplored.

So, the structural and ultrastructural data presented above indicate that the mammary gland of a woman by the third trimester of pregnancy, i.e., long before the birth of a child, reaches a development close to the state before childbirth. The secretory cells of the alveolus begin to form and withdraw into the cavity of the alveoli and ducts, a secret called colostrum. Additionally, the beginning of secretory activity of the breast is indicated by the results of studies on antenatal (prenatal) milk excretion (Cox 2006; Singh et al. 2009; Chapman et al. 2013a, b). The period from the beginning of secretory activity to the beginning of milk secretion after childbirth is called lactogenesis I (Hartmann 1973). Apparently, this development of the breast during pregnancy can be considered a kind of factor in the reliability of lactation function in the event that the birth for some reason was premature. That is, in this situation, the breast will need a relatively short period to achieve its final morphofunctional development and, accordingly, provide the child with full nutrition. Indeed, the literature data (Cregan et al. 2002a, b; Hill et al. 2005), as well as own clinical observations (we will discuss this in more detail in Chap. 3), show that in terms of productivity and composition of milk, the mammary glands of women with preterm birth in most cases "catch up" with these indicators for 10–15 days after delivery.

It is interesting to note that the amount of colostrum during antenatal pumping was usually in the range of 10–30 ml (Widdows et al. 1935), despite the fact that the concentration of prolactin by the middle of pregnancy exceeded eight to ten times its concentration at the end of the luteal phase of the cycle (Figs. 2.9 and 2.12). Studies have shown that a possible cause is the inhibitory effect of steroids mainly progesterone on the secretory activity of prolactin. This fact was first discovered in direct experiments on the mammary glands of pregnant rats (Kuhn 1969) and then on the mammary glands of other mammals (Neville et al. 2001). Exogenous progesterone blocked the synthesis of lactose and lipids in the mammary glands of pregnant rats that had their ovaries removed, which are the main source of progesterone in these animals during pregnancy (Kuhn 1969). There is no direct information about the effect of progesterone on the secretory function of women's mammary glands during pregnancy. However, data from the clinic (Neifert et al. 1981) on the inhibitory effect of placenta residues (the main source of progesterone in pregnant women) on milk secretion after childbirth indicate that progesterone is the main inhibitor of milk formation in women as well. In addition, it is believed that inhibitory effect on the secretion of milk in women has an estrogen. Thus, sex steroids in women do not affect the proliferating and differentiating effect of prolactin and placental lactogen on the breast during pregnancy, but inhibit their synthetic activity in the secretory cells of the alveoli. In addition to volume, the composition of antenatal colostrum also differed from that of postpartum milk (Kulski and Hartmann 1981; Allen et al. 1991) (Table 2.1). Thus, prenatal colostrum contains relatively high concentrations of sodium ions, chlorine, and proteins that perform a protective function, such as immunoglobulins and lactoferrin. The concentration of lactose is reduced (Kulski and Hartmann 1981; Allen et al. 1991). The unusual composition of colostrum and, in particular, the high concentration of sodium and chlorine ions and the low content of lactose explain the open state of dense contacts between secretory cells. On diagrams in various textbooks and reviews (e.g., Lawrence and Lawrence 1999, 2011; Neville et al. 2001; Riordan 2005), this is presented as mechanical gaps between secretory cells, which may mislead the reader. As mentioned above, the apical intercellular complex, which includes dense contacts and desmosomes, provides contact between cells, and isolates the alveolar cavity from the intralobular space throughout pregnancy, remains unchanged (Ferguson and Anderson 1983). Gaps in the literal sense of the word between secretory cells in the areas of dense contacts on electronograms were not observed. It should be noted here that significant material has been accumulated to date on the structure and function of dense contacts of various epithelial tissues (Johnson 2005), including the mammary alveoli

Table 2.1 Composition of antenatal human milk for 20.21 ± 12.18 days before delivery

Milk components	Concentration	±SD (number of women)
Fat	%	2.07 ± 0.98 (11)
Lactose	mM	79.78 ± 21.68 (9)
Protein	g/100 mL	5.44 ± 1.7 (8)
Glucose	mM	0.35 ± 0.16 (8)
Sodium	mM	61.26 ± 25.82 (10)
Potassium	mM	18.30 ± 5.67 (10)
Chloride	mM	62.21 ± 17.44 (10)
Calcium	g/100 mL	25.35 ± 8.48 (10)
Magnesium	g/100 mL	5.64 ± 1.44 (10)
Citrate	mM	0.40 ± 0.17 (8)
Phosphate	g/100 mL	2.32 ± 0.70 (9)
Ionized Calcium	mM	3.25 ± 0.84 (6)
pH		6.83 ± 0.18 (6)
Urea	g/100 mL	14.87 ± 2.40 (9)
Creatinine	g/100 mL	1.47 ± 0.35 (9)

of mice (Nguyen and Neville 1998; Nguyen et al. 2001; Markov et al. 2012). It turned out that tight contacts with the help of special proteins can very effectively regulate the intercellular permeability to various substances and ions. Dense contacts contain pores through which various substances diffuse. The pores can open or block under the influence of various biological and pathological factors. In the case of women's breast, a slight difference in the concentration of lactose in colostrum and in the intercellular fluid during pregnancy allows us to assume that the pores through which there is a diffusion of lactose through dense contacts between secretory cells are open. However, there is a significant difference between the concentration of sodium, potassium, and chloride ions in prenatal colostrum (70–80 mM, 10 mM, and 60–70 mM, respectively) (Kulski and Hartmann 1981; Allen et al. 1991) and intercellular fluid (136–145 mM, 3.5–5 mM, and 98–106 mM, respectively), whereas with open ionic pores in dense contacts between secretory cells, the concentration of ions in colostrum and intercellular fluid should be about the same, as in the case of lactose. Therefore, it can be assumed that the permeability of dense contacts to sodium and chloride ions is increased, and for potassium ions, it is reduced, but not to the maximum levels. Studies on the mammary glands of mice during pregnancy (Nguyen et al. 2001) found that the high permeability of dense contacts to sucrose depends on the presence of progesterone in the blood. Removing it by ovariectomy caused a block in the permeability of dense contacts to sucrose. It is possible that in the mammary gland of women, progesterone also determines the state of permeability of dense contacts. It should be noted that there is no natural "demand" for antenatal colostrum during term of delivery. The formation of large amounts of colostrum during pregnancy will cause discomfort in the state of the mammary glands in women. Therefore, it is logical to inhibit the secretion of milk during pregnancy. However, the relatively small volume of antenatal colostrum that fills the alveolar-ductal system of the breast immediately before delivery seems to be important for nutrition and immune protection of the newborn in the first 48 h after birth.

One more important fact must be pointed out here. It turned out that, as well as during the luteal phase of the menstrual cycle, mechanical stimulation of the nipple and areola of the breast during various periods of pregnancy, according to a number of authors, is accompanied by an increase in the concentration of oxytocin in the blood of women (Leake et al. 1984a; Finley et al. 1986). This is consistent with data (Cetin et al. 2014) that women, without any consequences (miscarriage, premature birth) for a significant time of pregnancy with the second child, for example, up to 32 weeks (Merchant et al. 1990), can breastfeed their first child. Since oxytocin is an essential hormone for removing milk from the alveolar-ductal system of the breast (for more information, the structure and function of oxytocin will be presented in Chap. 3) without its release into the bloodstream from the central nervous system, women would not be able to feed a child during the next pregnancy. The guarantee of the absence of undesirable consequences for pregnancy during this time is apparently a reduced concentration of oxytocin receptors in the endometrium of the uterus. In the last weeks of pregnancy and during childbirth (Christensson et al. 1989; Hatjis et al. 1989; Seoud et al. 1993), oxytocin levels increase significantly, despite mechanical stimulation of the nipple-areolar area of the breast. By this time, the content of oxytocin receptors in the endometrium of the uterus increases.

References

Alexander DP, Britton HG, Buttle HL, Nixon DA (1973) Impermeability of the sheep placenta towards endogenous prolactin. Res Vet Sci 14:271–272

Alipov VI, Kolodina LN, Korkhov VV, Radzinsky VE (1988) Woman lactation. "Ylym", Ashgabat. (in Russian)

Allen JC, Keller RP, Archer P, Neville MC (1991) Studies in human lactation: milk composition and daily secretion rates of macronutrients in the first year of lactation. Am J Clin Nutr 54 (1):69–80

Amico JA, Finley BE (1986) Breast stimulation in cycling women, pregnant women and a woman with induced lactation: pattern of release of oxytocin, prolactin and luteinizing hormone. Clin Endocrinol 25(2):97–106

Amico JA, Seif SM, Robinson AG (1981) Elevation of oxytocin and the oxytocin-associated neurophysin in the plasma of normal women during midcycle. J Clin Endocrinol Metab 53:1229–1232

Arscott GD, Craig HR, Gabay L (2001) Failure of bromocriptine therapy to control juvenile mammary hypertrophy. Br J Plast Surg 54:720–723

Atkinsson LE, Hotchkiss J, Surve AH, Neil JD, Knobil E (1975) Circulation levels of steroid and chronic gonadotropin during pregnancy in the rhesus monkey, with special attention to the rescue corpus luteum in early pregnancy. Biol Reprod 12:335–345

Aubert ML, Grumbach MM, Kaplan SL (1975) The ontogenesis of human fetal hormones. III. Prolactin. J Clin Invest 56:155–164

Backstrom CT, McNeeilly AS, Leask RM, Baird DT (1982) Pulsatile secretion of LH, FSH, prolactin, oestradiol and progesterone during the human menstrual cycle. Clin Endocrinol 17:29–42

Bamshad M, Lin RC, Jaw DJ, Wakins WS, Krakowiak PA, Moore ME, Franceschini P, Lala R, Holmes LB, Gebuhr TC, Bruneau BG, Schinzel A, Seidman CE, Jorde LB (1997) Mutations in human TBX3 alter limb, apocrine and genital development in ulnar-mammary syndrome. Nat Genet 16:311–315

Barber VC (1974) Cilia in sense organs. In: Sleigh MA (ed) Cilia and Flagella. Academic Press, London, pp 403–433

Battersby S, Anderson TJ (1988) Proliferative and secretory activity in the pregnant and lactating human breast. Virchows Arch A Pathol Anat Histopathol 413:189–196

Blackburn DG (1991) Evolutionary origins of the mammary gland. Mammal Rev 21:81–98

Bluestein DD, Wall GH (1963) Persistent neonatal breast hypertrophy. Am J Dis Child 105:292–294

Bosu WT, Johansson ED, Gemzell C (1973) Ovarian steroid patterns in peripheral plasma during the menstrual cycle in the rhesus monkey. Folia Primatol(Basel) 19:218–234

Braunstein GD, Rasor J, Danzer H, Adler D, Wade ME (1976) Serum human chorionic gonadotropin levels throughout normal pregnancy. Am J Obstet Gynecol 126:678–681

Braund W, Roeger DC, Judd SJ (1984) Synchronous secretion of luteinizing hormone and prolactin in the human luteal phase: neuroendocrine mechanisms. J Clin Endocrinol Metab 58:293–297

Brouha L (1905) Recherches sur les diverses phases du development et de l'activite de la mamelle. Arch Biol 21:459–603

Bussolati G, Cassoni P, Ghisolfi G, Negro F, Sapino A (1996) Immunolocalization and gene expression of oxytocin receptors in carcinomas and non-neoplastic tissues of the breast. Am J Pathol 148:1895–1903

Cardiff RD, Wellings SR (1999) The comparative pathology of human and mouse mammary gland. J Mam Gland Biol Neoplasia 4(1):103–122

Cetin I, Assandro P, Massari M, Sagone A, Gennaretti R, Donzelli G, Knowles A, Monasta L, Davanzo R (2014) Breastfeeding during pregnancy: position paper of the italian society of perinatal medicine and the task force on breastfeeding. J Hum Lact 30(1):20–27

Chapman T, Pincombe J, Harris M, Fereday J (2013a) Antenatal breast expression: exploration and extent of teaching practices amongst International Board Certified Lactation Consultant midwives across Australia. Women Birth 26(1):41–48

Chapman T, Pincombe J, Harris M (2013b) Antenatal breast expression: a critical review of the literature. Midwifery 29(3):203–210

Cheng G, Li Y, Wang Y, Berg T, Nord M, Vihko P, Warner M, Piao Y-S, Gustafsson J-A (2005) Differential regulation of estrogen receptor (ER) α and ERβ in primate mammary gland. J Clin End Metab 90:435–444

Christensson K, Nilsson BA, Stock S, Matthiesen AS, Uvnäs-Moberg K (1989) Effect of nipple stimulation on uterine activity and on plasma levels of oxytocin in full term, healthy, pregnant women. Acta Obstet Gynecol Scand 68:205–210

Clarke RB (2005) Isolation and characterization of human mammary stem cells. Cell Prolif 38:375–386

Cline MJ (2007) Assessing the mammary gland of nonhuman primates: effects of endogenous hormones and exogenous hormonal agents and growth factors. Birth Defects Res (Part B) 80:126–146

Cole EN, Sellwood RA, England PC, Griffiths K (1977) Serum prolactin concentrations in benign breast disease throughout the menstrual cycle. Eur J Cancer 13:597–603

Cox SG (2006) Expressing and storing colostrum antenatally for use in the newborn period. Breastfeed Rev 14:11–16

Cox DB, Owens RA, Hartmann PE (1996) Blood and milk prolactin and the rate of milk synthesis in women. Exp Physiol 81:1007–1020

Cregan MD, Mitoulas LR, Hartmann PE (2002a) Milk prolactin, feed volume and duration between feeds in women breastfeeding their full-term infants over a 24 h period. Exp Physiol 81:1007–1020
Cregan MD, De Mello TR, Kershaw D, McDougall K, Hartmann PE (2002b) Initiation of lactation in women after preterm delivery. Acta Obstet Gynecol Scand 81:870–877
Cregan MD, Fan Y, Appelbee A, Brown ML, Klopcic B, Koppen J, Mitoulas LR, Piper KM, Choolani MA, Chong YS, Hartmann PE (2007) Identification of nestin-positive putative mammary stem cells in human breastmilk. Cell Tissue Res 329:129–136
Cunha GR, Cooke PS, Kurita T (2004) Role of stromal-epithelial interactions in hormonal responses. Arch Histol Cytol 67(5):417–434
Dada OA, Ladipo OA, Osinusi BO, Osotimehin BO, Nduka EU (1981) Circulation blood level of gonadotropin s and prolactin in normal cycle. Int J Gynaecol Obstet 19:291–294
Dancey A, Khan M, Dawson J, Peart F (2008) Gigantomastia – a classification and review of the literature. J Plast Rec Aesth Surg 61:493–502
Davis WL, Moncrieff A (1938) Composition of the milk from the breasts of newly-born infants. Biochem J 32(7):1238–1240
Dawood MY, Raghavan KS, Pociask C, Fuchs F (1978) Oxytocin in human pregnancy and parturition. Obstet Gynecol 51:138–143
De Geest K, Thiery M, Piron-Possuyt G, Vanden Driessche R (1985) Plasma oxytocin in human pregnancy and parturition. J Perinat Med 13:3–13
Dimitrakakis C, Zhou J, Wang J, Belanger A, LaBrie F, Cheng C, Powell D, Bondy C (2003) A physiologic role for testosterone in limiting estrogenic stimulation of the breas. Menopause 10:292–298
Djonov V, Andres AC, Ziemiecki A (2001) Vascular remodelling during the normal and malignant life cycle of the mammary gland. Microsc Res Tech 52:182–189
Dossett JA (1960) The nature of breast secretion in infancy. J Pathol Bacteriol 80:93–99
Durnberger H, Kratochwil K (1980) Specificity of tissue interaction and origin of mesenchymal cells in the androgen response of the embryonic mammary gland. Cell 19:465–471
Eriksson M, Lindh B, Uvnas-Moberg K, Hokfelt T (1996) Distribution and origin of peptide-containing nerve fibres in the rat and human mammary gland. Neuroscience 70:227–245
Eyden BP, Watson RJ, Harris M, Howell A (1986) Intralobular stromal fibroblasts in the resting human mammary gland: ultrastructural properties and intercellular relationships. J Submicrosc Cytol 18:397–408
Fanger H, Ree HJ (1974) Cyclic changes of human mammary gland epithelium in relation to the menstrual cycle – an ultrastructural study. Cancer 34:574–585
Ferguson DJ (1985) Intraepithelial lymphocytes and macrophages in the normal breast. Virchows Arch A Pathol Anat Histopathol 407:369–378
Ferguson DJ, Anderson TJ (1983) An ultrastructural study of lactation in the human breast. Anat Embryol (Berl) 168(3):349–359
Finley BE, Amico J, Castillo M, Seitchik J (1986) Oxytocin and prolactin responses associated with nipple stimulation contraction stress tests. Obstet Gynecol 67:836–839
Fowler PA, Casey CE, Cameron GG, Foster MA, Knight CH (1990) Cyclic changes in composition and volume of the breast during the menstrual cycle, measured by magnetic resonance imaging. Br J Obstet Gynecol 97:595–602
Freeman ME, Kanyicska B, Lerant A, Nagy G (2000) Prolactin: structure, function, and regulation of secretion. Physiol Rev 80(4):1523–1631
Galantsev VP, Gulyaeva EP (1987) Evolution of lactation. L., "Science" (in Russian)
Gherghiceanu M, Popescu LM (2005) Interstitial Cajal-like cells (ICLC) in human resting mammary gland stroma. Transmission electron microscope (TEM) identification. J Cell Mol Med 4:893–910
Gill S, Peston D, Vonderhaar BK, Shousha S (2001) Expression of prolactin receptors in normal, benign, and malignant breast tissue: an immunohistological study. J Clin Pathol 54:956–960

Gimpl G, Reitz J, Brauer S, Trossen C (2008) Oxytocin receptors: ligand binding, signalling and cholesterol dependence. Prog Brain Res 170:193–204
Going JJ, Anderson TJ, Battersby S, MacIntyre CC (1988) Proliferative and secretory activity in human breast during natural and artificial menstrual cycles. Am J Pathol 130(1):193–204
Grachev II, Galantsev VP (1973) Physiology of lactation. General and comparative. L., Science (in Russian)
Guyda HJ, Friesen HG (1973) Serum prolactin levels in humans from birth to adult life. Pediat Res 7:534–540
Halban J (1905) Arch. Gynak 75:353 (from Mayer G, Klein M (1961) Histology and cytology of the mammary gland). In: Kon SK, Cowie AT (eds) Milk. The mammary gland and its secretion, vol 1. Academic, London, pp 47–126
Hartmann PE (1973) Changes in composition and yield of the mammary secretion of cows during the initiation of lactation. J Endocrinol 59:231–247
Hassiotou F, Beltran A, Chetwynd E, Stuebe AM, Twigger AJ, Metzger P, Trengove N, Tat LC, Filgueira L, Blancafort P, Hartmann PE (2012) Breastmilk is a novel source of stem cells with multi-lineage differentiation potential. Stem Cells 30:2164–2174
Hatjis CG, Morris M, Rose JC, Kofinas AD, Penry M, Swain M (1989) Oxytocin, vasopressin, and prolactin responses associated with nipple stimulation. South Med J82:193–196
Healy DL, Rattigan S, Hartmann PE, Herington AC, Burger HG (1980) Prolactin in human milk: correlation with lactose, total protein, and alpha-lactalbumin levels. Am J Physiol 238(1):E83–E86
Hens JR, Dann P, Zhang J-P, Harris S, Robinson GW, Wysolmerski J (2007) BMP4 and PTHrP interact to stimulate ductal outgrowth during embryonic mammary development and to inhibit hair follicle induction. Development 134:1221–1230
Hiba J, Del Pozo E, Genazzani A, Pusterla E, Lancranjan I, Sidoropoulos D, Gunti J (1977) Hormonal mechanisms of milk secretion in the newborn. J Clin Endocrinol Metab 44:973–976
Hill PD, Aldag JC, Chatterton RT, Zinaman M (2005) Comparison of milk output between mothers of preterm and term infants: the first 6 weeks after birth. J Hum Lact 21:22–30
Hovey RC, Trott JF, Vonderhaar BK (2002) Establishing a framework for the functional mammary gland: from endocrinology to morphology. J Mammary Gland Biol Neoplasia 7(1):17–38
Howard BA, Gusterson B (2000) Human breast development. J Mam Gland Biol Neoplasia 5 (2):119–137
Hughes ESR (1950) The development of the mammary gland. Ann R Coll Surg Engl 6:99–119
Johnson LG (2005) Applications of imaging techniques to studies of epithelial tight junctions. Adv Drug Deliv Rev 57:111–121
Johnson MC (2010) Anatomy and physiology of the breast. In: Management of the breast diseases. Springer, pp 1–36
Johnson RE, Murad H (2009) Gynecomastia: pathophysiology, evaluation and management. Mayo Clin Proc 84:1010–1015
Josimovich JBG, Weiss G, Hutchinson DL (1974) Sources and disposition of pituitary prolactin in maternal circulation, amniotic fluid, fetus and placenta in the pregnant rhesus monkey. Endocrinology 94:1364–1371
Kajava Y (1915) The proportions of supernumerary nipples in the Finnish population. Duodecim 31:143–170
Kaplan SL, Edgar JC, Ford EG, Adgent MA, Schall JI, Kelly A, Umbach DM, Rogan WJ, Stallings VA, Darge K (2016) Size of testes, ovaries, uterus and breast buds by ultrasound in healthy full-term neonates ages 0–3 days. Pediatr Radiol 46(13):1837–1847
Keeling JW, King G, Walker F (1998) Oestrogen and progesterone receptors in fetal, infant and childhood mammary tissue. J Pathol 186:21–28
Kellokumpu-Lehtinen P, Santi R, Pelliniemi LJ (1979) Early cytodifferentiation of human fetal prostatic urethra and Leydig cells. Anat Rec 194:429–444
Kellokumpu-Lehtinen P, Santi R, Pelliniemi LJ (1980) Correlation of early cytodifferentiation of human fetal prostate and Leydig cells. Anat Rec 196:263–273

Kellokumpu-Lehtinen P, Johansson R-M, Pelliniemi LJ (1987) Ultrastructure of human fetal mammary gland. Anat Rec 218:66–72
Khamadyanov UR, Laricheva IP, Smirnova LK (1980) Content of prolactin, follicle-stimulating, lutenizing hormones, estradiol and progesterone in the blood serum of women with a normal menstrual cycle. Problems Endocrinol 26(2):32–37. (in Russian)
Khizhnyakova KI (1965) Cytology of breast secretions in normal and certain diseases. Moscow (in Russian)
Kimura T, Ito Y, Einspanier A, Tohya K, Nobunaga T, Tokugawa Y, Takemura M, Kubota Y, Ivell R, Matsuura N, Saji F, Murata Y (1998) Expression and immunolocalization of the oxytocin receptor in human lactating and non-lactating mammary glands. Hum Reprod 13:2645–2653
Kratochwil K (1971) In vitro analysis of the hormonal basis for the sexual dimorphism in the embryonic development of the mouse mammary gland. J Embryol Exp Morphol 25:141–153
Kratochwil K (1977) Development and loss of androgen responsiveness in the embryonic rudiment of the mouse mammary gland. Dev Biol 61:358–365
Kratochwil K, Schwartz P (1976) Tissue interaction in androgen response of embryonic mammary rudiment of mouse: identification of target tissue of testosterone. PNAS 73:4041–4044
Kuhn NJ (1969) Progesterone withdrawal as lactogenesis trigger in the rat. J Endocrinol 44:39–54
Kukhtinova RA (1963) Development of the intra-organ lymphatic system of the human breast in embryogenesis. Uch Zap Stavropol State University Med Institute 10:32–35
Kulski JK, Hartmann PE (1981) Changes in human milk composition during the initiation of lactation. Aust J Exp Biol Med Sci 59:101–114
Lawrence RA, Lawrence RM (1999) Breastfeeding: a guide for medical profession, 5th edn. Mosby
Lawrence RA, Lawrence RM (2011) Breastfeeding: a guide for medical profession, 7th edn. Mosby
Leake RD, Waters CB, Rubin RT, Buster JE, Fisher DA (1983) Oxytocin and prolactin responses in long-term breast-feeding. Obstet Gynecol 62:565–568
Leake RD, Fisher DA, Ross M, Buster JE (1984a) Oxytocin secretory response to breast stimulation in pregnant women. Am J Obstet Gynecol 148:259–262
Leake RD, Buster JE, Fisher DA (1984b) The oxytocin secretory response to breast stimulation in women during the menstrual cycle. Am J Obstet Gynecol 148:457–460
Lee PA, Xenakis T, Winer J, Matsenbaugh S (1976) Puberty in girls: correlation of serum levels of gonadotropins, prolactin, androgens, estrogens, and progestins with physical changes. J Clin Endocrinol Metab 43:775–784
Lefèvre CM, Sharp JA, Nicholas KR (2010) Evolution of lactation: ancient origin and extreme adaptations of the lactation system. Annu Rev Genomics Hum Genet 11:219–238
Levine A, Zagoory-Sharon O, Feldman R, Weller A (2007) Oxytocin during pregnancy and early postpartum: individual patterns and maternal-fetal attachment. Peptides 28:1162–1169
Li S, Han B, Liu G, Li S, Ouellet J, Labrie F, Pelletier G (2010) Immunocytochemical localization of sex steroid hormone receptors in human mammary gland. J Histochem Cytochem 58:509–515
Lin LS, Roberts VJ, Yan SS (1995) Expression of human gonadotropin-releasing hormone receptor gene in the placenta and its functional relationship to human chorionic gonadotropin secretion. J Clin Endocrinol Metab 80:580–585
Linden E, Williams R, King J, Blair E, Kini U (2009) Ulnar mammary syndrome and TBX3: expanding the phenotype. Am J Med Gen 149A:2809–2812
Lloyd RV, Anagnostou D, Cano M, Barkan AL, Chandler WF (1988) Analysis of mammosomatotropic cells in normal and neoplastic human pituitary tissues by the reverse hemolytic plaque assay and immunocytochemistry. J Clin Endocrinol Metab 66:1103–1110
Lucas A, Cole TJ (1990) Breast milk and neonatal necrotising enterocolitis. Lancet 336 (8730):1519–1523
Macpherson EE, Montagna W (1974) Proceedings: The mammary glands of rhesus monkeys. J Invest Dermatol 63(1):17–18

Madjar H, Vetter M, Prompeler HJ, Wieacker P, Schillinger H (1992) The normal vascularization of the female breast in Doppler ultrasound. Ultrashall Med 13:171–177
Madlon-Kay DJ (1986) "Witch's milk". Galactorrhea in the newborn. Am J Dis Child 140 (3):252–253
Markov AG, Kruglova NM, Fomina YA, Fromm M, Amasheh S (2012) Altered expression of tight junction proteins in mammary epithelium after discontinued suckling in mice. Pflugers Arch 463:391–398
Matulka LA, Tpiplett AA, Wagner KU (2007) Parity-induced mammary epithelial cells are multipotent and express cell surface markers associated with stem cells. Dev Biol 303:29–44
McDermott KM, Liu BY, Tisty TD, Pazour GJ (2010) Primary cilia regulate branching morphogenesis during mammary gland development. Curr Biol 20:731–737
McKiernan JF (1984) Postnatal breast development of preterm infants. Arch Dis Child 59:1090–1092
McKiernan JF, Hull D (1981a) Breast development in the newborn. Arch Dis Child 56:525–529
McKiernan JF, Hull D (1981b) Prolactin, maternal oestrogens, and breast development in the newborn. Arch Dis Child 56:770–774
McKiernan JF, Hull D (1982) The constituents of neonatal milk. Pediatr Res 16(1):60–64
McKiernan JF, Coyne J, Cahalane S (1988) Histology of breast development in early life. Arch Dis Child 63:136–139
Merchant K, Martorell R, Haas JD (1990) Consequences for maternal nutrition of reproductive stress across consecutive pregnancies. Am J Clin Nutr 52(4):616–620
Milligan D, Drife JO, Short RV (1975) Changes in breast volume during normal menstrual cycle and after oral contraceptives. Br Med J 4:494–496
Morris M, Stevens SW, Adams MR (1980) Plasma oxytocin during pregnancy and lactation in the cynomolgus monkey. Biol Reprod 23:782–787
Mitchell MD, Haynes PJ, Anderson AB, Turnbull AC (1981) Plasma oxytocin concentrations during the menstrual cycle. Eur J Obstet Gynecol Reprod Biol 12:195–200
Naccarato AG, Vicava P, Vignati S, Fanelli G, Bonadio AG, Montruccolli G, Bevilacqua G (2000) Bio-morphological events in the development of the human female mammary gland from fetal age to puberty. Wirchows Arch 436:431–438
Nathan B, Anbazhagan R, Clarkson P, Bartkova J, Gusterson B (1994) Expression of BCL-2 in the developing human fetal and infant breast. Histopathology 24:73–76
Navarrete MA, Maier CM, Falzoni R, Quadros LG, Lima GR, Baracat EC, Nazário AC (2005) Assessment of the proliferative, apoptotic and cellular renovation indices of the human mammary epithelium during the follicular and luteal phases of the menstrual cycles. Breast Cancer Res 7(3):306–313
Neifert MR, McDonough SL, Neville MC (1981) Failure of lactogenesis associated with placental retention. Am J Obstet Gynecol 140:477–478
Nevalainen MT, Xie J, Bubendorf L, Wagner K-U, Rui H (2002) Basal activation of transcription factor signal transducer and activator of transcription (Stat 5) in nonpregnant mouse and human breast epithelium. Mol Endocrinol 16:1108–1124
Neville MC (1990) The physiological basis of milk secretion. Part I. Basic physiology. Ann N Y Acad Sci 586:1–11
Neville MC, Morton J, Umemura S (2001) Lactogenesis. The transition from pregnancy to lactation. Pediatr Clin N Am 48:35–52
Neville MC, Anderson SM, McManaman JL, Badger TM et al (2012) Lactation and neonatal nutrition: defining and refining the critical questions. J Mammary Gland Biol Neoplasia 17:167–188
Nguyen DA, Neville MC (1998) Tight junction regulation in the mammary gland. J Mammary Gland Biol Neoplasia 3:233–246
Nguyen DA, Parlow AF, Neville M (2001) Hormonal regulation of tight junction closure in the mouse mammary epithelium during the transition from pregnancy to lactation. J Endocrinol 170:347–356

Oftedal OT (2002a) The mammary gland and its origin during synapsid evolution. J Mammary Gland Biol Neoplasia 7:225–252
Oftedal OT (2002b) The origin of lactation as a water source for parchment-shelled eggs. J Mammary Gland Biol Neoplasia 7:253–265
Oftedal OT (2012) The evolution of milk secretion and its ancient origins. Animal 6(3):355–368
Oftedal OT, Dhouailly D (2013) Evo-devo of the mammary gland. J Mammary Gland Biol Neoplasia 18(2):105–120
Ognev IF (1915) On the question of the active state of the breast. Bull Soc Imp Nat Moscou Nouv Ser XXIX:13–44. (in Russian)
Osin PP, Anbazhagan R, Bartkova J, Nathan B, Gusterson F (1998) Breast development gives insights into breast disease. Histopathology 33:275–283
Otsuki Y, Yamaji K, Fujita M, Takagi T, Tanizawa O (1983) Serial plasma oxytocin levels during pregnancy and labor. Acta Obstet Gynecol Scand 62:15–18
Ozzello L (1974) Electron microscopic study of functional and dysfunctional human mammary glands. J Invest Dermatol 63:19–26
Parmar H, Cunha GR (2004) Epithelial-stromal interactions in the mouse and human mammary gland in vivo. Endocr Relat Cancer 11:437–458
Perlman M, Shenker J, Glassman M, Ben-David M (1978) Prolonged hyperprolactinemia in preterm infants. J Clin Endocrinol Metab 47(4):894–897
Peters F, Diemer P, Mecks O, Behnken LLJ (2003) Severity of mastalgia in relation to milk duct dilatation. Obstet Gynecol 101:54–60
Pittard WB, Geddes KM, Pepkowitz SH, Carr R (1988) The immunologic composition of neonatal milk: cellular components. Clin Immunol Immunopathol 46(2):294–298
Potten CS, Watson RJ, Williams GT, Tickle S, Roberts SA, Harris M, Howell A (1988) The effect of age and menstrual cycle upon proliferative activity of the normal human breast. Br J Cancer 58(2):163–170
Ramakrishnan R, Khan SA, Badve S (2002) Morphological changes in breast tissue with menstrual cycle. Mod Pathol 15(12):1348–1356
Ramsay DT, Kent JC, Hartmann RA, Hartmann PE (2005) Anatomy of the lactating human breast redefined with ultrasound imaging. J Anat 206(6):525–534
Raynaud A (1961) Morphogenesis of the mammary gland. In: Kon SR, Cowie AT (eds) Milk: the mammary gland and its secretion, vol 1. Academic, New York, pp 3–47
Reversi A, Cassoni P, Chini B (2005) Oxytocin receptor signaling in myoepithelial and cancer cells. J Mammary Gland Biol Neoplasia 10:221–229
Rigg LA, Lein A, Yen SS (1977) Pattern of increase in circulating prolactin levels during human gestation. Am J Obstet Gynecol 129:454–456
Robinson GW, Karpf ABC, Kratochwil K (1999) Regulation of mammary gland development by tissue interaction. J Mammary Gland Biol Neoplasia 4:9–19
Romanò N, Yip SH, Hodson DJ, Guillou A, Parnaudeau S, Kirk S, Tronche F, Bonnefont X, Le Tissier P, Bunn SJ, Grattan DR, Mollard P, Martin AO (2013) Plasticity of hypothalamic dopamine neurons during lactation results in dissociation of electrical activity and release. J Neurosci 33(10):4424–4433
Rudland PS (1991) Histochemical organization and cellular composition of ductal buds in developing human breast: evidence of cytochemical intermediates between epithelial and myoepithelial cells. J Histochem Cytochem 39:1471–1481
Russo J, Russo IH (2004a) Chapter 2. The breast as a developing organ. In: Molecular basis of breast cancer. Prevention and treatment. Springer, Berlin, pp 1–48
Russo J, Russo I (2004b) Development of the human breast. Maturitas 49:2–15
Russo J, Ao X, Grill C, Russo IH (1999) Pattern of distribution of cells positive for estrogen receptor alpha and progesterone receptor in relation to proliferating cells in the mammary gland. Breast Cancer Res Treat 53(3):217–227

Salonia A, Nappi RE, Pontillo M, Daverio R, Smeraldi A, Briganti A, Fabbri F, Zanni G, Rigatti P, Montorsi F (2005) Menstrual cycle-related changes in plasma oxytocin are relevant to normal sexual function in healthy women. Horm Behav 47:164–169

Sapino A, Macrì L, Tonda L, Bussolati G (1993) Oxytocin enhances myoepithelial cell differentiation and proliferation in the mouse mammary gland. Endocrinology 133:838–842

Selenkow HA, Saxena BN, Dana CL (1969) Measurement and pathophysiologic significance of human placental lactogen. In: Pecile A, Finzi C (eds) The Feto-Placental Unit. Excerpta Medica, Amsterdam, 340 p

Seoud MA, Sayigh R, Khayat H, Abou Ali L, Azoury RS (1993) Electrical breast stimulation. Oxytocin, prolactin and uterine response. J Reprod Med 38:438–442

Shaw JA, Udokang K, Mosquera J-M, Chauhan H, Jones JL, Walker R (2002) Oestrogen receptors alpha and beta differ in normal human breast and breast carcinomas. J Pathol 198:450–457

Shennan DB (1994) Regulation of water and solute transport across mammalian plasma cell membranes by prolactin. J Dairy Res 61:155–166

Short RV (1984) Estral and menstrual cycles. In: Austin CR, Short R (eds) Hormonal control of reproduction (Reproduction in Mammals Series) (Book 3), pp 145–192

Shukovski L, Healy DL, Findlay JK (1989) Circulating immunoreactive oxytocin during the human menstrual cycle comes from the pituitary and is estradiol dependent. J Clin Endocrinol Metab 68:455–460

Shutt DA, Smith ID, Sherman RP (1974) Oestrone, oestradiol-17β,and oestriol levels in human foetal plasma during gestation and at term. J Endocrinol 60:333–341

Singh G, Chouhan R, Sidhu K (2009) Effect of antenatal expression of breast milk at term in reducing breast feeding failures. Med J Armed Forces India 65(2):131–133

Smith RL, Pruthi S, Fitzpatrick LA (2004) Mayo Clin Proc 79:353–372

Soderqvist G (1998) Effect of sex steroids on proliferation in normal mammary tissue. Ann Med 30:511–524

Speirs V, Skliris GP, Burdall SE, Carder PJ (2002) Distinct expression patterns of ER alpha and ER beta in normal human mammary gland. J Clin Pathol 55(5):371–374

Sternlicht MD, Korous-Mehr H, Lu P, Werb Z (2006) Hormonal and local control of mammary branching morphogenesis. Differentiation 74(7):365–381

Stingl J, Raouf A, Emerman JT, Eaves CJ (2005) Epithelial progenitors in the normal human mammary gland. J Mammary Gland Biol Neoplasia 10:49–59

Stirling JW, Chandler JA (1976a) The fine structure of normal, resting terminal ductal-lobular unit of the female breast. Virchows Arch A Path Anat Histol 372:205–226

Stirling JW, Chandler JA (1976b) Ultrastructural studies of the female breast: I. 9 + 0 cilia in myoepithelial cells. Anat Rec 186(3):413–416

Stirling JW, Chandler JA (1977) The fine structure of ducts and subareolar ducts in the resting gland of the female breast. Virchows Arch A Path Anat Histol 373:119–132

Stock S, Bremme K, Uvnäs-Moberg K (1991) Plasma levels of oxytocin during the menstrual cycle, pregnancy and following treatment with HMG. Hum Reprod 6:1056–1062

Stoeckelhuber M, Stumpf P, Hoefter EA, Welsch U (2002) Proteoglycan-collagen associations in the non lactating human breast connective tissue during menstrual cycle. Histochem Cell Biol 118:221–230

Suzuki R, Atherton AJ, O'Hare MJ, Entwistle A, Lakhani SR, Clarke C (2000) Proliferation and differentiation in the human breast during pregnancy. Differentiation 66:106–115

Tanner MJ, Hardlow NC, Wardrop R (2011) Variation of female prolactin levels with menopausal status and phase of menstrual cycle. Aust New Zeal J Obst Gynecol 51:321–324

Taylor D, Pearce CL, Hovanessian-Larsen L, Downey S, Spicer DV, Bartow S, Pike MC, Wu AH, Hawes D (2009) Progesterone and estrogen receptors in pregnant and premenopausal non-pregnant normal human breast. Breast Cancer Res Treat 118:161–168

Thoresen M, Wesche J (1988) Doppler measurements of changes in human mammary and uterine blood flow during pregnancy and lactation. Acta Obstet Gynecol Scand 67:741–745

Thorner MO, Round J, Jones A, Fahmy D, Groom GV, Butcher S, Thompson K (1977) Serum prolactin and oestradiol levels at different stages of puberty. Clin Endocrinol 7:463–468
Tobon H, Salazar H (1974) Ultrastructure of the human mammary gland. 1. Development of the fetal gland throughout gestation. J Clin Endocrinol Metab 39:443–456
Tsuchiya S, Li F (2005) Electron microscopic findings for diagnosis of breast lesions. Med Mol Morphol 38:216–224
Tulchinsky D, Hobel CJ, Yeager E, Marshall JR (1972) Plasma estrone, estradiol, estriol, progesterone, and 17-hydroxyprogesterone in human pregnancy. I. Normal pregnancy. Am J Obstet Gynecol 112:1095–1100
Ueda EK, Huang K, Nguyen V, Ferreira M, Andre S, Walker AM (2011) Distribution of prolactin receptors suggests an intraductal role for prolactin in the mouse and human mammary gland, a finding supported by analysis of signaling in polarized monolayer cultures. Cell Tissue Res 346:175–189
Vekemans M, Delvoye P, L'Hermite M, Robyn C (1977) Serum prolactin levels during the menstrual cycle. J Clin Endocrinol Metab 44:989–993
Villadsen R, Fridriksdottir AJ, Rønnov-Jessen L, Gudjonsson T, Rank F, LaBarge MA, Bissell MJ, Petersen OW (2007) Evidence for a stem cell hierarchy in the adult human breast. J Cell Biol 177:87–101
Vogel PM, Georgiade NG, Fetter BF, Vogel FS, McCarty KS Jr (1981) The correlation of histologic changes in the human breast with the menstrual cycle. Am J Pathol 104:23–34
von Schoultz B (2007) Androgens and the breast. Maturitas 57:47–49
Wagner KU, Smith GH (2005) Pregnancy and stem cell behavior. J Mammary Gland Biol Neoplasia 10:25–36
Watson RJ, Eyden BP, Howell A, Sellwood RA (1988) Ultrastructural observation on basal lamina in the normal human breast. J Anat 156:1–10
Weinstein SP, Conant EF, Sehgal CM, Woo IP, Patton JA (2005) Hormonal variation in the vascularity of breast tissue. J Ultrasound Med 24:67–72
Weitzman RE, Leake RD, Rubin RT, Fisher DA (1980) The effect of nursing on neurohypophyseal hormone and prolactin secretion in human subjects. J Clin Endocrinol Metab 51:836–839
Welsch U, Oppermann T, Mortezza M, Höfter E, Unterberger P (2007) Secretory phenomena in the non-lactating human mammary gland. Ann Anat 189(2):131–141
Widdows ST, Lowenfeld MF, Bond M, Shiskin C, Taylor EI (1935) A study of the antenatal secretion of the human mammary gland and a comparison between this and the secretion obtained directly after birth. Biochem J 29:1145–1166
Yap PL, Mirtle CL, Harvie A, McClelland DBL (1980) Milk protein concentrations in neonatal milk (witch's milk). Clin Exp Immunol 39:695–697
Yap PL, McKiernan J, Mirtle CL, McClelland DBL (1981) The development of mammary secretory immunity in the human newborn. Acta Paediatr Scand 70(4):459–465
Yasugi T, Kaido T, Uehara Y (1989) Changes in density and architecture of microvessels of the rat mammary gland during pregnancy and lactation. Arch Histol Cytol 52(2):115–122
Zhou J, Siu NG, Adesanya-Famuiya O, Anderson K, Bondy CA (2000) Testosterone inhibits estrogen-induced mammary epithelial proliferation and suppresses estrogen receptor expression. FASEB J 14:1725–1730

3 The Structure of the Lactating Mammary Gland of a Woman

Abstract

Data on structure and ultrastructure of secretory, myoepithelial, and stem cells of mammary gland alveoli are presented. Cell layer forming walls of ducts and alveoli are completely covered with a basal shell consisting of light and dark layers. The alveolar-ductal system is surrounded by a number of connective tissues and cellular structures—the stroma. During lactation, the volume of the stroma is significantly reduced, mainly due to the replacement of glandular connective tissue and fat components. Data obtained on blood supply, lymph supply, and innervation of the mammary gland belong to the level that is designated as "gross anatomy," i.e., the work describes mainly the general course and branching of blood and lymphatic vessels, as well as nerve stems visible to the naked eye. The internal structures of the gland are innervated by nerve fibers of the vagus nerve (parasympathetic system) and spinal nerve fibers (sympathetic system).

In the previous chapter, when presenting the material on the development and function of the breast, data on the structure of the gland in the corresponding period were simultaneously presented. Given that the structure of a woman's breast reaches its maximum development after the birth of a child for a better understanding of the function of the gland, it is advisable to present in a special chapter more detailed information about the structure and ultrastructure of a woman's lactating gland.

In 1975, in the introduction to their article, G. Tobon and G. Salazar (1975) noted the lack of data in the literature on the ultrastructure of a woman's lactating breast. The situation has not changed much to date. Against the background of a fairly large number of studies regarding the structure and ultrastructure of the mammals of various mammals during lactation, single works are devoted to the structure of the woman's breast during this period not only at the ultrastructural, but also at the light-microscopic levels (Ferguson and Anderson 1983). However, according to the

N. P. Alekseev, *Physiology of Human Female Lactation*,
https://doi.org/10.1007/978-3-030-66364-3_3

experimental data available to date, there were no significant morphological changes in the external and internal structures of a woman's breast compared to the end of pregnancy (Toker 1967; Tobon and Salazar 1975; Battersby and Anderson 1988; Russo and Russo 2004a, b; Hassiotou and Geddes 2013).

As follows from the previous sections, the largest amount of material on the structure of the breast refers to glandular tissue. However, in addition to glandular tissue, other components of the breast that are necessary for its normal functioning during lactation develop and change. Unfortunately, there is relatively little information about blood supply, lymph supply, innervation of internal and surface areas of the breast, as well as data on connective tissue formations of a woman's lactating breast. However, for a better understanding of the work of a woman's breast in the main period of its functioning—lactation in the postpartum period—it is advisable to present together the available data on the structure and ultrastructure of the lactating breast of a woman.

3.1 Structure of the Alveolar-Ductal System

After childbirth, the glandular tissue reaches a significant development and it is dominated by the milk lobes of type 3 (Fig. 2.11). The growth of the alveolar-ductal system slows down, but what is very important is the differentiation and proliferation of the alveolar-ductal system does not end completely, but continues throughout lactation (Russo and Russo 2004a, b). Therefore, along with type 3 lobes, type 2 and type 1 lobes are marked in glandular tissue.

3.1.1 The Alveolus

The main structural and functional unit of glandular tissue is the alveola. From the currently available morphological works on the light or ultrastructural levels, it is difficult to choose a photo illustrating the general structure of the alveola and the ducts of the mammary gland of a woman. Therefore, in Fig. 3.1, the structure of the alveolus and primary duct of the lactating mammary gland of a woman is given in a schematic form on an appropriate scale. It should be noted here that a detailed comparative study of the general structure of the mammary alveoli of women and rats showed a great similarity in their size and location of secretory and myoepithelial cells in the alveolus (Emerman and Vogl 1986). In this regard, in addition to a better understanding of the structure of the alveolus, Fig. 3.1a shows a photo of the alveolus of the mammary gland of a lactating rat obtained by electronic scanning.

In the lactating mammary gland (Fig. 3.1), the alveola is a sphere formed by two layers of cells—secretory (epithelial) and myoepithelial. According to morphological data of various authors, the diameter of the alveoli in the lobes of the mammary glands of women during this period can vary within 50–140 microns and mainly depends on the degree of filling of the alveolar cavity with secret. The alveoli are

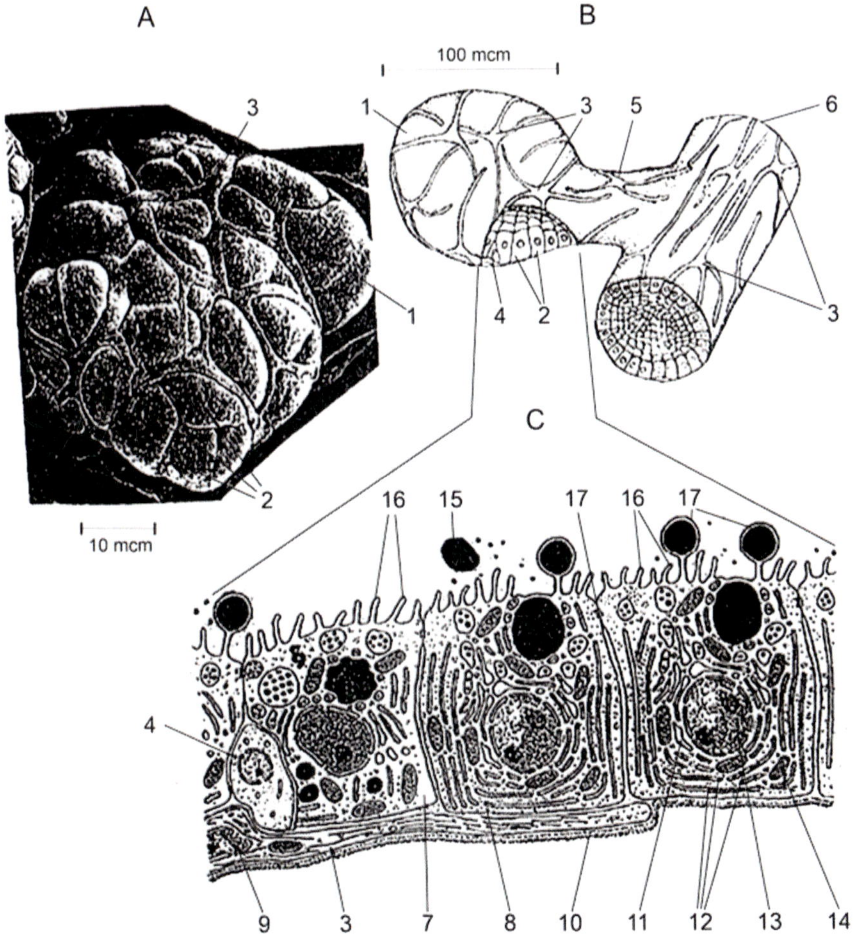

Fig. 3.1 Structure of the alveolus of the lactating breast of a woman. (**a**) Scanning photo of the alveolus of the lactating mammary gland (from Nagato et al. 1980). *al* alveola, *sc* secretory cells, *mc* myoepithelial cells. (**b**) Alveola, primary and intralobular duct. The drawing is made on the basis of electron microscopic data of the alveoli of the lactating breast. *al* alveola, *mc* myoepithelial cell, *sc* secretory cell, *stc* stem cell, *ad* alveolar duct, *ild* intralobular duct. (**c**) Drawing of alveolar wall cells. *mc* myoepithelial cell, *mcn* myoepithelial cell nucleus, *lsc* light secretory cell, *dsc* dark secretory cell, *stc* stem cell, *bs* basal shell, Golgi apparatus, *rr* rough reticulum, *scn* secretory cell nucleus, *mt* mitochondria, *mv* microvilli, *f* fat drop, *tc* intercellular tight contact

connected by short primary ducts with a diameter of 40–70 microns with a wider common duct of the lobule. The inner cavity of the alveolus and primary duct is lined with a layer of secretory cells—lactocytes. As already mentioned in the previous chapter, using a system of tight junctions, the lateral sections of secretory cells located near the alveolar cavity are bound together throughout the entire length so that the contents of the alveolar cavity are largely isolated from the surrounding intercellular environment (Fig. 3.1, as well as Fig. 4.2). This is done using special proteins. The main proteins of tight junction of the mammary glands are claudins, occludins, and adhesive proteins—JAM (junction adhesion molecule) (Itoh and Bissell 2003).

In addition to tight junction, the mechanical integration of cells that make up the alveolus involves contacts called desmosomes (Fig. 3.1). The function of desmosomes is mainly to provide mechanical communication between cells (Andrews et al. 2012). In the desmosome region, the membranes of the lateral surfaces of secretory cells in limited areas (up to 0.5 microns in diameter) are tightly adjacent to each other. In this area between the cell membranes there are cell adhesion proteins from the cadherin family and connecting proteins that combine them with intermediate filaments located in the cytoplasm of the cell and form a network that has a high tensile strength. Cell adhesion proteins include desmoglein and desmocollin. These transmembrane proteins have five extracellular domains and are calcium-binding. They provide a homophilic connection of cells, i.e., two identical protein molecules are connected to each other. The intracellular protein desmoplakin (with the participation of two more proteins, plakophilin and plakoglobin) connects desmoglein domains with intermediate filaments.

Another type of contact was found between the lateral surfaces of secretory cells. Just as in the case of desmosomes in limited areas, special protein formations pass through the cell membranes of neighboring cells, which are hexagonal subunits—connexons—with a distance between them of 8–10 nm. Each connexon consists of six connexins of a polypeptide nature, constructed so that they create a channel, as if surrounding it. Passing through the bilayers of the membranes of each of the two neighboring cells, the connexons exit into the intercellular gap, where they connect with each other and form a contact structure in the form of a channel between the cytoplasm of two neighboring cells. As a result of the fact that the single connexins of each connexon have the ability to shift relative to each other, the central channel of the connexon can be opened or closed so that water, small molecules, and ions, as well as an electric current freely pass from one cell to another (in both directions). This structure is called the gap junction. Studies of the distribution and structure of slit contacts in the alveoli of a woman's breast have found that slit contacts between secretory cells are quite rare (Monaghan and Moss 1996).

The second layer of alveola cells is formed by myoepithelial cells, which by their processes braid the layer of secretory cells of the alveola and duct (Fig. 3.1). On the ducts, myoepithelial cells are located more densely than on the alveoli and their processes are localized mainly along the longitudinal axis of the duct. The processes of myoepithelial cells form slotted contacts between themselves, and their density is much greater than between secretory cells. No slotted contacts were found between secretory and myoepithelial cells in the alveoli and ducts of the woman breast (Monaghan and Moss 1996). The processes of myoepithelial cells form desmosomal contacts with the basal membrane, which are designated as hemidesmosomes.

In addition to secretory and myoepithelial cells, stem cells are localized in the walls of the alveoli and ducts (Fig. 3.1) and progenitor cells (Smith et al. 1984; Smith and Chepko 2001), the number of which increases in the lactating mammary gland (Hassiotou et al. 2012). The number of stem cells is greater in the ducts than in the alveoli (Villadsen et al. 2007). Along with the cell types mentioned above that are constantly present in the alveoli and ducts, lymphocytes and macrophages are found migrating in an unknown way from the inside and between the lobules of the gland. Just as in the glands of non-pregnant and pregnant women, the lactating mammary gland is dominated by lymphocytes. The total percentage of lymphocytes and macrophages relative to the entire cell population of alveoli and ducts is similar to that in the mammary gland of pregnant women and is about 10% (Ferguson 1985). The alveoli and ducts are separated from the interlobular stroma by a basal shell. Let's take a closer look at the structure of cells that form the alveolus and the primary duct of the lactating breast.

The greatest changes during lactation of a woman occur in the structure of secretory cells (Tobon and Salazar 1975) (Fig. 3.1c) in lactocytes, the number of organelles increases significantly, which indicates an increase in their metabolic activity (Fig. 3.1c). In particular, the number of free ribosomes and polyribosomes increases. The rough endoplasmic reticulum is located in the cell in the form of ladders. Enlarged cisterns of the rough endoplasmic reticulum contain granular dense material and occupy a significant volume of the cell. As a result, the cytoplasm of lactocytes on the sections becomes dark. The Golgi apparatus is mainly concentrated in the apical regions of secretory cells. Its expanded cisterns and vesicles contain osmiophilic material organized as sinuous bead-like structures. Numerous elongated or oval mitochondria with clearly defined crists are present in the cytoplasm of cells. Lipid droplets of various sizes are observed in the cytoplasm of cells. The nuclei of secretory cells are rounded, often have wavy contours, sometimes with deep invaginations. Droplets of chromatin fill the inner volume of the nuclei. The surface of secretory cells in contact with the alveolar cavity has short thin outgrowths—microvilli, between which secretory and pinocytotic vesicles are observed. In addition, there are larger outgrowths containing lipid droplets entering the alveolar cavity by the type of apocrine secretion (Fig. 3.1c).

In addition to secretory cells with increased metabolic activity (their overwhelming number in the alveoli), cells are observed to be "at rest" (Tobon and Salazar 1975) (Fig. 3.1c). Their cytoplasm contains oval or slightly elongated mitochondria, as well as free ribosomes and thin parallel profiles of a rough reticulum, in the cisterns of which there is a rare fine-grained granular material. The capacitance system of the Golgi apparatus is slightly expanded and the content of granular material is minimal. In the cytoplasm of cells, fat droplets are noted, but their number and size are less than in actively secreting cells. In the images, these cells look lighter than actively secreting cells. Microvilli are also present on the luminal surface of "resting" secretory cells, but there were no cases of secretions (Fig. 3.1c). The shape and structure of the nuclei of "resting" cells did not differ from those of actively secreting cells.

The second layer of the wall of the alveoli and ducts consists of myoepithelial cells. Unlike secretory (epithelial) cells, they form a mesh structure, entwining the alveoli and ducts with their cytoplasmic processes, between which desmosomes are observed. On the alveoli, they form plexuses shaped like baskets (Fig. 3.1b). Hence another name for myoepithelial cells—basket cells. On the walls of the ducts, the

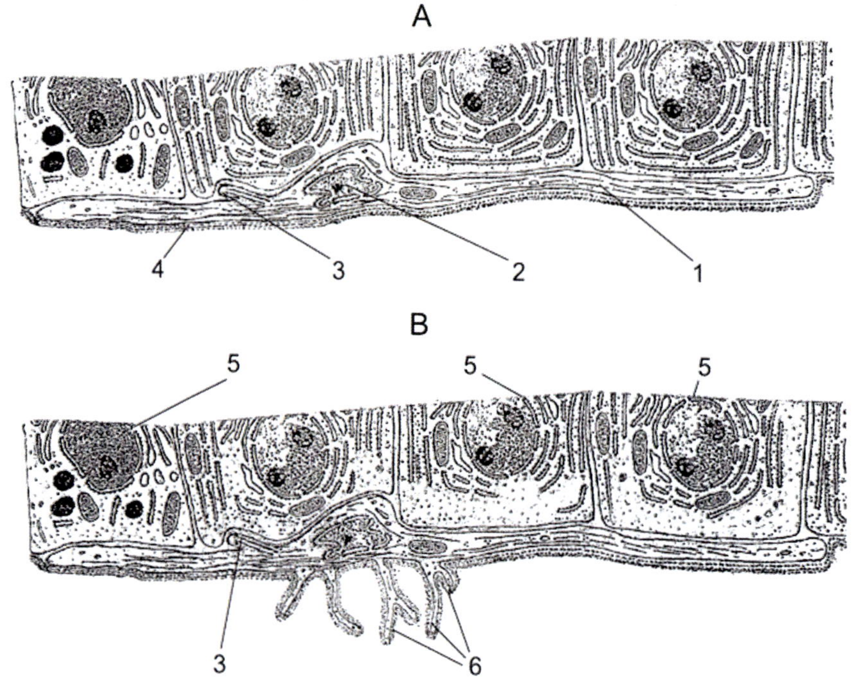

Fig. 3.2 Myoepithelial cells of the breast. The figures show only the basal parts of secretory cells. (**a**) Myoepithelial cells of alveoli, intralobular, and thin interlobular ducts. (**b**) Myoepithelial cells of the milk ducts located under the areolar region. *mc* myoepithelial cell, *mcn* myoepithelial cell nucleus, *mcc* myoepithelial cell cilia, *bs* basal shell, *mcp* myoepithelial cell processes, *sc* secretory cells

processes of myoepithelial cells are located mainly along the longitudinal axis of the duct. Myoepithelial cells are characterized by the presence of a multi-lobed nucleus. The Golgi apparatus and rough reticulum are less developed than secretory cells and never show secretory activity. In addition, the cells contain rare small oval-shaped mitochondria and glycogen droplets. A distinctive feature of the internal structure of myoepithelial cells is the presence in the cytoplasmic processes of a large number of filaments, which are similar in structure to the myofilaments of smooth muscle cells and also in smooth muscle cells give color to cytokeratin 14 (Hassiotou et al. 2013). Another feature of myoepithelial cells is the presence of special cilia hairs (Stirling and Chandler 1976, 1977) (Fig. 3.2b). There are 9 pairs of fibrils located inside the cilia along the periphery, and there is no central pair of fibrils. Cilia penetrate the cell and are associated with the basal body. Cilia are localized on the surface of the myoepithelial cell facing the basal sides of the epithelial cells. The diameter of the cilia is about 250 nm, and the maximum length is 830 nm. Cilia were observed in myoepithelial cells located in all parts of the ductal system of the breast. The significance of cilia for the morphogenesis of the ductal system of the mammary glands has already been discussed in the previous chapter. Additionally, it is assumed that in the lactating mammary gland, cilia can participate in coordinating the functioning of myoepithelial cells during milk excretion. This function of cilia will be discussed in more detail in the next chapter. Myoepithelial cells had a noticeable difference in shape depending on their location. Thus, myoepithelial cells located on the ducts closer to the areolar region and under the areolar region often formed quite long processes immersed in the stroma. While on the alveoli and thin ducts, the contact of myoepithelial cells with the stroma was flat form (Fig. 3.2a) (Tobon and Salazar 1975; Stirling and Chandler 1977; Tsuchiya and Li 2005).

The third type of cells constantly present in the alveoli and ducts of non-lactating and lactating mammary glands is stem cells. Systematic studies of the structure and function of breast stem cells in women began in the 1980s of the last century (Smith et al. 1984), but the most intensive study was conducted in the first decade of this century (Villadsen et al. 2007; Patki et al. 2010; Hassiotou et al. 2012, 2013, Hassiotou and Geddes 2013; Hosseini et al. 2014). Some of the data on stem cells were presented in the previous chapter in the section on breast development during puberty. It should be noted that sometimes other cell types are identified as stem cells. Therefore, to date, there are a number of criteria by which cells are defined as stem cells (Johnson 2010). In particular stem cells should: (1) be multipotential, (2) produce their own copies, (3) do not react to markers of mature cells, (4) have low metabolic activity (be in a "resting" state), and (5) provide long-term regeneration of tissue sites in which they are localized. In recent experiments using cytochemical and tissue culture techniques, cells that meet the criteria were found in the alveoli and ducts of a non-lactating woman breast (Villadsen et al. 2007). Experimental confirmation that the lactating mammary gland of a woman contains stem cells is the information obtained using cytochemical methods and tissue culture methods about their presence in woman milk (Cregan et al. 2007; Hassiotou et al. 2012). Stem cells get there from the walls of the alveoli and ducts in some unknown way. Data on the structure and localization of stem cells in the walls of the alveoli and ducts at the ultrastructural level of a woman's lactating gland are not available in the literature. However, a comparison of the results of ultrastructural studies of non-lactating mammary glands in women (Stirling and Chandler 1976, 1977; Smith et al. 1984) and data obtained on lactating mammary glands in rats (Smith and Chepko 2001) suggests that they are localized between the lateral surfaces of secretory cells (Fig. 3.1c) or located between the basal areas of lactocytes and myoepithelial cells. Probably, stem cells do not communicate with any part of their surface with the cavity of the alveolus or duct. Unlike leukocytes and macrophages, stem cells contact secretory and myoepithelial cells via desmosomes (Smith et al. 1984). Stem cells are usually smaller than secretory cells and have an amoeboid shape. The number of intracellular organelles in them is less than in secretory cells with increased metabolic activity. Therefore, the cytoplasm of stem cells is light. It should be noted that the number of stem cells in the glandular tissue of lactating rats is approximately 3% (Smith and Chepko 2001). It is possible that the lactating woman gland contains stem cells of the same order. Undoubtedly, the low content of stem cells makes their morphological study difficult. Differentiation of stem cells into epithelial (secretory) and myoepithelial cells occurs not directly, but through an intermediate link—progenitor cells (precursors), which are present in the walls of the alveoli and ducts. Three types of progenitors were identified using appropriate markers in the woman breast (Stingl et al. 2005). The first type is bipotent, resulting in differentiation of progenitors for secretory and progenitors for myoepithelial cells, respectively. The second type is a progenitor only for secretory cells and the third type, respectively, for myoepithelial cells. It is interesting to note that when breast stem cells are cultured in a special environment, they can differentiate into neurons, oligodendrocytes, and astrocytes (Hosseini et al. 2014).

It should be noted that information about progenitors was obtained by histochemical methods in tissue culture. Unfortunately, accurate data on their structure and localization at the ultrastructural level in tissue culture, as well as for stem cells in the alveoli and ducts of the mammary gland of women are not available. It is important to note that at all stages of breast development: the fetal mammary glands before birth, during puberty, the mammary glands of women with an ovulatory cycle, and the mammary glands during pregnancy (see in the previous chapter), cells with reduced metabolic activity are marked-light cells that can be stem cells or progenitors.

The cell layer that forms the walls of the ducts and alveoli of the lactating gland is completely covered by the basal membrane or rather the basal shell (Figs. 2.1 and 3.2). The basal shell consists of two layers, called light (lamina lucida) with a thickness of 15–30 nm and dark (lamina densa) with a thickness of 40–80 nm (Tsuchiya and Li 2005). The components of the basal lining of the breast include type IV collagen and a number of glycoproteins: laminin, nidogen type 1 and 2, perlecan, and fibronectin. All these glycoproteins are found in the basal shell covering the alveoli and various parts of the milk ducts (Johnson 2010). Myoepithelial cells in contact with basal membrane form hemidesmosome. Areas of epithelial cells are also suitable close to the basal shell, but the formation of hemidesmosome was not observed.

It is important to note that due to the content of various glycoproteins, the basal shell is of great importance for the normal functioning of secretory and myoepithelial cells (Johnson 2010).

3.1.2 Ductal System of the Breast

Alveoli, connecting with the help of their ducts with a wider common duct, form lobules (Fig. 2.11). In turn, lobules form lobes by combining their ducts. As the lobes join the common duct of the lobe, its diameter increases and reaches the maximum value in the subareolar region when approaching the nipple. Then the milk duct at the entrance to the nipple narrows several times in diameter and forms a milk channel with its exit at the tip of the nipple. Externally, the ductal system of a woman's breast lobe resembles a river with numerous tributaries. For illustration, Fig. 3.3a shows in a very visual form the duct system of one lobe, obtained using the radiography technique by D. B. Kopans (1989). Various manuals on women's lactation and breastfeeding, as well as anatomical atlases, indicate that there may be 15–25 such

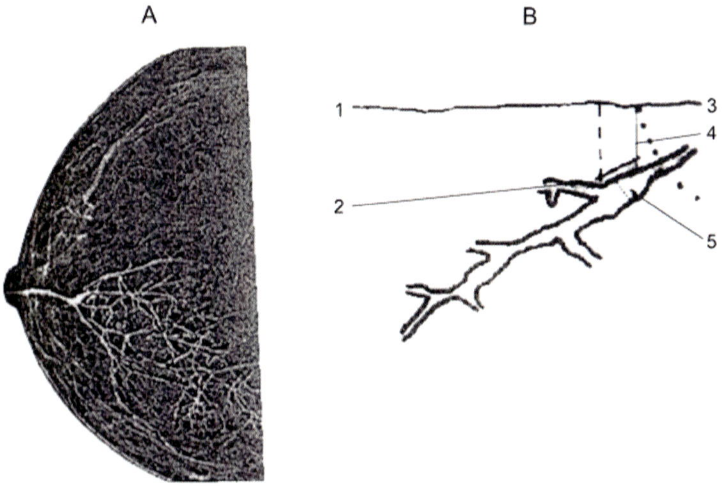

Fig. 3.3 Ductal system of the breast. (**a**) Ductal system of one lobe of the breast (from Kopans 1989). White branching lines—milk ducts. In the image of Kopans D. B., only the contours of the nipple are guessed, so the nipple is retouched for clarity and it shows the milk flow in scale. (**b**) Location of the breast ducts when they approach the breast nipple (from Ramsay et al. 2005a). 1—the skin surface of the breast, 2—the first branch of the duct, 3—the base of the nipple, 4—the depth of the main duct when it passes into the milk duct of the nipple (indicated by a thin solid line), 5—the main duct. Thick solid line is the length of the non-branching part of the main milk duct; thin dotted line is the diameter of the main duct; thick dotted line indicates the depth of the first branch of the main channel; dotted thick line is the base of the nipple

"milk rivers" (Grachev and Galantsev 1973; Alipov et al. 1988; Akre 1989; Vorontsov et al. 1993; Lawrence and Lawrence 1999, 2011, Koyama et al. 2013, etc.). However, experimentally confirmed data indicate that the number of lobes in a woman's breast and common ducts that fit to the nipple varies between 5 and 12 in different women (King and Love 2006). For comparison, in the mammary glands of mice and rats, the lobules are united in a single milk duct. For the first time, a detailed study of the ductal system of the lactating breast of a woman, which can be called a classic, was carried out by E. Cooper (1840). Through various milk ducts of the nipple of the breast of a woman who died during lactation, wax of different colors was introduced, filling the ductal system of various lobes. Then, after appropriate treatment, the breast tissue was removed, and a wax "cast" remained. Color drawing of the "cast" of the ductal system of the breast made a strong impression on experts so that it was actively used to illustrate the structure of the breast of women in various anatomical atlases and textbooks. Further studies of the ductal system of the breast of lactating and non-lactating women with the use of other more modern methods (radiography, ultrasound) in the main features confirmed the data of E. Cooper. These methods have made additional adjustments to the overall structure (gross anatomy) of the ductal system of the woman breast (Fig. 3.3a) (Kopans 1989; Hou et al. 2001; Ramsay et al. 2005a). In particular, when approaching the base of the nipple, the milk duct of the lobe had the largest diameter, but bag-like extensions (sinuses) as it is shown in the drawings of A. Cooper in the lactating and non-lactating gland were not observed. It is believed that the appearance of milk sinuses on wax casts in A. Cooper was associated with a fairly intense increase in pressure and accordingly, "inflating" the sections of ducts near the base of the nipple when the wax was introduced (Ramsay et al. 2005a; Gooding et al. 2010).

In A. Cooper's drawings, the lobe ducts radiate from the base of the nipple. However, as shown by radiography and ultrasonography (Kopans 1989; Hou et al. 2001; Ramsay et al. 2005a; Gooding et al. 2010), this pattern is not observed in lactating and non-lactating mammary glands. The milk ducts in the areolar region can intersect, passing over each other in a variety of ways. Here it is necessary to point out that A. Cooper in his work noted that for greater clarity of the structure of the ducts, he "untangled" their final parts and arranged them in a circle (Cooper 1840). Unfortunately, this idealized drawing was later used without reservation in anatomical atlases and textbooks as an illustration of the structure of a woman's normal breast.

A feature of the ductal system of a woman's breast was the lack of uniformity in increasing the diameter of the ducts as they merge and approach the areolar region. So, in some places where the lobe ducts join at the periphery of the lobe, the ducts may expand to the diameter of the ducts in the areolar region. But in the future, as you move to the nipple, the expansion decreases, which is clearly seen in the "waxograms" of A. Cooper (1840) and was later confirmed using the radiography technique (Kopans 1989; Hou et al. 2001) (Fig. 3.3a). The distribution of alveolar tissue belonging to one lobe varies significantly from lobe to lobe (Kopans 1989). So in some cases, the proportion is relatively compact located in any part of the breast. In other cases, part of the glandular tissue of the lobe may be localized at a significant distance from the main location of the lobe. In addition, by volume (by the degree of branching) and therefore by the amount of milk contained in them, the shares may differ markedly. For example, in experiments on cadavers during reconstruction of the ductal system of the breast (Going and Moffat 2004), it was found that the volume of the capacitive system belonging to one main duct can be equal to about 23% of the volume of the gland, and the volume of the three largest ducts—about 75% of the volume of the gland. At the same time, the volume of the other four ducts can be about 1.6% of the volume of the gland.

The use of an ultrasound non-invasive technique allowed examining the ductal system of the breast in lactating women not only in the periods between feedings, but also during the feeding of the child (Ramsay et al. 2005a, b; Geddes 2009a, b; Prime et al. 2009). However, the authors note that the resolution of the ultrasound method has certain limitations. Thus, ducts smaller than 0.5 mm were not clearly visible on ultrasound images. In addition, the course of the ducts could not be traced in the thickness of glandular tissue. Therefore, the examination was limited to the subareolar region, i.e., the location of the end of the large lobe ducts and their transition to the milk passages of the nipple.

According to Ramsay and colleagues (Ramsay et al. 2005a, b), the number of main ducts that fit to the nipple of the left breast was in the range of 6–12 and the right 4–14. The diameter of the main ducts varied between 1.0 and 4.4 mm and 1.0

and 4.0 mm, respectively, and did not differ from the diameter of the mammary ducts of non-lactating women (Mendelson 1998). The number of main ducts and their diameter was not related to nipple diameter, areola radius of the gland, and milk productivity of women. In addition, there was no correlation between the number of main ducts and their diameter. The depth of the main milk ducts relative to the skin surface of the areola was 4.5 ± 1.98 mm for the left and 4.74 ± 1.59 mm for the right breast. The distance from the base of the nipple to the beginning of branching of the main ducts for the left breast was 8.21 ± 3.96 mm and 7.20 ± 2.43 mm for the right (Fig. 3.3b). It should be noted that in clinical studies during breast ductoscopy (insertion of a catheter into the milk ducts of the nipple and the ducts of the gland), the presence of sphincters in the milk ducts is indicated (Zografos et al. 2009). According to the authors, sphincters, structures that regulate the movement of milk, are located approximately 10 mm from the base of the nipple, i.e., according to Ramsay et al. (2005a, b) in those places where the main ducts begin to branch. However, in studies conducted using radiography (Kopans 1989; Hou et al. 2001), ultrasound (Ramsay et al. 2005a, b; Geddes 2009a, b; Prime et al. 2009), and morphological methods: light (Giacometti and Mjntagna 1962; Montagna 1970) and electron microscopic (Stirling and Chandler 1977) do not note the presence of structures in the milk ducts that could be attributed to sphincters.

The ductal system of glandular tissue passes into the milk ducts—channels of the nipple of the breast. When passing the ducts of the glandular tissue in the area of the base of the nipple, the main ducts are significantly narrowed. The milk channels of the nipple open mainly on the surface of the tip of the nipple in the depressions between the furrows of the epidermis (Fig. 3.4). Unfortunately, due to the high density of the nipple tissue and the small diameter of the nipple ducts, it is impossible to apply ultrasound techniques and, accordingly, measure the diameter, number, and track the course of the milk channels along the entire length of the nipple of lactating women. It should be noted that the question of the number of milk channels in the nipple resembles the situation of the number of lobes in the breast of a woman and the common ducts of the lobes that fit to the nipple. In various reviews, manuals, monographs, and anatomical atlases devoted to the mammary gland of women, it is indicated without reference to experimental confirmation that the presence of milk channels in the nipples significantly exceeds the number of ducts that fit to the base of the nipple. The number 15–25 is often mentioned (Grachev and Galantsev 1973; Akre 1989; Lawrence and Lawrence 1999, 2011; Koyama et al. 2013, etc.). Moreover, some histological studies of recent years indicate the possibility of the presence of 28 (Rusby et al. 2007), 30 (Taneri et al. 2006), or even 48 (Going and Moffat 2004) milk passages in the nipple. It is important to note that for the normal functioning of the gland, it is necessary that the milk ducts of the lobes through the milk passages of the nipple open outwards. Detailed analysis of data on the localization of milk ducts obtained using histological and immunohistochemical methods on mastectomized nipples (Going and Moffat 2004; Love and Barsky 2004), as well as using transareolar injection of paint to the base of the nipple in vitro and in vivo in lactating women to detect the exit of the nipples and direct observation of the exit of the nipple milk channels in lactating women (Love and

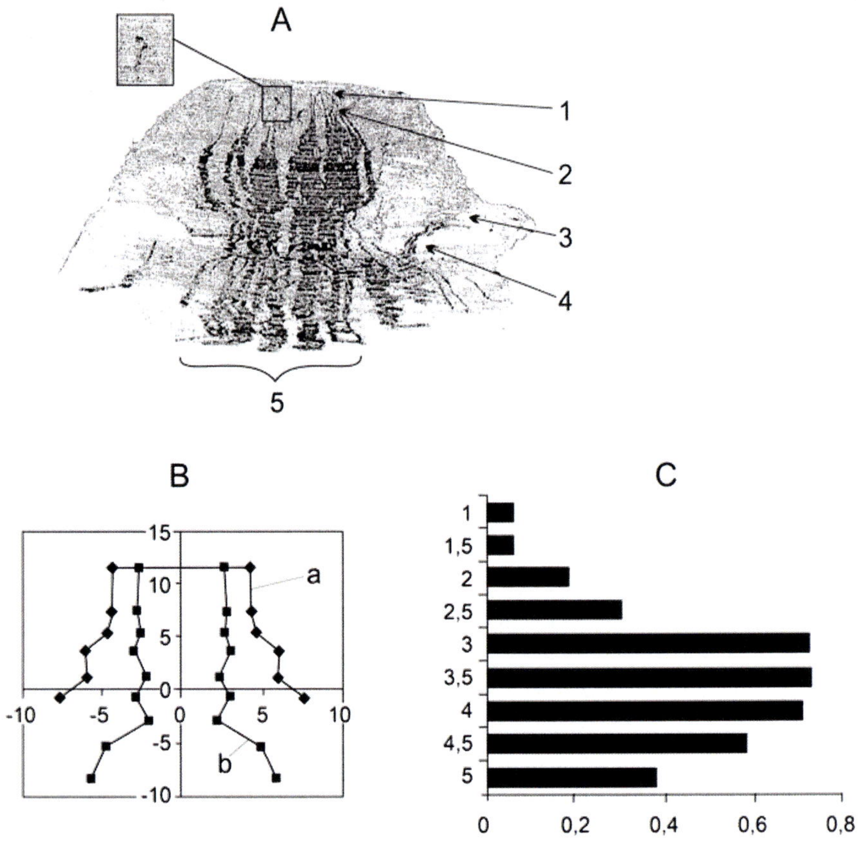

Fig. 3.4 Milk ducts of the breast nipple (from Rusby et al. 2007). (**A**) Three-dimensional reconstruction based on a series of 5 μm slices. a—mammary ducts of the nipple with a common output, b—reduction of the diameter of the ducts on the tip of the nipple, in the region of the areola, the rectangle shows an enlarged increase in the diameter of the milk duct when exiting the nipple. (**B**) Graphic illustration of a sagittal section through a schematically represented nipple. The horizontal line (as the abscissus axis) indicates the level of the base of the nipple and, accordingly, its radius in mm. The vertical (ordinate axis) represents the length of the nipple. GS—the outer part of the nipple, GA—the border of the bundle of milk passages passing inside the nipple. (**C**) Change in the diameter of the milk ducts along the course of the nipple. On the ordinate axis—the length of the nipple (increase in length from top to bottom) (in mm), on the abscissus axis—the diameter of the milk ducts (in mm)

Barsky 2004) allowed us to conclude that the number of milk passages opening at the tip of the nipple is 5–9. The remaining ducts, or rather tubular structures in the nipple, may be ducts belonging to the sebaceous and sweat glands passing among the milk ducts of the nipple. This was pointed out in early histological works by W. Montana (Giacometti and Mjntagna 1962; Montagna 1970). Here it is interesting to note that the ducts of some sebaceous glands sometimes open into the milk ducts of the nipple (Giacometti and Mjntagna 1962). The increased number of milk ducts may be due to their branching in the nipple (Love and Barsky 2004; Going and Moffat 2004; Taneri et al. 2006). In this case, one branch may open at the tip of the nipple, and the other does not reach the surface, but on a histological section, for example, made in the middle of the nipple, two ducts will be found. In addition, it is believed that there may be two populations of ducts in the nipple. One population is true milk ducts, the other consists of tubular structures originating from the outgrowth of the nipple skin (Going and Moffat 2004). Thus, the analysis of experimental data from various authors allowed us to conclude that the number of milk ducts passing through the nipple and opening out is within the range of 5–12 (King and Love 2006) and "joins" with the number of milk ducts of the lobes that fit to the base of the nipple.

The walls of the milk ducts, which are a continuation of the milk ducts of the lobes, are formed by a single-layer cuboid epithelium and covered with a basal shell. There were no myoepithelial cells in the walls of the nipple ducts. The milk ducts are grouped and pass in the central part of the nipple in the form of a bundle (Fig. 3.4). When the duct exits, the epithelium of the milk ducts passes into the epithelium of the skin's epidermis. The diameter of the milk ducts in the same nipple varies markedly. This can be easily detected by the thickness of the milk streams coming out of the nipple when the areola is compressed, where the end parts of the lobe ducts are located. An approximate estimate from our research shows that the thickness of the streams can vary from 0.1 to 0.6 mm. In addition, as has been shown in histological studies, the diameter of the milk ducts along the course of the nipple also changes. However, experiments were performed on mastectomized nipples of non-lactating women (Going and Moffat 2004; Rusby et al. 2007) (Fig. 3.4). In these experiments, it was found that when approaching the tip of the nipple (1 mm from the tip), the diameter of the ducts is the smallest and is about 0.1 mm. To the outlet, the ducts in most cases expand in the form of a funnel with an increase in diameter by 1.5–2 times. When moving deeper into the nipple (3.5–4 mm from the tip), the diameter of the ducts increases and reaches 0.4–0.8 mm. Then the diameter decreases again to 0.4 mm (Going and Moffat 2004; Rusby et al. 2007). At the exit, the milk passages on the cross-section have a round or oval shape, but when passing inside the nipple they take a star shape (Fig. 3.5) (Going and Moffat 2004; Taneri et al. 2006). Probably, the folding is due to the lack of secret in the milk passages. When filling the milk ducts with secret, the folds are straightened and the ducts take a circular shape in cross-section. It can be assumed that the profile and diameter of the nipple ducts along the course of the nipple are preserved in lactating women. This is supported by data from D. Geddes (2009b). In particular, when milk enters from the main ducts of the gland during the reflex of milk ejection to the milk ducts of the nipple, the diameter of the latter increases so that, at least in the middle part of the nipple, they become visible on ultrasound images.

When exiting the nipple surface, the presence of keratin plugs inside the ducts is noted (Fig. 3.5) (Giacometti and Mjntagna 1962; Montagna 1970; Going and Moffat 2004). It was found that keratin, a waxy substance, serves as a barrier to the path of microorganisms in the milk cistern of the teat and udder of cows. It is believed that a lack of keratin may be the cause of inflammation (mastitis) of the udder in cows.

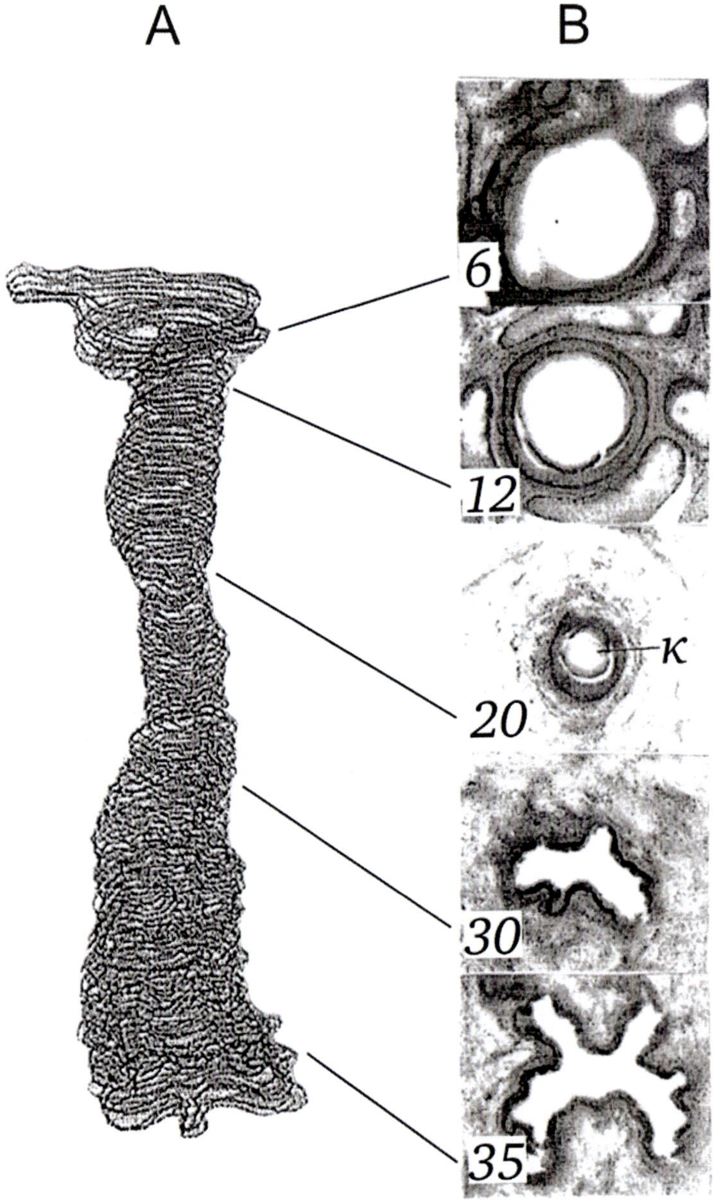

Fig. 3.5 Milk duct of the breast nipple, which has an outlet at the tip of the nipple. (**a**) Three-dimensional reconstruction of the milk channel based on serial sections (from Going and Mohan 2004). (**b**) Electron microscopic series of sections at different levels of the nipple duct. The numbers on the images are the number of the cut, and the lines indicate the approximate alignment of the cut along the milk duct of the nipple. On a 20 slice k-keratin plug in the duct (from Going and Moffat 2004)

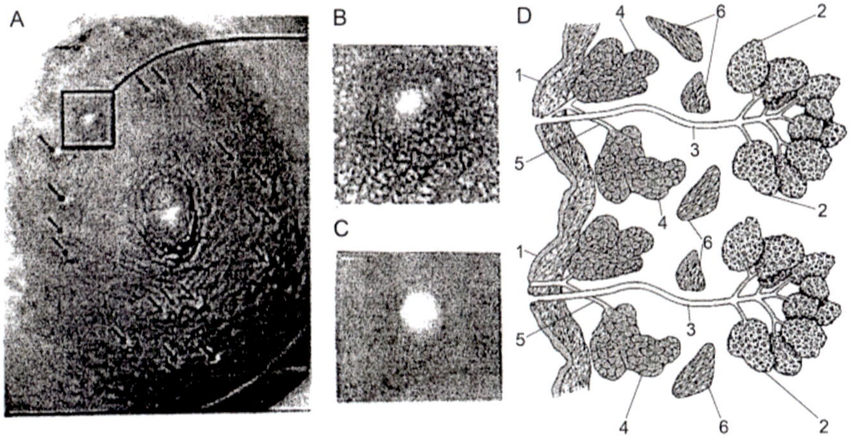

Fig. 3.6 Montgomery's glands on the breast areola. (**a**) Photo of the breast areola (left gland, 3 days after birth). Arrows mark the Morgagni-Montgomery tubercles on the surface of the areola (0.71 of the original size). (**b**) Photo of the Morgani-Montgomery hillock enclosed in a square on A is enlarged by 2.5 times. (**c**) Photo of a drop of secretions on the top of the hillock, released from the Montgomery gland enlarged 3 times. (**a, b**) From Doucet et al. (2012). (**d**) Schematic representation of the Morgani-Montgomery hillock and its internal structures (from Lawrence and Lawrence 1999). 1—tubercle on the skin of the areola, 2—milk lobules (Montgomery glands), 3—ducts of the mammary glands, 4—sebaceous glands, 5—ducts of the sebaceous glands, 6—smooth muscles of the areola

Perhaps the same function is performed by keratin plugs in the breast of women, closing the entry of infection into the ducts of the lactating and non-lactating glands.

In addition to the tip of the nipple on the skin surface of the areola and especially the lactating breast, there are tubercles, from which drops of milk are very often released. These structures on the surface of the areola were first described by Morgagni in 1719. Further light-microscopic studies of the tubercles, which were initiated by V. Montgomery (Montgomery 1837; Montagna and Yun 1972; Smith Jr et al. 1982; Rusby et al. 2007) and were, respectively, called Morgani—Montgomery tubercles, found that the duct of the mammary lobe exited at the top of the tubercle. In addition, a duct of the sebaceous gland opened nearby, which could sometimes be combined with the duct of the breast (Montagna and Yun 1972) (Fig. 3.6). The glandular lobes whose ducts exit at the top of the areola tubercles of the mammary gland are called the Montgomery glands. The number of Morgani—Montgomery tubercles on the areola of the lactating gland can vary between 0 and 40 (Doucet et al. 2009, 2012). When oxytocin is reflexively entered into the bloodstream of the breast (for more information, see Chap. 4) milk drops are released from the tubercles so that at a high density of the tubercles, the areola looks like the milky field of lower mammals (Grachev and Galantsev 1973). In connection with the peculiar "milk atavism" there was a question about the functional significance of the Montgomery glands. In the literature on lactation, a number of provisions concerning the structure and function of the Montgomery glands have appeared at various times. In particular, it was assumed that these glands differ in structure from the "normal" milk lobes and produce a secret that also differs in composition from normal milk. However, in morphological studies, at least at the light-microscopic level (Montagna and Yun 1972; Smith Jr et al. 1982), it was found that the Montgomery glands do not differ in structure from the other lobes, whose ducts pass inside the nipple and open at its tip. The outlets of the mammary ducts of the glands on the surface of the tubercle have a funnel-shaped shape, the diameter of which can reach 0.5 mm. The ducts pass inside the gland at a distance of about 4 mm and begin to branch so that each branch is a duct of a separate glandular lobule. Tubercles (Fig. 3.6d). The walls of the milk ducts are formed by a single-layer cuboid epithelium, on top of which is a layer of myoepithelial cells, covered with a basal shell. In the area where the duct exits to the surface, the cuboid epithelium is replaced with a special epithelium that can secrete keratin. Just like the outlets of the milk ducts at the tip of the nipple, the outlets of the ducts of the Montgomery glands are closed by keratin plugs (Montagna and Yun 1972). The segments of the Montgomery glands are located near the skin surface of the areola. Their volume is significantly less than the volume of the "normal" lobes located in deeper areas of the breast. The composition of Montgomery gland milk has not been studied. Since the volume of milk in the glands of Montgomery is small, it is unlikely that it is included in the provision of nutritional needs of the child. However, recent work has appeared (Doucet et al. 2009, 2012; Schal 2010), the results of which indicate that probably volatile components of the milk of the Montgomery glands provide more effective contact of the child with the mother and the formation of the process of removing milk from the gland. These issues will be discussed in more detail in the next chapter.

Here we note that in the skin of various parts of the gland, in addition to the sebaceous glands, there are sweat glands. Both types of glands do not differ in structure from the sweat and sebaceous glands located in the skin of other parts of the body of a woman.

3.1.3 Stromal Environment of the Alveolar Duct System

Recall that the alveolar-ductal system of lobes is surrounded by a number of connective tissue and cellular structures united by the common name stroma. The stroma is divided into intra- and interlobular parts. The composition of the stroma of the lactating breast does not differ from that of the non-lactating breast of women with a normal menstrual cycle and in women during pregnancy. Connective tissue structures of the stroma include bundles of various types of collagen and elastin fibers. The structure of stroma cell structures includes fibroblasts, fat cells, macrophages, lymphocytes, mast cells, Cajal cells, or cells similar to Cajal cells. During lactation, the volume of the stroma is significantly reduced due to the replacement of glandular tissue mainly its connective tissue and fat components.

After the basal shell, the alveolar-ductal system of the breast is surrounded by a layer of collagen fibers, which lie loosely in the intralobular part and pass in various directions. The thickness of the collagen fibrils is approximately 45 nm. Then, through the gap filled with collagen fibers, a layer of flattened fibroblasts adjoins the basal shell. Using contacts between their cellular processes, they form a mesh structure around the alveoli and ducts, resembling a basket layer of myoepithelial cells (Eyden et al. 1986; Howard and Gusterson 2000). As a result, the walls of the alveoli and ducts acquire another cell layer and are "reinforced" with a layer of collagen fibers. Here it should be noted that during lactation, in the intervals between feeding the child, the alveolar-ductal system is filled with milk, and therefore, the pressure inside it increases. Even more pressure increases during the formation of the milk withdrawal reflex and can reach 25 mm Hg (Cobo 1993; Alekseev et al. 1994, 1998) (see more in Chap. 4). Fibroblasts and collagen fibers increase the strength properties of the alveolar-ductal system and prevent damage to the walls of the alveoli and ducts. Fibroblasts of the mammary gland have characteristics typical of synthetically active cells. They show a well-developed Golgi apparatus and a rough reticulum (Eyden et al. 1986). Based on coloration for the enzyme dipeptidyl peptidase IV (an enzyme involved in the formation of breast cancer metastases), two populations of fibroblasts were found in the breast stroma. In the intralobular region of the gland, fibroblasts had a negative staining on this enzyme, and in the interlobular part a positive reaction (Atherton et al. 1992). Fibroblasts are of great importance for the functioning of the secretory cells of the mammary gland. In particular, in in vitro and in vivo experiments on the mammary glands of women and mice, it was found that fibroblasts stimulate the proliferation of secretory cells (Parmar and Cunha 2004). As already mentioned in the previous chapter, cells resembling Cajal cells (interstitial Cajal-like cell (ICLC)) have been found and relatively recently identified in the non-lactating, mature mammary gland in the intralobular space (Gherghiceanu and Popescu 2005). Detailed features of their structure and ultrastructure are presented in the previous chapter. A characteristic feature of ICLC was the presence of 2–3 fairly long, several tens of microns of cytoplasmic processes. Their thickness is 0.1–0.5 microns and they had extensions along the way. Undoubtedly, these cells are present in the lactating gland, but, unfortunately, there is no data on this.

Lymphocytes, macrophages, and mast cells are observed in the intralobular space. Macrophages and lymphocytes have the ability in some unknown way to penetrate the network of fibroblasts and basal membrane into the wall of the alveoli and ducts. In turn, passing also in an unknown way through the dense contacts of secretory cells, they end up in the milk. Mast cells contain several types of inflammatory mediators, including histamine, proteinases, and cytokines. However, the exact significance of these cells for breast function remains unknown (Dabiri et al. 2004).

The interlobular stroma of the lactating gland contains a large number of collagen fibers, but in comparison with the intralobular stroma, it is "poor" in cellular structures. The collagen fibrils of the interlobular stroma are thicker (52 nm in diameter) than the intralobular stroma and are arranged more orderly, parallel to each other.

The main cell type of the interlobular stroma is fat cells, which are localized between the lobes of the layers covered by the basement membrane (Rønnov-Jessen et al. 1996). In the lactating mammary gland of a woman, adipose tissue makes up the largest percentage of connective tissue and cellular structures of the stroma. A detailed study of the distribution of adipose tissue in the lactating mammary gland was performed using the ultrasonic method (Ramsay et al. 2005a, b). In addition to the fat layers located between the milk lobes, layers of fat cells surround the glandular tissue on all sides in the form of a fat "cover." The ratio between the volume of adipose tissue to the total volume of adipose and glandular tissue in different women varied significantly and amounted to 37 $\pm$ 9% for the left gland (limits of change 16–51%) and 35 $\pm$ 12% for the right (limits of change 9–54%). The largest percentage of adipose tissue was located under the skin surface of the gland, which was equal to 24 $\pm$ 7% for the left and 22 $\pm$ 7% for the right gland. Inside the glandular tissue, the fat layers were 7 $\pm$ 5% for the left and 6 $\pm$ 4% for the right gland. In the areas between the chest muscles and the milk lobes, the fat layer was 7 $\pm$ 3% for the left and 9 $\pm$ 3% for the right breast. The thickness of the subcutaneous fat layer was the smallest in the area of the base of the nipple and was, respectively, 5.5% $\pm$ 3.4% and 6.3% $\pm$ 4.2% of the total size of the cross-section of the gland in this place in the left gland. The thickness of the fat layer gradually increased in the direction of the chest and at a distance of 30 mm from the base of the nipple reached 17.0 $\pm$ 5.3% of the total size of the cross-section of the gland in this place for the left and 18.0 $\pm$ 7.6 for the right breast. The fat layer between the pectoral muscles and the glandular lobes was approximately the same thickness throughout. The percentage of glandular tissue relative to the total volume of adipose and glandular tissue was 63 $\pm$ 9% (limits of change 46–83%) for the left and 65 $\pm$ 11% (limits of change 45–83%) for the right breast. A large number of blood vessels (arterioles, capillaries, venules) are localized in the intra- and interlobular stroma. In addition, there are also lymphatic vessels, as well as nerve fibers and afferent nerve endings formed by them. These structures will be discussed in detail in the following sections of the chapter.

It should be noted that the milk lobes with layers of fat, connective tissue stroma with blood, and lymph vessels located in it, as well as nerve fibers are quite soft pliable structures and can easily shift in the event of a change in the position of the gland. In this case, there may be "twisting" and compression of the milk ducts of blood and lymphatic vessels, which will disrupt the excretion and formation of milk. However, this does not happen due to the fact that the glandular lobes with interlobular fat, stroma with blood, lymphatic vessels, and nerve fibers are located in a kind of connective tissue "pockets," which were first described by E. Cooper (1840) (Fig. 3.7). Connective tissue "pockets," called Cooper's ligaments are attached by processes (Fig. 3.7) to the deep layers of the dermis covering the mammary gland so that with the movement of the gland in any direction, the glandular tissue and stroma also shift, without changing their position in the internal volume of the gland.

In addition to Cooper's ligaments, there are a number of connective tissue structures in the inner volume of the gland that provide passage of blood vessels

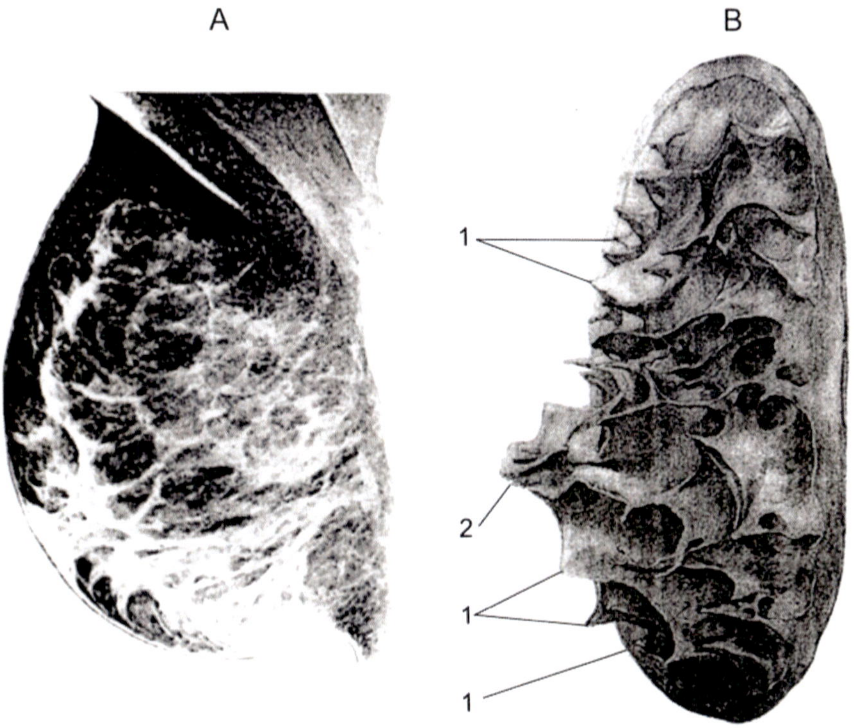

Fig. 3.7 Cooper's ligaments. (**a**) Cooper's ligaments in the mammary gland look like light lines of various curvature (from: Kopans 1989). (**b**) Drawing by A. Cooper (1840) of connective tissue ligaments in the mammary gland. lC—places of attachment of Cooper's ligaments to the skin of the gland, n—breast nipple

and nerve fibers to the glandular tissue and the external integument of the breast, as well as supporting the mammary gland in the necessary position. First of all, the horizontal connective tissue partition should be noted (Würinger et al. 1998; Wueringer and Tschabitscher 2002). It originates from the thoracic fascia at the level of the fifth rib, passes through the entire volume of the gland to the areolar-nipple part of the gland, forming a connection on all sides with the inner part of the skin of the gland (Fig. 3.8), and accordingly divides the internal volume of the gland into two parts. At the edges, the horizontal septum is rounded up, forming vertical medial and lateral ligaments that attach the gland to the sternum and the lateral edge of the pectoralis minor muscle (Fig. 3.8). In general, the horizontal fibrous sheath and ligaments form a suspension that functions as a distinctive "internal bra."

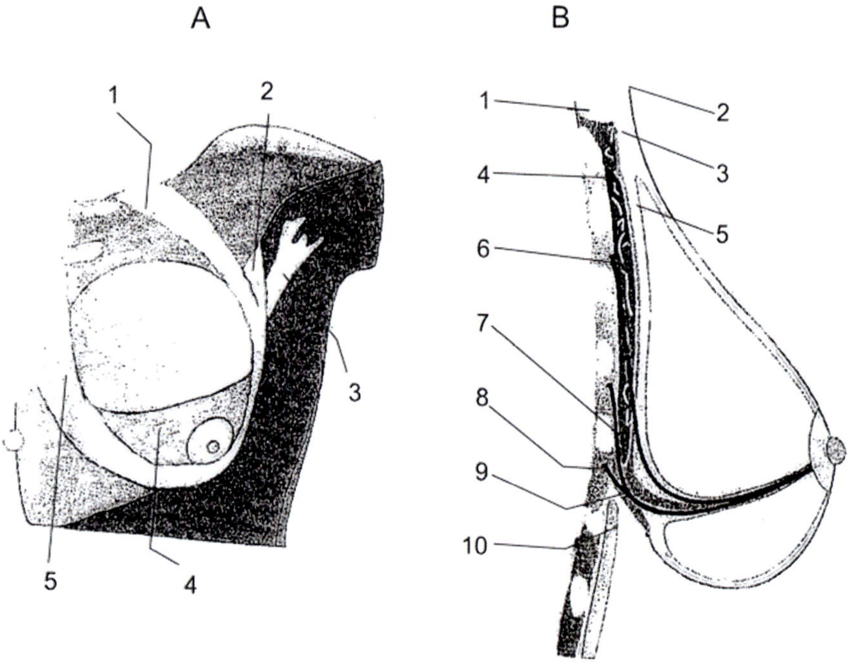

Fig. 3.8 Internal connective tissue ligaments of the breast (from: Wueringer et al. 1998). (**a**) Left breast and its supporting ligaments. 1—internal cranial ligament, 2—internal lateral ligament, 3—superficial lateral ligament, 4—horizontal connective tissue sheath, 5—internal medial ligament. (**b**) Longitudinal section of the breast. 1—clavicle, 2—skin surface, 3—thoracic ligament, 4—pectoralis major muscle, 5—retromammary space, 6—torocoacromial artery, 7—fourth intercostal artery, 8—fifth intercostal artery, 9—horizontal connective tissue sheath, 10—intragastric fold ligament

3.2 The Circulatory System of the Breast

The development and functioning of the breast is largely determined by its blood supply. With the blood flow, milk precursors, hormones, and water are delivered to the gland. Systematic studies of blood supply to the breast have been conducted since the middle of the last century. The intensification of these studies, as well as works on the study of lymph supply and innervation of women's mammary glands, occurred at the end of the last century due to a significant increase in the number of surgical operations to correct the volume and shape of the breast of women (Hamdi and Rasheed 2012). Here it should be noted that the data obtained on blood supply, lymph supply, and innervation of the mammary gland belong to the level that is designated as "gross anatomy," i.e., the work describes mainly the general course and branching of blood and lymphatic vessels, as well as nerve stems visible to the naked eye when they approach the skin of the gland, on their inner surface or intravascular space. There is very little information on the "microscopic level"—the

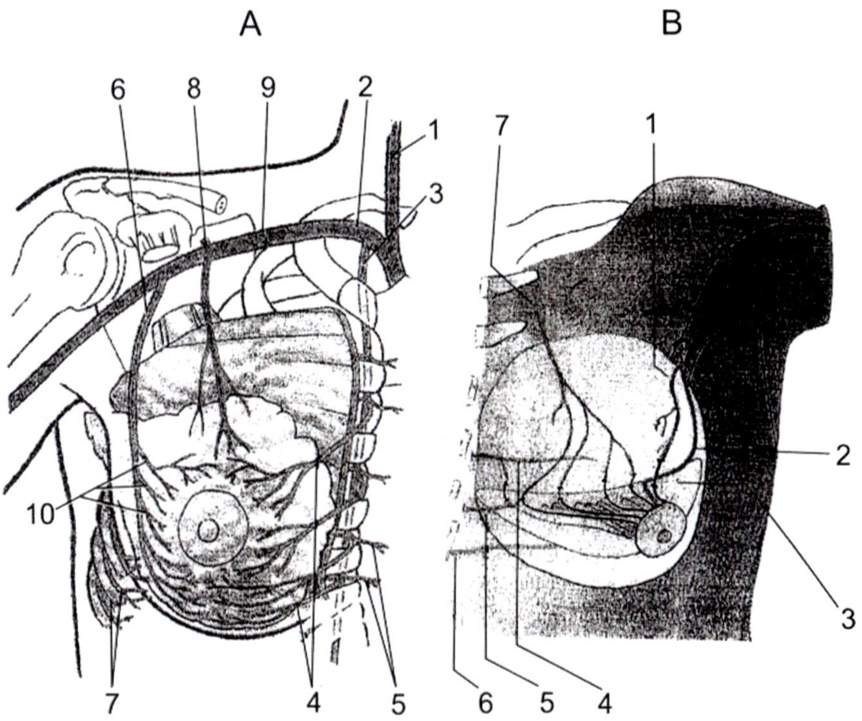

Fig. 3.9 Blood supply of the breast. (**a**) Schematic representation of the main arterial vessels. 1—right carotid artery, 2—right subclavian artery, 3—intra-thoracic artery, 4—medial branches of the internal thoracic artery, 5—sternal branches, 6—lateral thoracic artery, 7—lateral branches of intercostal arteries, 8—thorocoacromial artery, 9—axillary artery, 10—lateral branches of the thoracic artery. (**b**) Location of blood vessels and nerves on the horizontal connective tissue membrane (from: Wueringer et al. 1998). 1—lateral thoracic artery, 2—branch 4 of the intercostal nerve, 3—horizontal connective tissue sheath, 4–6—respectively, the fourth, fifth, and sixth intercostal arteries, 7—torocoacrimal artery

distribution of blood and lymphatic microvessels and nerve fibers in the intralobular space and in the skin surface of the breast of a woman.

To date, the main blood vessels supplying the mammary glands of women have been established. Here it should be noted that the location and number of the main vessels of the mammary glands of women are the same in non-pregnant, pregnant, and lactating women. Noticeable changes in the branching and thickness of blood vessels occur mainly in the peripheral areas of the main blood vessels during pregnancy and lactation. Blood supply to the internal structures of the breast, as well as its external skin is primarily due to large arterial vessels, which are branches of the subclavian and axillary arteries, as well as branches of the intercostal arteries (mainly the second, third, and fourth pairs). Figure 3.9a shows a simplified scheme of the main arterial vessels supplying various parts of the breast. For clarity, the arteries and their branches are depicted on the surface of the gland. However, in fact, in most cases, the vessels pass through (perforate) the muscles and glandular tissue.

The medial and central part of the breast receives blood from the perforant branches of the internal thoracic artery, which pass in 1–4 intercostals lateral to the sternum through the thickness of the pectoralis major muscle. The lateral thoracic artery departs from the axillary artery, goes down, along the lateral edge of the pectoralis minor muscle, to the anterior dentate muscle. The external branches of this artery, as well as the lateral perforant branches of the intercostal arteries, provide blood supply to the lateral parts of the breast. Thoracic artery departs from thoracoacromial artery and has a downward direction, passing between the great and small pectoral muscles. At the expense of the thoracic artery, the blood supply to the posterior parts of the mammary glands is carried out. The most vascularized areas of the external space of the breast are the area of the areolar-nipple complex. Branches of the internal thoracic artery and lateral thoracic artery approach the areola and nipple, forming anastomoses. Additionally, the anastomoses may be formed with branches of the intercostal arteries. Due to the abundant vascularization in the postpartum period, the intensity of the color of the nipple and areola increases. A more detailed study (Würinger et al. 1998; Wueringer and Tschabitscher 2002) of breast vascularization found that blood vessels inside the gland pass in most cases as part of connective tissue ligaments. Thus, on the horizontally located connective tissue ligament (Figs. 3.8a and 3.9b) on its cranial (i.e., upper surface) are located branches of the thoracoacromial artery, which exit through the pectoralis major muscle at the level of 4 ribs. Branches of the lateral thoracic artery also pass on the cranial surface. On the lower (caudal surface) pass mainly perforated branches from anastomoses 4 and 5 intercostal arteries. In some cases, branches from the intercostal artery 6 were observed. In the composition of the medial and lateral ligaments are also blood vessels. Medial ligaments contain perforated branches from the anastomoses of the internal thoracic artery, passing in the region of 2–4 vertebrae. The lateral ligaments contain branches of the lateral thoracic artery, extending from it at the level of 2–4 vertebrae. Arterial branches in the ligaments go toward the nipple (Würinger et al. 1998; Wueringer and Tschabitscher 2002; Ryssel et al. 2010). It should be noted that when describing the general blood supply to the mammary glands, the location of mainly arterial vessels is given in detail (Grachev and Galantsev 1973; Lawrence and Lawrence 1999, 2011; Johnson 2010, etc.) since the venous vessels are located parallel to the arterial vessels. Veins provide outflow of blood from breast tissue located in connective tissue ligaments, muscles, and intercostal spaces, as well as arterial vessels. The breast veins flow into the internal thoracic and axillary veins. Some veins may reach the external jugular vein. The network of superficial veins of the breast presents a fairly variable picture. The skin veins form a kind of loop, anastomosing with each other, and in the areolar region form a circular plexus around the base of the nipple, which is called the Haller's circle. It should be noted that the location of vessels in the external space and in the internal structures of the mammary glands of different women is very variable.

There are no data on the localization of microvessels inside the lobes of a woman's lactating gland. There is only information about the distribution of capillaries in normal glandular tissue of women during the ovulatory phase of the menstrual cycle (Naccarato et al. 2000). It turned out that the microvascularization of the ducts and lobes is different. Thus, the ducts are surrounded by a large number of "typical capillaries." In lobules, the number of capillaries is smaller, but they are wider in size and have a sinusoidal shape. It is assumed that the shape and size of the capillaries in the lobules are of great functional significance. In particular, an increase in the diameter and "sinuosity" of capillaries will slow down the blood flow near the alveoli, contributing to a longer contact of hormones and milk precursors with the cells forming the alveola (Naccarato et al. 2000). It is possible that this ratio in microcirculation is preserved in the lactating mammary gland.

3.3 The Lymphatic System of the Breast

The lymphatic system together with the venous system is a system of tissue outflow and ensures the maintenance of a constant volume and composition of the tissue fluid of the breast. The lymphatic system participates in the immunological reactions of the breast by delivering lymphocytes, plasma cells, and antibodies from the lymphoid organs. Systematic studies of the lymphatic system, as well as the circulatory system of the breast, have been conducted since the middle of the last century (Tanis et al. 2001; Suami et al. 2009). The study of intra-organ lymphatic vessels and lymph nodes found that the mammary lymph vessels form a very complex highly branching plexus system. These include the superficial skin plexus, the areolar-nipple plexus (Sappey plexus), and the plexus inside the glandular tissue (Fig. 3.10). The initial link of the lymphatic system in the plexus of the breast is the lymphatic capillaries (diameter 10–50 microns), which merge with each other, form larger collector vessels that pass in the connective tissue partitions, or next to the blood vessels and flow into the lymph nodes (Tanis et al. 2001). It should be noted that from various parts of the gland, the lymph flow does not go to strictly defined regional lymph nodes, but in any direction (Grachev and Galantsev 1973). However, by now it can be considered established that the main outflow of lymph from the breast is directed to the axillary lymph nodes and to the lymph nodes located along the internal thoracic artery inside the chest cavity. Other points of lymph outflow are the thoracic lymph nodes located between the pectoralis major and minor muscles and the subclavian neck nodes (Fig. 3.10). Lymph nodes may also be present inside the glandular tissue (Tanis et al. 2001). Additionally, lymph vessels from the breast can pass to the liver and abdominal lymph nodes. In addition, there are lymph pathways between the right and left mammary glands (Fig. 3.10).

The distribution of lymph vessels within glandular tissue is of great interest. There is evidence that the initial lymphatic capillaries in the breast during lactation are located in the stroma inside the lobes and form a dense network (Kukhtinova 1963). However, there are no detailed data on the distribution of lymphatic capillaries inside the lobes of a woman's breast.

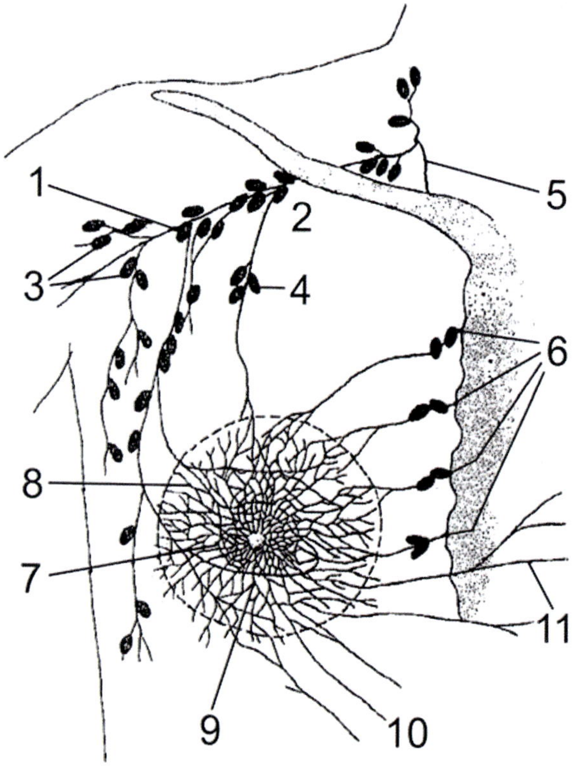

Fig. 3.10 Lymphatic drainage of the breast. Major drainage in toward axilla (by: Lawrence and Lawrence 1999). 1—central axillary lymph node, 2—deep axillary lymph nodes, 3—subscapular lymph nodes, 4—intermuscular lymph nodes (Rotter nodes), 5—subclavian lymph nodes, 6—lymph nodes located along the internal thoracic artery, 7—areolar-nipple lymph plexus (Sappey plexus), 8—lateral collector lymph vessel, 9—medial collector lymph vessel, 10—lymphatic pathways to lymph nodes of the liver and abdominal cavity, 11—lymphatic pathways between the mammary glands

3.4 Innervation of the Breast

The process of the secretion and ejection of milk from the woman breast occurs with the active participation of the nervous system. Innervation of the skin surface of the breast is carried out by the somatic nervous system and internal structures by the autonomous nervous system. The study of breast innervation as well as blood and lymph supply began in the last century. The intensification of these studies, as already noted, occurred at the end of the last century due to an increase in the number of surgical operations to correct the volume and shape of the breast of women (Hamdi and Rasheed 2012). When studying the innervation of the mammary glands at the macro level, it was found that the innervation of the skin of the breast is carried out by somatic nerve fibers, which are mainly part of 2–6 pairs of intercostal thoracic nerves. In addition, 3,4 subclavian nerves participate in innervation (Fig. 3.11) (Sarhadi et al. 1997). After entering in the parenchyma of the breast, the nerve branches pass in the connective tissue ligaments in the depth of the parenchyma at a constant level, at a distance of about two-thirds of their total length. Then they rise to the inner surface of the skin and repeatedly branch out further so that the areas of the skin surface, innervated by neighboring intercostal nerves,

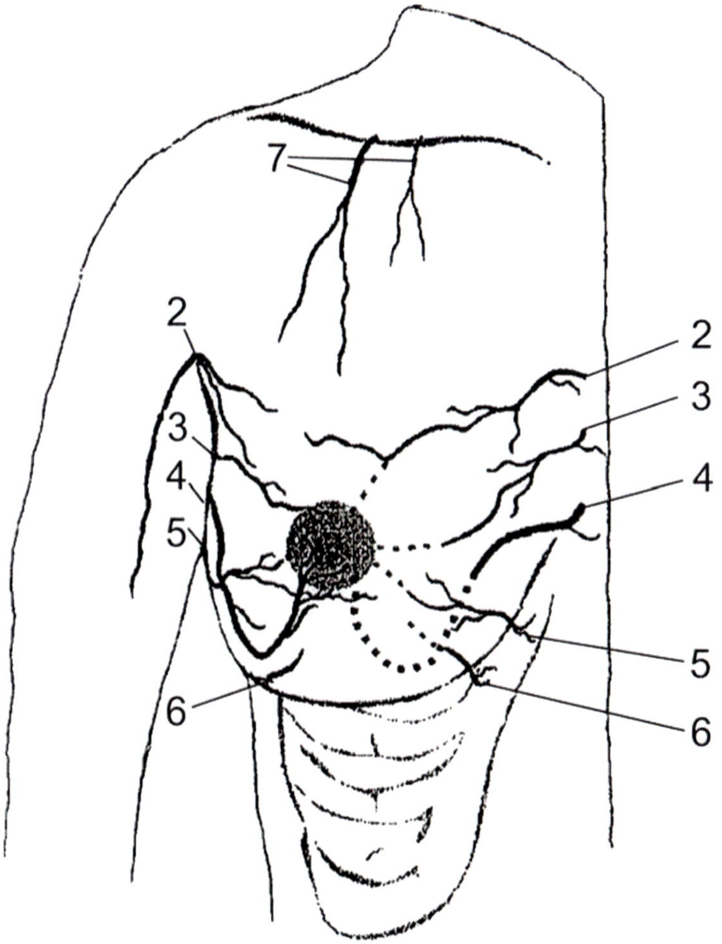

Fig. 3.11 Innervation of the breast (from Sarhadi et al. 1997). 2–6—intercostal nerves, 7—subclavian nerves

partially overlap with each other (Jaspars et al. 1997; Sarhadi et al. 1997). Along the path of the nerves in the parenchyma of the gland, there were no nerve branches ending in the depth of the breast (Jaspars et al. 1997). It should be noted that the branching and course of the nerve branches is extremely variable and differs in different women. Therefore, as noted by all researchers, it is extremely difficult to identify common features of branching and give a general scheme of branching even relatively thick nerve branches.

Special attention was paid to the innervation of the surface and internal areas of the areolar-nipple area of the breast. This is due to the method of moving the areola with the nipple on the parenchymal or dermal-parenchymal "feeding legs" during the formation of the areolar-nipple complex of the breast during surgical operations

(Hamdi and Rasheed 2012; Blondeel et al. 2003). It turned out that the areolar-nipple area of the breast is supplied mainly with a deep branch of the intercostal nerve 4 and sometimes partially with branches of the intercostal nerve 5. The nerve trunks exit at the edge of the pectoralis major muscle, pass through the retromammary space, and then follow the caudal surface of the horizontal connective tissue septum (Fig. 3.9b) to the nipple-areolar part of the gland.

The internal structures of the gland are innervated by nerve fibers of the autonomic nervous system: branches of the vagus nerve (parasympathetic system) and spinal nerve fibers (sympathetic system) (Eriksson et al. 1996). The course of parasympathetic nerve branches of the autonomous nervous system and their distribution in the mammary gland of a woman is unknown. With respect to the sympathetic branches, it can be assumed that they pass as part of the intercostal nerves.

3.4.1 Afferent Receptors of the External Integument of the Breast

The skin surface covering the internal structures of the breast can be divided into two areas. The first, hairless part is located on the nipple and areola, the second, containing hair, is located behind the areola and covers the main part of the breast. Studies (Giacometti and Mjntagna 1962; Montagna 1970; Koyama et al. 2013) have shown that the hairy skin surface located behind the areola does not differ in structure from the skin surface of the adjacent areas of the woman body, while the skin on the nipple and areola has noticeable differences. A thick, highly pigmented epidermis covers the nipple and areola (Fig. 3.12). On the surface of the nipple and areola, the epidermis forms numerous grooves. In addition, there are folds on the skin of the nipple and areola. Inside the skin of the nipple and areola, the epidermis forms outgrowths. At the tip of the nipple, the epidermal outgrowths go to a greater depth than on the sidewalls and areola. Numerous bundles of elastic fibers are observed in the skin of the nipple and areola. Moreover, their greatest density is noted under the epidermis at the tip of the nipple. In some cases, elastic fibers completely fill the space between the epidermal outgrowths (Montagna 1970). It should be noted that during the removal of milk by the child, the nipple and areola are subjected to intense mechanical influences. In addition, the nipple and part of the areola during feeding in the baby's mouth are surrounded by saliva, which is a biologically active liquid and contains digestive enzymes. Undoubtedly, the special structure of the skin of the nipple and areola provides a protective function in the process of feeding the child and, first of all, high strength mechanical characteristics. Psychophysical examinations of the sensory sensitivity of the skin surface of the breast of a woman indicate that it has receptors that perceive mechanical, cold, heat, and pain stimuli. Morphological research of the skin surface receptors of lactating and non-lactating mammary glands of women is devoted to a small number of works and only at the light-microscopic level. At the same time, numerous studies of receptors of various modalities of other areas of the skin surface of humans and animals have found a great similarity in their morphological characteristics at the light-microscopic and ultrastructural levels (Johansson and Vallbo 1983; Macefield

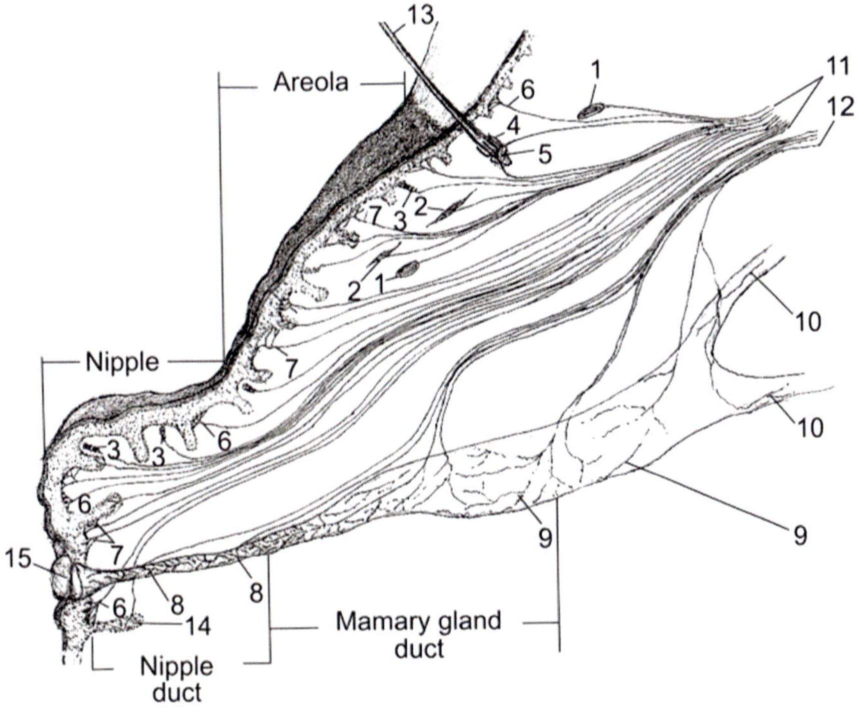

Fig. 3.12 Distribution of afferent nerve endings in the mammary gland of a woman. 1—Pacini's body, 2—Ruffini's body, 3—Meissner's body, 4—palisade of lanceolate nerve endings of the fluffy hair follicle, 5—circular nerve endings of the hair follicle, 6—Merkel complex, 7—free nerve endings that can be mechanoreceptors, pain, cold, or heat receptors, 8—free nerve endings located next to the walls of the milk duct, which can be mechanoreceptors and pain receptors, 9, 10—free nerve endings located next to the walls of the milk duct, which can be mechanoreceptors and pain receptors, 11—free nerve endings located next to the walls of the milk duct, which can be mechanoreceptors and pain receptors—nerve fibers of the somatic nervous system, 12—nerve fibers of the autonomous nervous system, 13—downy hair, 14—outgrowth of the epidermis, 15—exit of the milk duct

2005; Roudaut et al. 2012; Abraira and Ginty 2013). Therefore, with a high degree of probability, it is possible to extrapolate these data to the structure and ultrastructure of receptors located in the skin surface of a woman's breast.

The mechanoreceptor function of hairless and hair-covered skin of humans and other mammals is provided by receptors formed by branching nerve fibers belonging to nerve cells localized in the posterior spinal ganglia. All skin sensory neurons can be classified as Aß, Aδ, or C neurons based on their cell body size, axon diameter, degree of myelination, and velocity of conduction along the afferent axon. C-type of sensory neurons is the smallest in size and most commonly found in the ganglia. It has an unmyelinated afferent axon and the lowest carrying velocity of 0.2–2 m/s. Aß and Aδ neurons have average and largest sizes with varying degrees of myelination. For Aß, the velocity of conduction along the axon is in the range of 16–100 m/s, respectively, for Aδ 5–30 m/s. In a number of mechanoreceptors, branching nerve fibers are surrounded by rather complex accessory structures-capsules. The branched nerve endings of other mechanoreceptors do not have special accessory structures and are "freely" located in various areas of the skin or around the hair roots. These receptors are called free nerve endings.

Encapsulated Mechanoreceptors Among the mechanoreceptors with special accessory structures, four types have been well studied: the Meissner corpuscle, the Merkel complex, the Ruffini corpuscle, and the Pacini corpuscle. These receptors belong to the group of low-threshold mechanoreceptors. It should be noted that all these mechanoreceptors were found in the skin covering the mammary gland of non-lactating and lactating women (Cathcart et al. 1948; Morozova 1954; Ganul 1956; Miller and Kasahara 1959; Giacometti and Mjntagna 1962; Montagna 1970). Morphofunctional properties of various types of human skin mechanoreceptors have been reviewed in detail (Johansson and Vallbo 1983; Macefield 2005; Roudaut et al. 2012; Abraira and Ginty 2013). Here we will briefly discuss the morphological characteristics of encapsulated mechanoreceptors found in human skin, including the skin of a woman's breast.

The nerve component of the **Meissner body** is formed by a myelinated Aß nerve fiber. The afferent axon loses myelin and forms a stack of flattened sections that curl and spread over each other. Each section is also covered on both sides by flattened Schwann cells. With the help of thin finger-like processes, sections of nerve endings contact the wall of the connective tissue capsule, which covers a column of flattened nerve endings and Schwann cells. In turn, the connective tissue capsule contacts the basement membrane of the epidermis through collagen fibers. Here it is important to note that in addition to the "classic" Meissner bodies in the skin of humans and other mammals, often found receptor structures that resemble the structure of the Meissner body, called the Krause body, genital bodies or bodies that do not have a name—Meissner-like bodies. In addition to the similarity in morphological characteristics, these bodies have similar functional properties. These Meissner-like bodies were found in the dermis of the nipple, areola, and in areas of the breast skin located behind the areola (Cathcart et al. 1948; Miller and Kasahara 1959; Giacometti and Mjntagna 1962; Montagna 1970) (Fig. 3.12).

The Merkel complex consists of special Merkel cells adjacent on one side to flattened single non-myelinated nerve endings formed by branching Aß myelinated fiber. Nerve endings contain a large number of mitochondria. Conversely, Merkel cells are in contact with the basement membrane of epidermal cells by means of thin processes. A single nerve fiber can form up to 150 nerve endings, each of which contacts the Merkel cell. Merkel cells are characterized by a multi-lobed nucleus and a large number of granulated vesicles. Merkel cells have desmosomal and synaptic contacts with nerve endings near which there is a cluster of granulated vesicles. Merkel complexes are located in the basal layer of the epidermis of the areolar and hairy parts of the breast skin or near the follicles of downy hairs (Miller and Kasahara 1959; Fradette et al. 1995) (Fig. 3.12).

The Ruffini body is a fusiform structure with a thin lamellar capsule filled with liquid. The myelinated axon (Aß) enters the capsule, loses myelin, and branches inside it into numerous terminals that are in close contact with the collagen fibers. These fibers pass through the central part of the body, exit at its poles, and end in the skin and subcutaneous layers surrounding the Ruffini body. This type of receptor was found in areas of the skin surface of the breast located in the distal areas of the areolar region (Miller and Kasahara 1959; Lassman 1964) (Fig. 3.12).

The Pacini body is perhaps one of the most studied encapsulated skin mechanoreceptors. This is due to the fact that Pacini bodies are relatively large and relatively easily available for morphological and physiological research. The Pacini body has an elongated shape and consists of a nerve ending enclosed in a capsule. The myelinated axon (Aß) loses myelin when entering the capsule, and the non-myelinated nerve endings usually do not branch. In the distal part, it usually ends in a thickening, and sometimes dichotomous branching. The nerve end contains many small mitochondria. Flattened capsule cells are organized in such a way that they form two zones—central and peripheral. The peripheral zone consists of 30–40 concentric closed elastic shells. The distance between them decreases as you approach the central zone. The central zone consists of closely adjacent lamellar layers (about 60 of them), separated by a gap and surrounding the nerve end. Pacini corpuscles are the most rarely encountered encapsulated mechanoreceptor in the skin of a woman's breast. The presence of Pacini corpuscles in the skin of the areola was noted only in one work (Giacometti and Mjntagna 1962) (Fig. 3.12).

Non-encapsulated Mechanoreceptors If researchers often disagree on the presence and localization of various types of encapsulated mechanoreceptors in the breast skin, the authors' data are the same for non-encapsulated mechanoreceptors (Cathcart et al. 1948; Morozova 1954; Ganul 1956; Miller and Kasahara 1959; Giacometti and Mjntagna 1962; Lassman 1964; Montagna 1970; Grachev and Alekseev 1980; Eriksson et al. 1996; McGrouther and Ahmad 1998). In the skin of the breast of a woman, there is a large number of non-encapsulated nerve endings of varying complexity and degree of branching that perform a mechanoreceptor function, called free nerve endings.

It should be noted that free nerve endings are the most common type of skin receptors (Roudaut et al. 2012; Abraira and Ginty 2013). Free nerve endings are located in the epidermis and the skin itself, as well as around the follicles of skin hair. They are formed mainly by non-myelinated nerve fibers (C–fibers), which make up about 80% of skin afferents. Free nerve endings can also form thin (Aδ) and medium (Aß) myelinated nerve fibers, whose diameter is usually less than 6 microns. Strictly speaking, the term "free nerve endings" is not entirely successful, since these endings are more or less surrounded by a shell of Schwann cells and necessarily basal membrane. However, of all afferent nerve endings, they are the most free from accessory structures (Andres and During 1973).

In the hairy part of the skin covering the mammary gland of a woman, free nerve endings are localized around the hair follicles and in the epidermis. The skin, which is located behind the areola, contains downy hairs. Innervation of downy hair follicles in mammals is carried out by two types of free nerve endings. The first type includes lanceolate (palisade) endings located along the major axis of the hair follicle and surrounding it from all sides. The second type, lanceolate endings, localizes around the circumference of the hair follicle (Roudaut et al. 2012; Abraira and Ginty 2013).

Studies at the light level have shown that in the skin of the nipple and the areola of the breast of a woman, free endings are localized mainly in the skin itself and only a small number of them occur in the deep layers of the epidermis (Cathcart et al. 1948; Miller and Kasahara 1959; Giacometti and Mjntagna 1962; Lassman 1964; Montagna 1970; Grachev and Alekseev 1980; Eriksson et al. 1996). A particularly high concentration of nerve endings was found in the skin of the tip of the nipple (Miller and Kasahara 1959). Histochemical study of the innervation of the nipple and the areola of the breast found that the nerve endings in these areas contain calcitonin gene-related peptide (CGRP) and substance P (SP), which are markers of afferent nerve fibers that form sensory receptors (Eriksson et al. 1996).

Thus, morphological information about the composition and location of mechanoreceptors in the skin of the nipple and areola of the breast of a woman suggests that this area of the skin has special "sensory" characteristics. In particular, the "deficit" of low-threshold encapsulated mechanoreceptors, which are thought to provide high tactile resolution of hairless skin, such as the skin of a woman's palm (Johansson and Vallbo 1983), will increase the threshold of spatial mechanical sensitivity. Recall that this threshold is defined as the minimum distance between two point touches, at which these effects are perceived separately. The location of mechanoreceptors in the skin itself and in the deep layers of the epidermis will reduce the overall mechanical sensitivity. Indeed, the results of early psychophysical examinations of the skin sensitivity of a woman's breast (Sherrington 1900; Wood-Jones and Turner 1931; Belonoschkin 1933) indicated a reduced sensitivity of the skin of the nipple and areola compared to other areas of the skin to mechanical tactile stimuli. Surveys C. Weinstein (1968) threshold of spatial mechanical sensitivity of various parts of the breast skin found that the lowest resolution is the skin of the nipple and areola. The results of these studies confirmed the data of more detailed psychophysical examinations of skin sensitivity of various parts of the breast in subsequent years (Terzis et al. 1987; Tairych et al. 1998). In addition, similar information was obtained in connection with the study of the restoration of skin sensitivity after surgical operations to correct the volume and shape of the breast of women. In particular, preoperative examination of the mechanical sensitivity of the skin of the nipple and areola, as well as the skin of the gland adjacent to the areola, found that the skin of the nipple and areola have a lower mechanical sensitivity. Moreover, the areolar part had the highest threshold of mechanosensitivity (Chiari Jr et al. 2012; Longo et al. 2014). Thus, the available data indicate that the mechanoreceptor apparatus of the skin surface of a woman's breast is tuned to the perception of undifferentiated influences and according to the Hed classification has protopathic sensitivity compared to, for example, the skin of the face and hands.

The skin surface of a woman's breast as well as the skin of other parts of the body has a well-expressed sensitivity to heat and cold stimuli. To date, it has been established that, in contrast to mechanoreceptors, the skin cold and heat receptors of humans and other mammals are represented only by free nerve endings (Schepers and Ringkamp 2009). Cold receptor endings are formed by thin myelinated Aδ and non-myelinated C fibers and are located mainly at the border of the skin and epidermis (Hensel et al. 1974) or in some cases enter the epidermis (Dhaka et al. 2008). Heat receptors are branches of non-myelinated C fibers and are located in the upper layer of the skin. It should be noted here that it is very difficult to localize and investigate the structure and ultrastructure of skin encapsulated mechanoreceptors. However, these difficulties increase significantly when studying the terminal sites of receptors that are free nerve endings that receive mechanical, cold, and heat stimuli. Morphological studies have found a great similarity in the structure and ultrastructure of nerve terminals that receive mechanical, cold, and heat stimuli. The end sections of the nerve branches contain a large number of small mitochondria, glycogen granules, and microtubules inside. However, using methods of immunohistochemistry, molecular genetics, and tissue culture, a fundamental difference was established between free nerve endings that receive stimuli of various modes. It turned out that the membrane of nerve terminals contains special ion channels called transient receptor potential (TRP) channels that change their permeability under the influence of mechanical, heat, and cold stimuli (Belmonte and Viana 2008; Schepers and Ringkamp 2009). In particular, TRPM8 channels provide the perception of stimuli within 35 °C–12 °C as moderately cold and TRPV3 and TRPV4 within 35 °C–43 °C as moderately warm stimuli (Nomoto et al. 2004; Belmonte and Viana 2008; Schepers and Ringkamp 2009).

Unfortunately, there is no data on the location and structure of thermoreceptors in the areola and nipple of a woman's breast. However, by analogy with cold and heat receptors of other areas of the skin of a woman, it is highly likely that part of the free nerve endings of the skin of the nipple and areola, in addition to the mechanoreceptor, perform a thermoreceptor function. And accordingly, the reception of cold stimuli is carried out TRPM8 ion channels, and warm stimuli is carried out TRPV3 and TRPV4 ion channels located in the membrane of these endings. Special psychophysical examinations of the heat sensitivity of various areas of the skin surface of the breast were not carried out. In the literature, there are only preoperative data on the heat and cold sensitivity of the skin of the nipple and areola in connection with the study of the restoration of skin sensitivity after surgical operations to correct the volume and shape of the breast of women. It turned out that in contrast to mechanosensitivity, heat and cold sensitivity are the same in different areas of the skin surface of the breast (Gahm et al. 2013).

To date, it is an established fact that the pain receptors of the skin of mammals and including humans, as well as thermoreceptors, are free nerve endings formed by Aδ and C nerve fibers (Haggard et al. 2013). Pain nerve terminals are located in the same areas of the skin as thermoreceptors. The available morphological data do not allow distinguishing pain receptor endings from mechanoreceptor, cold and heat nerve terminals at the light and ultrastructural levels. Special mechanical pain receptors were found in the skin that respond to strong mechanical stimuli, as well as heat pain receptors that perceive cold and heat stimuli below 12 °C and above 43 °C, respectively. In addition, there are multimodal pain receptors. Using methods of immunohistochemistry, molecular genetics, and tissue culture, it was found that the membrane of mechanical pain receptors contains high-threshold mechano-activated MS channels, and in the membrane of cold and heat pain receptors, TRPA1 and TRPV2 channels are localized, respectively. TRPV1 channels were found in the membrane of multimodal pain receptors, individual subunits of which are tuned to receive strong mechanical, heat, and chemical stimuli (Belmonte and Viana 2008). Morphological studies of pain receptors in the mammary gland of women were not conducted. However, by analogy with thermoreceptors, it can be assumed that pain receptors are also free nerve endings located in the nipple and areola at the border of the skin and epidermis. With a high degree of probability, it can be assumed that the reception of pain stimuli is carried out by the same ion channels (MS, TRPV1, TRPA1, TRPV2) found in the membrane of pain terminals, localized in various parts of the skin of humans and other mammals.

Quantitative psychophysical studies of pain reception of various parts of the breast of a woman were not conducted. However, our own clinical examinations have found that the breast nipple is the most painful part of the breast. Very often, with a sufficiently intense mechanical compression of the area of the areola or the hairy part of the skin of the gland, the woman did not feel pain, while approximately the same compression of the nipple caused pain.

3.4.2 Afferent Receptors of Internal Structures of the Breast

A large experimental material shows that the sensory systems that ensure the functioning of internal organs-visceral sensory systems have noticeable differences from the sensory systems that supply the body with information about changes in the external environment-extrasensory systems. Among the main ones, we can distinguish a low density of afferent receptors in the innervated organ and a less precise "blurred" localization of the action of stimuli in the internal structures of the body (Alekseev 2009). The internal structures of a woman's breast are innervated by nerve fibers of the autonomic nervous system: branches of the vagus nerve and spinal nerve fibers (Eriksson et al. 1996). The structure of the nerves includes afferent and efferent fibers, but the detailed characteristics of the nerve fibers that form visceral nerves and their course inside the breast of a woman are unknown. In early works devoted to the study of innervation of internal breast structures of a woman, the high density of nerve fibers (terminals) located among the mammary ducts and glandular lobes of the breast was indicated (an overview of these studies is presented in the work of Grachev I.I. and Alekseev N.P. 1980). However, these results were not confirmed in the future. It turned out that the density of innervation of the parenchyma is significantly inferior to that of the skin of the nipple and areola.

Comparative histochemical studies of innervation of internal structures of mammary glands in lactating rats and non-lactating women using afferent nerve fiber markers and their CGRP and SP terminals were conducted by M. Erikson and colleagues (Eriksson et al. 1996). CGRP and SP—immunoreactive nerve fibers and terminals were found in both cases in the breast parenchyma. CGRP-positive nerve fibers were observed in the stroma surrounding the lobules of the gland, their ducts, as well as in the walls of arteries and veins localized in glandular tissue. The density of nerve fibers increased along the course of the milk ducts and reached the maximum value in the environment of the walls of the milk ducts of the nipple (Fig. 3.12). SP-positive nerve fibers were present in the same areas as CGRP-positive nerve fibers. However, the number of SP-positive nerve fibers was noticeably less than CGRP-positive nerve fibers. Data from histochemical studies on the distribution of nerve fibers and their endings in the alveolar-ductal system of the woman breast are consistent with the results of previous studies, in which other histological dyes were used to color the nerve structures inside the gland (Miller and Kasahara 1959; Giacometti and Mjntagna 1962; Lassman 1964; Montagna 1970; Montagna and Macpherson 1974). Here it should be noted that a special impression is made by preparations that present the innervation of the milk ducts of the nipple (Miller and Kasahara 1959; Montagna 1970; Montagna and Macpherson 1974). The walls of the milk ducts inside the nipple are covered with a dense network of nerve endings, which in their course form bead-like extensions of various shapes and diameters and have a great similarity to the nerve endings located in the walls of other hollow visceral organs, such as the bladder, gastrointestinal tract, and blood vessels (Zagorodnyuk et al. 2010). Since the walls of the milk ducts inside the nipple are formed by a single-layer cuboid epithelium and covered with a basal shell, it can be considered that the nerve endings located on the walls of the milk channels of the nipple are afferent. It is interesting to note that the walls of the excretory ducts of the Montgomery glands are also covered with a dense network of nerve endings (Montagna and Yun 1972). Studies of the receptor function of the nerve terminals of the internal structures of the woman breast have not been conducted. However, as already mentioned, clinical data indicate the presence of mechanoreceptors and pain receptors in the parenchyma of the gland. Thus, when filling the ducts system with colostrum or milk, a woman feels "heaviness" in the gland, and the overflow of milk ducts in the case of an increase in the interval between feedings is often accompanied by pain. While feeding a baby, some women feel a tingling sensation when the hormone oxytocin reflexively enters the bloodstream of the breast and increases the pressure in the ductal system of the breast (more on this in the next chapter). It is believed that this is a reaction of pain receptors in blood vessels to the vasodilating effect of oxytocin and to the stretching of the walls of the ducts when the milk pressure increases in them. During ductoscopy, i.e., the introduction of catheters into the milk ducts of the nipple and, accordingly, their expansion, the woman feels pain.

3.5 Muscular System of the Breast

In addition to myoepithelial cells, the motor function of the breast is provided by smooth muscle fibers. In the glandular department, smooth muscles are located in the walls of blood and lymphatic vessels. A large number of smooth muscle fibers were found under the skin surface of the areola and in the nipple. Under the areolar skin surface, smooth muscle fibers are arranged as a layer in which muscle cells are localized in a radial and circular direction. Most often, smooth muscle fibers have a fusiform shape. However, there are cells that form processes that take a variety of forms (Cathcart et al. 1948; Gairns and Garven 1949). Montagna (1970), so that on histological preparations it is almost very difficult to trace the exact orientation of smooth muscle fibers in the areolar region. However, according to observations, most areola muscle fibers show convergence toward the nipple. In this case, there is a partial mixing of radially and circularly located muscle fibers. At the periphery of the areola, some smooth muscle fibers pass at an acute angle to the skin. This arrangement of muscle fibers in the areola suggests that the areola becomes wrinkled, the size of the areola decreases, and its stiffness increases (Cathcart et al. 1948). In the nipple, most of the smooth muscle fibers are localized parallel to the milk ducts of the nipple. Muscle fibers are surrounded by elastin fibers, which are a kind of "skeleton" and anchoring smooth muscle cells along the length of the nipple in the surrounding tissues (Montagna 1970). In addition, some of the muscle fibers are located circularly. It was assumed that the nipple sphincter is formed at the tip of the nipple using special circular and longitudinally arranged fibers (Giacometti and Montagna 1962). However, detailed morphological studies on this topic have not been conducted and the question of the presence of a sphincter in the nipple for each milk duct in the breast of a woman is still open.

When the skin of the nipple and areola of a woman's breast is affected by stimuli of various modalities: mechanical, temperature, pain, a reflex change in the shape and size of the nipple and areola occurs. The efferent link of the reflex can be humoral, nervous, or simultaneously neuro-humoral. Histochemical study of innervation of internal structures of the woman breast revealed immunoreactive nerve fibers and terminals with efferent markers (Eriksson et al. 1996). Part of the nerve fibers and their endings were immunoreactive for neuropeptide Y (NPY) and tyrosine hydroxylase (TH). The content of these peptides is characteristic of sympathetic nerve fibers and their efferent endings. In addition, nerve fibers and terminals were detected with immunoreaction to vasoactive intestinal polypeptide (VIP) and to the peptide histidine isoleucine (PHL). The reaction to these peptides is characteristic of parasympathetic nerve fibers. It turned out that the walls of the woman's breast veins contain nerve terminals that are immunoreactive for NPY and TN. In the adventitia of the mammary arteries, nerve fibers immunoreactive for NPY and TN are also noted. At the same time, near the arteries there are nerve fibers, with reaction for VIP and PHL. Nerve terminals located near smooth muscle cells in the nipple and areola contain VIP, PHL, NPY, and TH (Eriksson et al. 1996).

References

Abraira VE, Ginty DD (2013) The sensory neurons of touch. Neuron 79:618–639

Akre J (1989) Infant feeding. The physiological basis. Bull World Health Organ 67(Suppl):1–108

Alekseev NP (2009) Visceral sensory system. In: Altman JA, Kulikov GA (eds) Physiology of sensory systems and higher nervous activity, vol 1. Academy, pp 236–283. (in Russian)

Alekseev NP, Ilyin VI, Yaroslavsky VK, Gaidukov SN, Tikhonova TK, YuA S, Omelianyuk EV, Tkachenko NN (1994) Role of vacuum and tactile stimuli in the process of removing milk from a woman's breast. Physiol Zh SSSR 89:67–84. (in Russian)

Alekseev NP, Ilyin VI, Yaroslavski VK, Gaidukov SN, Tikhonova TK, Specivcev YA, Omelyanjuk EV, Tkachenko NN (1998) Compression stimuli increase the efficacy of breast pump function. Eur J Obstet Gynecol Reprod Biol 77(2):131–139

Alipov VI, Kolodina LN, Korkhov VV, Radzinsky VE (1988) Woman lactation. Ylym, Ashgabat. (in Russian)

Andres KH, During M (1973) Morphology of cutaneous receptors. In: Handbook of sensory physiology, vol 2. Springer, pp 1–23

Andrews JL, Kim AC, Hens JR (2012) The role and function of cadherins in the mammary gland. Breast Cancer Res 14:203–213

Atherton AJ, Monaghan P, Warburton MJ, Robertson D, Kenny AJ, Gusterson BA (1992) Dipeptidyl peptidase IV expression identifies a functional sub-population of breast fibroblasts. Int J Cancer 50:15–19

Battersby S, Anderson TJ (1988) Proliferative and secretory activity in the pregnant and lactating human breast. Virchows Arch A Pathol Anat Histopathol 413:189–196

Belmonte C, Viana F (2008) Molecular and cellular limits to somatosensory specificity. Mol Pain 4:4–14

Belonoschkin B (1933) Physiologishe- anatomische Unterssuchengen uber die Empfenger der Kaltemplindung. Z Zellforsch Mikrosk Anat 18:555–572

Blondeel PN, Hamdi M, Van de Sijpe KA, Van Landuyt KH, Thiessen FE, Monstrey SJ (2003) The latero-central glandular pedicle technique for breast reduction. Br J Plast Surg 56:348–359

Cathcart EP, Gairns FW, Garven HSD (1948–1949) The innervation of the human quiescent nipple, with notes on pigmentation, erection, and hyperneury. Trans Roy Soc Edinburgh 61:699–717

Chiari A Jr, Nunes TA, Grotting JC, Cotta FB, Gomes RC (2012) Breast sensitivity before and after the L short-scar mammaplasty. Aesthet Plast Surg 36:105–114

Cobo E (1993) Characteristics of the spontaneous milk ejecting activity occurring during human lactation. J Perinat Med 2:77–85

Cooper AP (1840) Anatomy of the breast. Longman, Orme, Green, Browne and Longmans, London

Cregan MD, Fan Y, Appelbee A, Brown ML, Klopcic B, Koppen J, Mitoulas LR, Piper KM, Choolani MA, Chong YS, Hartmann PE (2007) Identification of nestin-positive putative mammary stem cells in human breastmilk. Cell Tissue Res 329:129–136

Dabiri S, Huntsman D, Makretsov N, Cheang M, Gilks B, Bajdik C, Gelmon K, Chia S, Hayes M (2004) The presence of stromal mast cells identifies a subset of invasive breast cancers with a favorable prognosis. Mod Pathol 17:690–695

Dhaka A, Earley TJ, Watson J, Patapoutian A (2008) Visualizing cold spots: TRPM8-expressing sensory neurons and their projections. J Neurosci 28:566–575

Doucet S, Soussignan R, Sagot P, Schaal B (2009) The secretion of areolar (Montgomery's) glands from lactating women elicits selective, unconditional responses in neonates. PLoS 4:e7579

Doucet S, Soussignan R, Sagot P, Schaal B (2012) An overlooked aspect of the human breast: areolar glands in relation with breastfeeding pattern, neonatal weight gain, and the dynamics of lactation. Early Hum Dev 88:119–128

Emerman JT, Vogl AW (1986) Cell size and shape changes in the myoepithelium of the mammary gland during differentiation. Anat Rec 216:405–416

Eriksson M, Lindh B, Uvnas-Moberg K, Hokfelt T (1996) Distribution and origin of peptide-containing nerve fibres in the rat and human mammary gland. Neuroscience 70:227–245

Eyden BP, Watson RJ, Harris M, Howell A (1986) Intralobular stromal fibroblasts in the resting human mammary gland: ultrastructural properties and intercellular relationships. J Submicrosc Cytol 18:397–408

Ferguson DJ, Anderson TJ (1983) An ultrastructural study of lactation in the human breast. Anat Embryol (Berl) 168(3): 349–359

Ferguson DJ (1985) Intraepithelial lymphocytes and macrophages in the normal breast. Virchows Arch A Pathol Anat Histopathol 407:369–378

Fradette J, Godbout MJ, Michel M, Germain L (1995) Localization of Merkel cells at hairless and hairy human skin sites using keratin 18. Biochem Cell Biol 9–10:635–639

Gahm J, Hansson P, Brandberg Y, Wickman M (2013) J Plast Reconstr Aesthet Surg 66:1521–1527

Gairns FW, Garven HS (1949) The smooth muscle cell types and their associated elastic fibers in the female nipple. J Physiol 110:18

Ganul LA (1956) About the innervation of the female breast in the childbearing period. Obstet Gynecol 5:14–19. (in Russian)

Geddes DT (2009a) Ultrasound imaging of the lactating breast: methodology and application. Int Breastfeed J 4:1–17

Geddes DT (2009b) The use of ultrasound to identify milk ejection in women – tips and pitfalls. Int Breastfeed J 4(5):1–7

Gherghiceanu M, Popescu LM (2005) Interstitial Cajal-like cells (ICLC) in human resting mammary gland stroma. Transmission electron microscope (TEM) identification. J Cell Mol Med 4:893–910

Giacometti L, Mjntagna W (1962) The nipple and the areola of the human female breast. Anat Rec 144:191–197

Going JJ, Moffat DF (2004) Escaping from Flatland: clinical and biological aspects of human mammary duct anatomy in three dimensions. J Pathol 203:538–544

Gooding MJ, Finlay J, Shipley JA, Halliwell M, Duck F (2010) Three-dimensional ultrasound imaging of mammary ducts in lactating women: a feasibility study. J Ultrasound Med 29 (1):95–103

Grachev II, Alekseev NP (1980) Role of receptors in lactation regulation. L., Nauka

Grachev II, Galantsev VP (1973) Physiology of lactation. General and comparative. L., Science (in Russian)

Haggard P, Iannetti GD, Longo MR (2013) Spatial sensory organization and body representation in pain perception. Curr Biol 23:R164–R176

Hamdi M, Rasheed MZ (2012) Advances in autologous breast reconstruction with pedicled perforator flaps. Clin Plast Surg 39:477–490

Hassiotou F, Geddes D (2013) Anatomy of the human mammary gland: current status of knowledge. Clin Anat 26:29–48

Hassiotou F, Beltran A, Chetwynd E, Stuebe AM, Twigger AJ, Metzger P, Trengove N, Tat LC, Filgueira L, Blancafort P, Hartmann PE (2012) Breastmilk is a novel source of stem cells with multi-lineage differentiation potential. Stem Cells 30:2164–2174

Hassiotou F, Geddes DT, Hartmann PE (2013) Cells in human milk: state of the science. J Hum Lact 29:171–182

Hensel H, Andres KH, von Düring M (1974) Structure and function of cold receptors. Pflugers Arch 352:1–10

Hosseini SM, Talaei-Khozani T, Sani M, Owrangi B (2014) Differentiation of human breast-milk stem cells to neural stem cells and neurons. Neurol Res Int 2014:1–8

Hou MF, Huang TJ, Liu GC (2001) The diagnostic value of galactography in patients with nipple discharge. Clin Imaging 25:75–81

Howard BA, Gusterson B (2000) Human breast development. J Mam Gland Biol Neoplasia 5 (2):119–137

Itoh M, Bissell MJ (2003) The organization of tight junctions in epithelia: implications for mammary gland biology and breast tumorigenesis. J Mammary Gland Biol Neoplasia 8:449–462

Jaspars JJ, Posma AN, van Immerseel AA, Gittenberger-de Groot AC (1997) The cutaneous innervation of the female breast and nipple-areola complex: implications for surgery. Br J Plast Surg 50(4):249–259
Johansson RS, Vallbo AB (1983) Tactile sensory coding in the glabrous skin of the human hand. Trend Neusci 6:27–32
Johnson MC (2010) Anatomy and physiology of the breast. In: Management of the breast diseases. Springer, pp 1–36
King BL, Love SM (2006) The intraductal approach to the breast: raison d'être. Breast Cancer Res 8:206–221
Kopans DB (1989) Breast imaging. Anatomy, histology and physiology. JB Lippincott Co, Philadelphia, PA
Koyama S, Wu HJ, Easwaran T, Thopady S, Foley J (2013) The nipple: a simple intersection of mammary gland and integument, but focal point of organ function. J Mammary Gland Biol Neoplasia 18:121–131
Kukhtinova RA (1963) Development of the intra-organ lymphatic system of the human breast in embryogenesis. Uch Zap Stavropol State University Med Institute 10:32–35
Lassman G (1964) Beitrag zur innervation der menschlichen brustdruse. Acta Anat 58:131–159
Lawrence RA, Lawrence RM (1999) Breastfeeding: a guide for medical profession, 5th edn. Mosby
Lawrence RA, Lawrence RM (2011) Breastfeeding: a guide for medical profession, 7th edn. Mosby
Longo B, Campanale A, Santanelli di Pompeo F (2014) Nipple-areola complex cutaneous sensitivity: a systematic approach to classification and breast volume. J Plast Reconstr Aesthet Surg 67:1630–1636
Love SM, Barsky SH (2004) Anatomy of the nipple and breast ducts revisited. Cancer 101:1947–1957
Macefield VG (2005) Physiological characteristics of low-threshold mechanoreceptors in joints, muscle and skin in human subjects. Clin Exp Pharmacol Physiol 32:135–144
McGrouther DA, Ahmad FS (1998) A preliminary report: the changes in the neuropeptide containing epidermal innervation in response to inflammatory reactions elicited in human breast skin. J R Coll Surg Edinb 43:49–52
Mendelson EB (1998) The breast. In: Rumack CM, Wilson SR, Charboneau JW (eds) Diagnostic ultrasound, 2nd edn. Elsevier Mosby, St. Louis, pp 755–760
Miller MR, Kasahara M (1959) The cutaneous innervation of the human female breast. Anat Rec 135:153–167
Monaghan P, Moss D (1996) Connexin expression and gap junctions in the mammary gland. Cell Biol Int 20:121–125
Montagna W (1970) Histology and cytochemistry of human skin. XXXV. The nipple and areola. Br J Dermatol 83(Suppl):2–13
Montagna W, Yun JS (1972) The glands of Montgomery. Br J Dermatol 86:126–133
Montagna W, Macpherson EE (1974) Proceedings: some neglected aspects of the anatomy of human breasts. J Invest Dermatol 63:10–11
Montgomery WF (1837) It is quoted on: Doucet S. et al. 2009
Morozova MG (1954) On the morphology of the sensitive innervation of the mammary gland of women. Arch Anat Histol Embryo 31(1):50–55. (in Russian)
Naccarato AG, Vicava P, Vignati S, Fanelli G, Bonadio AG, Montruccolli G, Bevilacqua G (2000) Bio-morphological events in the development of the human female mammary gland from fetal age to puberty. Wirchows Arch 436:431–438
Nagato T, Yoshida H, Yoshida A, Uehara Y (1980) A scanning electron microscope study of myoepithelial cells in exocrine glands. Cell Tissue Res 209:1–10
Nomoto S, Shibata M, Iriki M, Riedel W (2004) Role of afferent pathways of heat and cold in body temperature regulation. Int J Biometeorol 49:67–85
Parmar H, Cunha GR (2004) Epithelial-stromal interactions in the mouse and human mammary gland in vivo. Endocr Relat Cancer 11:437–458

Patki S, Kadam S, Chandra V, Bhonde R (2010) Human breast milk is a rich source of multipotent mesenchymal stem cells. Hum Cell 23:35–40

Prime DK, Geddes DT, Spatz DL, Robert M, Trengove NJ, Hartmann PE (2009) Using milk flow rate to investigate milk ejection in the left and right breasts during simultaneous breast expression in women. Int Breastfeed J 4(10):1–10

Ramsay DT, Kent JC, Hartmann RA, Hartmann PE (2005a) Anatomy of the lactating human breast redefined with ultrasound imaging. J Anat 206(6):525–534

Ramsay DT, Mitoulas LR, Kent JC, Larsson M, Hartmann PE (2005b) The use of ultrasound to characterize milk ejection in women using an electric breast pump. J Hum Lact 21(4):421–428

Rønnov-Jessen L, Petersen OW, Bissell MJ (1996) Cellular changes involved in conversion of normal to malignant breast: importance of the stromal reaction. Physiol Rev 76:69–125

Roudaut Y, Lonigro A, Coste B, Hao J, Delmas P, Crest M (2012) Touch sense: functional organization and molecular determinants of mechanosensitive receptors. Channels (Austin) 6:234–245

Rusby JE, Brachtel EF, Michaelson JS, Koerner FC, Smith BL (2007) Breast duct anatomy in the human nipple: three-dimensional patterns and clinical implications. Breast Cancer Res Treat 106:171–179

Russo J, Russo IH (2004a) The breast as a developing organ, Chapter 2. In: Molecular basis of breast cancer. Prevention and treatment. Springer, pp 1–48

Russo J, Russo I (2004b) Development of the human breast. Maturitas 49:2–15

Ryssel H, Germann G, Reichenberger R (2010) Craniomedial pedicled mammaplasty based on Würinger's horizontal septum. Aesthet Plast Surg 34(4):494–501

Sarhadi NS, Shaw-Dunn J, Soutar DS (1997) Nerve supply of the breast with special reference to the nipple and areola: Sir Astley Cooper revisited. Clin Anat 10:283–288

Schal B (2010) Mammary odor cues and pheromones: mammalian infant-directed communication about maternal state, mammae, and milk. Vitam Horm 83:83–136

Schepers RJ, Ringkamp M (2009) Thermoreceptors and thermosensitive afferents. Neurosci Biobehav Rev 33:205–212

Sherrington CS (1900) Cutaneous sensation. In: Schafe N (ed) Textbook of physiology, vol 2. Edinburgh, Pentland, pp 929–1001

Smith GH, Chepko G (2001) Mammary epithelial stem cells. Microsc Res Tech 52:190–203

Smith DM Jr, Peters TG, Donegan WL (1982) Montgomery's areolar tubercle. A light microscopic study. Arch Pathol Lab Med 106:60–63

Smith CA, Monaghan P, Neville AM (1984) Basal clear cells of the normal human breast. Virchows Arch A Pathol Anat Histopathol 402:319–329

Stingl J, Raouf A, Emerman JT, Eaves CJ (2005) Epithelial progenitors in the normal human mammary gland. J Mammary Gland Biol Neoplasia 10:49–59

Stirling JW, Chandler JA (1976) Ultrastructural studies of the female breast: I. 9 + 0 cilia in myoepithelial cells. Anat Rec 186(3):413–416

Stirling JW, Chandler JA (1977) The fine structure of ducts and subareolar ducts in the resting gland of the female breast. Virchows Arch A Path Anat Histol 373:119–132

Suami H, Pan WR, Taylor GI (2009) Historical review of breast lymphatic studies. Clin Anat 22 (5):531–536

Tairych GV, Kuzbari R, Rigel S, Todoroff BP, Schneider B, Deutinger M (1998) Normal cutaneous sensibility of the breast. Plast Reconstr Surg 102(3):701–704

Taneri F, Kurukahvecioglu O, Akyurek N, Tekin EH, Ilhan MN, Cifter C, Bozkurt S, Dursun A, Bayram O, Onuk E (2006) Microanatomy of milk ducts in the nipple. Eur Surg Res 38:545–549

Tanis PJ, Nieweg OE, Valdés Olmos RA, Kroon BB (2001) Anatomy and physiology of lymphatic drainage of the breast from the perspective of sentinel node biopsy. J Am Coll Surg 192 (3):399–409

Terzis JK, Vincent MP, Wilkins LM, Rutledge K, Deane LM (1987) Breast sensibility: a neurophysiological appraisal in the normal breast. Ann Plast Surg 19(4):318–322

Tobon H, Salazar H (1975) Ultrastructure of the human mammary gland. II. Postpartum lactogenesis. J Clin Endocrinol Metab 40:834–844

Toker C (1967) Observations on the ultrastructure of a mammary ductule. J Ultrastruct Res 21:9–25

Tsuchiya S, Li F (2005) Electron microscopic findings for diagnosis of breast lesions. Med Mol Morphol 38:216–224

Villadsen R, Fridriksdottir AJ, Rønnov-Jessen L, Gudjonsson T, Rank F, LaBarge MA, Bissell MJ, Petersen OW (2007) Evidence for a stem cell hierarchy in the adult human breast. J Cell Biol 177:87–101

Vorontsov IM, Fateeva EM, Hasenson LB (1993) Breastfeeding children. Ppmi, SPb, p 198 (in Russian)

Weinstein S (1968) Intensive and extensive aspects of tactile sensitivity as a function body part, sex and laterality. In: Kenshalo DR (ed) The skin senses. Charles C. Thomas, Springfield, IL, pp 195–222

Wood-Jones F, Turner JB (1931) A note on the sensory characters of the nipple and areola. Med J Australia 1:778–779

Wueringer E, Tschabitscher M (2002) New aspects of the topographical anatomy of the mammary gland regarding its neurovascular supply along a regular ligamentous suspension. Eur J Morphol 40(3):181–189

Würinger E, Mader N, Posch E, Holle J (1998) Nerve and vessel supplying ligamentous suspension of the mammary gland. Plast Reconstr Surg 101(6):1486–1493

Zagorodnyuk VP, Brookes SJ, Spencer NJ (2010) Structure-function relationship of sensory endings in the gut and bladder. Auton Neurosci 153(1–2):3–11

Zografos GC, Zagouri F, Sergentanis TN, Oikonomou V, Fotou M, Nonni A, Patsouris E (2009) Ductal endoscopy of the breast: more painful at the luteal phase? Pain Pract 9:221–224

Functioning of a Woman's Breast in the Initial Period of Lactation

4

Abstract

During the period of established lactation, the process of milk formation and excretion in a lactating woman begins to depend on the active participation of the central nervous system. The chapter presents humoral and nervous mechanisms for the formation of reflexes of secretion and excretion of milk. Detailed data of amplitude–time characteristics of vacuum and compression stimuli affecting the sensory receptors of breast areola during milk ejection by child are presented. The release of oxytocin and prolactin from the neurohypophysis and adenohypophysis during the mother's milk excretion and secretion reflex is an impulse. It is believed that the impulse nature of this release is provided by special neuronal structures localized in the hypothalamus and the brain structures surrounding it. During secretion, the components of milk can enter the alveolar cavity through transcellular and intercellular pathways.

4.1 Beginning of Lactation. End of Lactogenesis I

After childbirth, the functioning of the breast of a woman is characterized by a peculiar "inertia," which is manifested in the fact that the amount and composition of secreted milk during the first 1.5–2 days, called postpartum colostrum, changes quite slowly, remaining in most cases at the antenatal level. The study of the content of prolactin, progesterone, and estrogen in the blood of lactating women at this time indicates that this is mainly due to the high content of the hormones progesterone and estrogen remaining after pregnancy (Fig. 4.1a, b). Similar to the mammary glands of animals, estrogen and progesterone are believed to inhibit the synthetic activity of prolactin in the secretory cells of the alveoli of the mammary glands of women (Neville et al. 2001, 2002). It is important to note that the experimental data available in the literature (West and McNeilly 1979; Martin et al. 1980) indicate that the change in the concentration of hormones in the first 2 days after birth did not depend

N. P. Alekseev, *Physiology of Human Female Lactation*,
https://doi.org/10.1007/978-3-030-66364-3_4

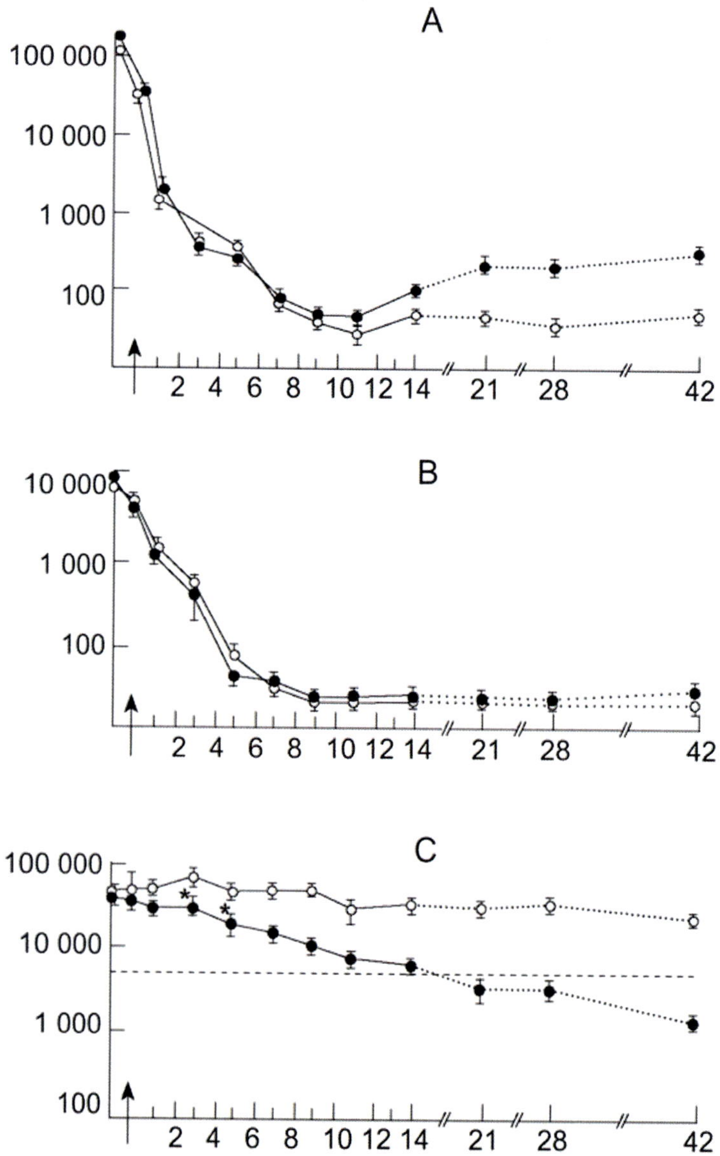

Fig. 4.1 Basal levels of 17-beta-estradiol (**a**), progesterone (**b**), and prolactin (**c**) in women who breastfeed in the postpartum period (light circles) and not breastfeed (dark circles) (from Martin et al. 1980). (**a**, **b**) Ordinate axis-respectively the concentration of 17-beta estradiol and progesterone in nmol/l, on the abscissas axis-time in days, the arrow indicates the time of delivery. (**c**) Ordinate axis—the concentration of prolactin miu/L: abscissas axis—the time in days, the arrow indicates the time of delivery. The dashed line shows the level of prolactin in nonpregnant women with a normal menstrual cycle

on whether the child was excreted colostrum during this period (Fig. 4.1). This indicates that the dynamics of the hormone content in the blood of women at the beginning of lactation seem to be genetically programmed.

Special surveys (Santoro Jr et al. 2010), devoted to the consumption of colostrum by a child on the first day, using a high-resolution scale (0.5 g), showed that the volume of colostrum excreted during 10 feedings was 15 ± 11 g. At the same time, as the authors note, the amount of colostrum excreted by the child in the first day was not associated with the further milk productivity of the woman and the duration of breastfeeding. On the second day, the amount of colostrum excreted by the child was not measured with such accuracy, but the available data indicate that the volume of colostrum was of the same order and was approximately 30 ml (Pang and Hartmann 2007). At the same time, in our surveys, it was found that the amount of colostrum excreted during this period by the child could differ markedly from the content of it in the alveolar-ductal system of the mother's breast. Thus, additional pumping of colostrum with the help of breast pump showed that the amount of colostrum could exceed the volume of colostrum excreted by the child by 2–3 times (Alekseev et al. 2010).

It is believed that the first 2 days for the child are the main transition period from continuous feeding and receiving oxygen through the umbilical cord during pregnancy to intermittent feeding from the mother's breast and pulmonary respiration. A newborn child enters a non-sterile world with an immature immune system (Newton 2004; Jackson and Nazar 2006). Extremely important at this time is the immune protection of the respiratory and gastrointestinal tract. Therefore, the child through a small amount of colostrum additionally receives a high concentration of immune components that provide a kind of initial "disinfection" of the entrance surface to the respiratory tract and gastrointestinal tract.

Due to the small amount of colostrum consumed, there is a question of compensation for energy costs during this period for a newly born child. It should be noted here that, first, colostrum has a high energy value, which is 128–150 kcal/kg and approximately 2–2.4 times higher than mature milk—62.5–75 kcal/kg (Vorontsov et al. 1993; Riordan 2005). Second, the food supply is compensated at first by the mobilization of energy from the fat reserves of the newborn, which exceed the fat content of the born cubs of other mammals (Pang and Hartmann 2007).

Reducing the concentration of progesterone and estrogen in the blood of women by 2–3 days after delivery to the level of their content in the blood of women with a normal menstrual cycle (West and McNeilly 1979) eliminates the block of synthetic prolactin function in the secretory cells of the alveoli. In addition, there is a decrease in the permeability of dense contacts between the secretory cells of the alveoli and thin ducts, as well as between the epithelial cells of the walls of the middle and thick milk ducts of the breast. Recall that in the mammary glands of mice during pregnancy (Nguyen et al. 2001), the high permeability of dense contacts, for example, to sucrose, depended on the presence of progesterone in the blood. Removing it by ovariectomy in mice caused a block in the permeability of dense contacts to sucrose. It is possible that in the mammary gland of women, the permeability of dense contacts is blocked for 2–3 days after birth due to a decrease in the concentration of progesterone. On the one hand, this will reduce the influence of the mineral and organic composition of the intercellular environment on the composition of the milk. On the other hand, it reduces the leakage of milk components from the alveolar-ductal system into the intercellular fluid.

As a result of hormonal changes, the synthetic and secretory function of the secretory cells of the breast alveoli is enhanced, and the pathways of milk components entering the alveolar-ductal capacitance system of the breast are activated. The experimental material available to date, obtained on the mammary glands of various animals, made it possible to identify several main routes of milk components entering the alveolar-ductal system (Neville 1990; Lawrence and Lawrence 1999; Shennan and Peaker 2000; Montalbetti et al. 2014) (Fig. 4.2). Unfortunately, there is no information about the peculiarities of the ways in which precursors of the milk and water enter the alveolar-ductal system of a woman's lactating breast due to methodological and ethical difficulties (Neville et al. 2012; Montalbetti et al. 2014). However, given the great similarity in the structural and functional characteristics of the cells that make up the alveoli, as well as the structure of the alveoli and ducts of the mammary glands of animals and humans, it can be assumed that this scheme (Fig. 4.2) in general can be applied to the breast of a woman.

The components of milk can enter the alveolar cavity through transcellular and intercellular pathways, among which there are five main pathways: four transcellular and one intercellular. In turn, transcellular pathways can be divided into pathways that transport milk components formed directly in secretory cells and transcellular pathways that transport milk components from blood serum or from stroma cells.

4.1.1 Transcellular Pathways

I. **Exocytosis**. This type of secretion in breast lactocytes has a great similarity to exocytosis in other types of cells in the body. The exocytosis pathway is the main one for entering the alveolar cavity of milk and lactose proteins, as well as oligosaccharides, phosphate, calcium, citrate, zinc, and water. These components are packed into vesicles in the Golgi apparatus of the secretory cells of the alveoli and then transported to the apical region of the cells. Here the vesicle membranes merge with the apical membrane, which is perforated. In this case, the contents of the vesicles are output to the alveolar lumen (Fig. 4.2).

II. **Transport of lipids**. Milk lipids, mainly triglycerides and lipoproteins, are also secreted into the alveolar cavity by a pathway common to all mammalian epithelial cells. Lipids are synthesized in a smooth plasma reticulum localized in the basal region of the cell from fatty acids and glycerol. Synthesized lipid molecules form fat droplets, which are covered by a protein shell. When approaching the apical part of the secretory cell, the droplets increase in size, merging with each other. In contact with the apical membrane, fat droplets are enveloped by the cell membrane, forming fat globules, and then separated, unlaced from the cell so that the integrity of the cell is not violated (Fig. 4.2).

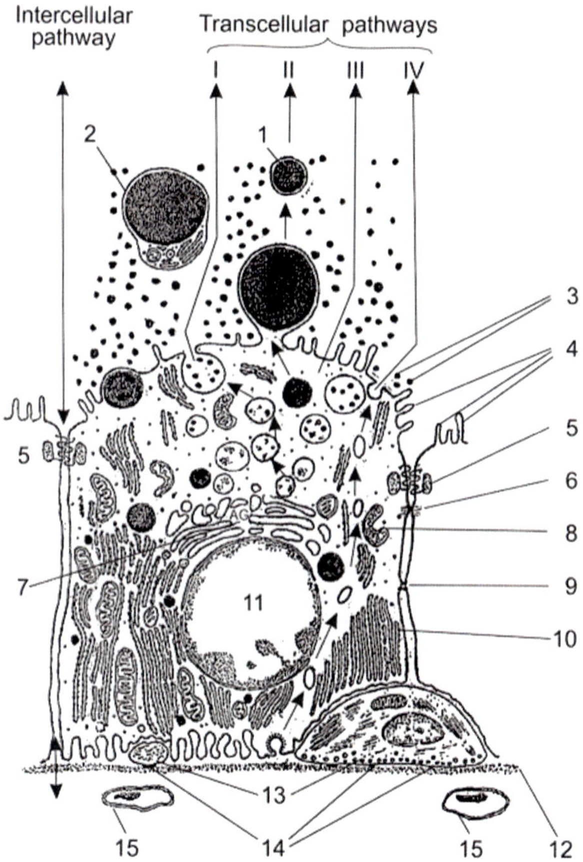

Fig. 4.2 Ways of secreting milk components into the alveolar cavity (from Shennan and Peaker 2000 with additions). Transcellular pathways I. Exocytosis of milk protein and lactose using vesicles formed in the Golgi apparatus. II. Secretion of lipids using a fat drops III. Secretion of ions and water through the apical membrane of the secretory cell. IV. Pinocytosis–exocytosis pathway for the transport of serum proteins: immunoglobulins, albumins, and transferrin; endocrine hormones: insulin, prolactin, and estrogen; secretory immunoglobulin A (sIgA), cytokines, and lipoprotein come from the stroma. Intercellular pathway for components of the intercellular environment, leukocytes, and stem cells. *fg* fat globule, *fgc* fat globule with a part of the cytoplasm of the secretory cell, *pg* protein granules, *mv* microvilli, *AG* Golgi apparatus, *tj* tight contacts, *d* desmosome, *mt* mitochondria, *sk* slit contact, *rr* rough reticulum, *n* nucleus, *bm* basal shell, *mp* processes of myoepithelial cells, *hd* hemidesmosomes, *k* capillaries

As a result of this process, a small volume of secretory cell cytoplasm is added to the fat globule in the form of a narrow strip. The membrane surrounding the globule prevents the formation of large fat droplets, which can hinder the movement and exit of milk from the alveolar-ductal system. In addition, the globule membrane is a source of phospholipids, as well as enzymes including oxidases, reductases, and hydrolases with high specific activity (McManaman and Neville 2003).

III. **A transmembrane pathway for the transport of ions and small organic molecules** such as glucose, amino acids, and water through the basal and apical part of the plasma membrane.

IV. **Along this path, exogenously formed milk components are transported to the alveolar cavity, including macromolecular substances, the source of which is blood serum.** These include the serum proteins immunoglobulins, albumin, and transferrin, and endocrine hormones such as insulin, prolactin, and estrogen. Secretory immunoglobulin A (sIgA), cytokines, and lipoprotein come from the stroma. These components are incorporated into vesicles formed by the basolateral membrane of the secretory cell and transported to the apical part of the cell. Here the vesicle membranes merge with the apical membrane, which is perforated, and the contents of the vesicles enter the alveolar lumen.

4.1.2 Intercellular Pathway

This pathway provides direct entry of components of blood serum and intercellular medium into milk. Transport through this pathway is regulated by the direct and indirect action of hormones and growth factors on the permeability of intercellular contacts. The transport of small hydrophilic molecules and the passage of macrophages, lymphocytes, and stem cells into the cavity of the alveolus and ducts is regulated by the proteins of dense contacts of the mammary glands.

Activation of milk intake routes and water components will manifest in changes in the quality and amount of secreted postpartum colostrum during approximately the first 4 days (continuation of lactogenesis I) and then during the next 4–10 days (lactogenesis II) of transition milk (Lawrence and Lawrence 1999). It should be noted that very precise methods of analysis of many components of a woman's milk have been developed. For a number of the main components of milk, it was possible to trace the dynamics of changes in their concentrations and the level of secretion during various periods of lactation formation. Figures 4.3 and 4.4 (Neville et al. 1991) present data on the dynamics of concentrations and levels of secretion of certain mineral and organic components of milk during lactogenesis I and II. The figures clearly show that the amount and composition of secreted during the first 1.5–2 days of postpartum colostrum changes quite slowly, remaining in most cases at the antenatal level (Table 2.1). The most rapid changes in the content of milk components and the level of their secretion occur in the next 2–2.5 days. In particular, starting from the end of the second day, along with a rapid increase in the volume of milk (colostrum), the concentration of citrate, free phosphorus, glucose, and lactose in milk increases to the same extent. The content of magnesium, sodium, potassium, chlorine, calcium, and ionized calcium decreases by the end of the second day. Then for magnesium, sodium and chlorine concentrations remain at a constant level, and for potassium, calcium and including ionized calcium concentrations begin to increase and stabilize by the fourth day.

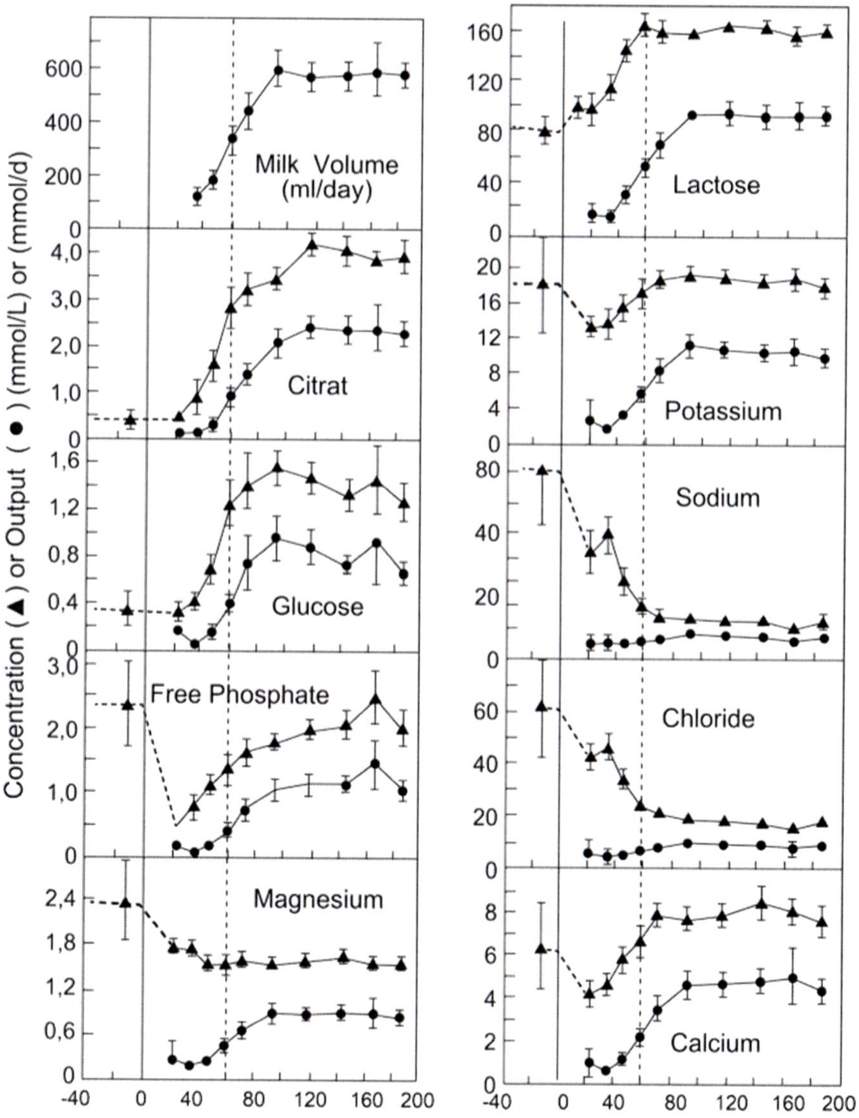

Fig. 4.3 Concentration (triangles) and secretion rate (circles). some mineral components of milk and lactose during lactogenesis I and II (from Neville et al. 1991). Ordinate axis: milk volume—ml/day; ion concentration—mmol/l. Mean values for milk composition were determined by pooling across individuals. The secretion rate of each component for these days was derived for each individual by multiplying her time-corrected volume by time-correcting concentration of that component. Mean secretion rate wMeas then derived by pooling across individuals. On the abscissa axis for all graphs: hours before and after childbirth. A vertical solid line with a zero mark on all graphs indicates the time of childbirth. A vertical dashed line in all graphs indicates the time when the lactose concentration reaches the maximum level and its reaching a constant level

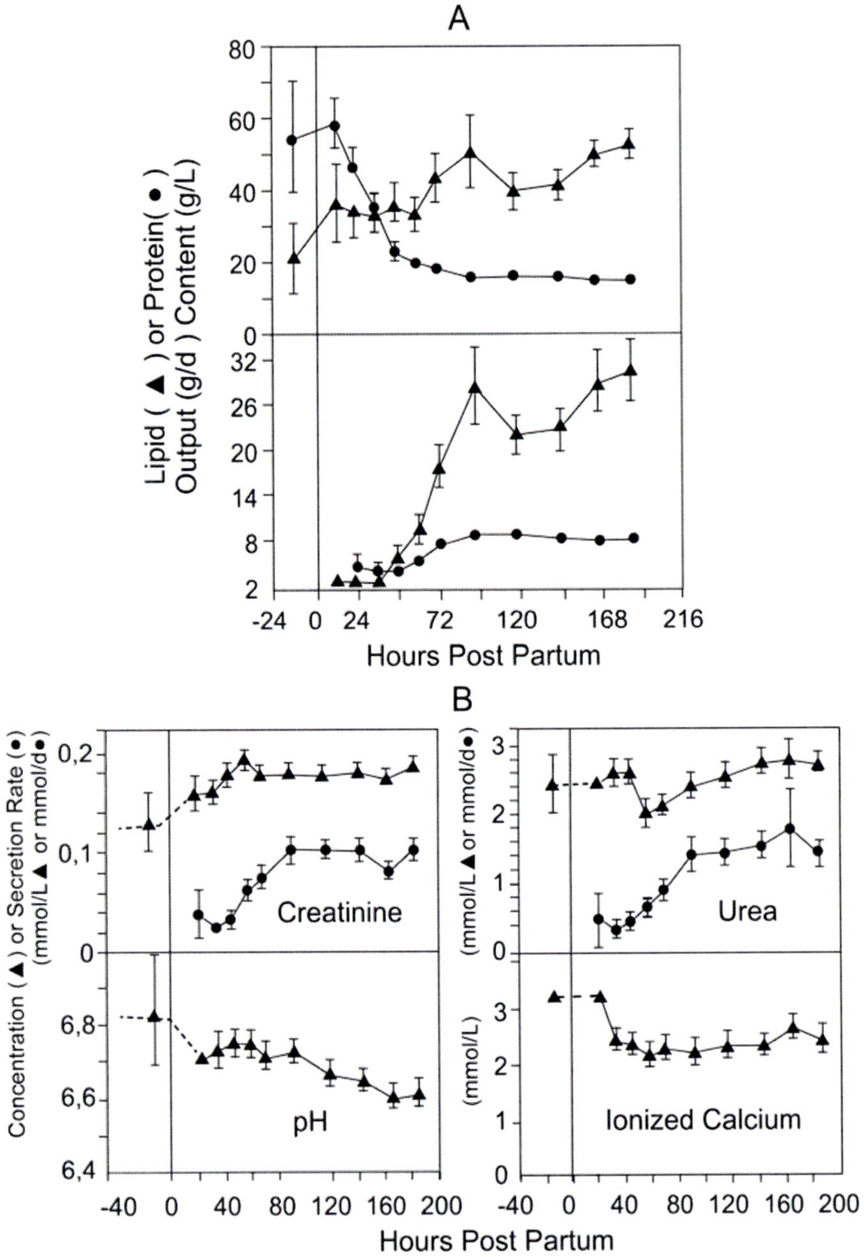

Fig. 4.4 Concentration (triangles) and rate of secretion (circles) of certain organic components of milk, and pH level during lactogenesis I and II (by Neville et al. 1991). (**a**) Lipids and proteins. Ordinate axis: concentration—g/l, secretion rate—g/day. On the abscissa axis for all graphs: hours before and after childbirth. (**b**) Creatinine and urea secretion concentration and rate, ionized calcium concentration, and pH level. On the ordinate axis: concentration in g/l, on the abscissa axis: hours before and after childbirth. A vertical solid line with a zero mark on all graphs (**a**, **b**) indicates the time of childbirth

The concentration of lipids in the colostrum increases immediately after delivery, while the concentration of protein, which is most important on the first day after birth, begins to decrease (Fig. 4.4a). The ratio of lipid and protein concentrations to colostrum volume allowed us to estimate the rate of secretion of these components. It turned out that the rate of secretion of proteins and lipids for 1.5 days after delivery was at a constant level, and then began to increase linearly and reached the maximum value by 4 days. Creatinine and urea concentrations did not change significantly in the first 4 days after delivery. At the same time, the rate of their secretion increased at this time, pH values decreased in the initial postpartum period and stabilized in 7–8 days.

The total volume of secreted milk, starting from the second day, increases rapidly. As a result, in 3–4 days, the alveolar-ductal system of the gland is filled with a fairly large amount of milk, called transitional milk. At the same time, the blood flow inside and in the skin of the gland increases. The glands become dense, and the temperature often rises. This condition is called "postpartum engorgement" of the mammary glands. It is interesting to note here that the onset of postpartum engorgement did not depend on whether the mother breastfed the child or expressed milk manually or with a breast pump (West and McNeilly 1979).

Recall that a similar phenomenon of "engorgement" was observed on days 4–6 in the mammary glands of newborns—"neonatal engorgement," as well as in women with a normal menstrual cycle at the end of the luteal phase—"premenstrual engorgement." In all cases, breast enlargement was associated with a decrease in the concentration of steroid hormones and an increase in the concentration of prolactin. At the same time, there was a growth of glandular and stromal tissue of the breast, filling with secret cavities of the alveolar buds (alveoli) and ducts, the appearance of edema in the interlobular stroma, an increase in vascularization and blood flow intensity. Note that at all stages of development of the mammary glands, engorgement is a normal physiological process, and the degree of engorgement has significant individual fluctuations. Postpartum lactation engorgement of the mammary glands is a signal of the end of lactogenesis I and the onset of lactogenesis II.

4.2 The Establishment of Lactation. The Lactogenesis II

In contrast to early lactation, when hormonal changes blood lactating women, the composition and amount of milk in her mammary glands a little depended on, were withdrawn or no colostrum from the breast, the establishment and implementation of the lactational function of the breastfeeding woman begins to depend on stimulation of sensory receptors of the breast and excretion of milk. The processes of secretion and excretion of milk occur with the active participation of the central nervous system (CNS) of the mother and are largely provided by the maternal reflexes and by the reflexes of the newborn baby. The main reflexes of the mother are a reflex excretion and reflex secretion of milk, and baby the rooting reflex and milk ejection reflex (sucking reflex). Recall that the reflex is defined as the response of the body to any impact, which is implemented in the form of a sequential awakening of the chain of elements that make up the reflex arc. The reflex arc includes the sensory receptor, afferent pathway, central link, efferent pathway, and effector (working organ). Reflexes are divided into two types: unconditional and conditional. Unconditioned reflexes are innate, inherited reflexes that are inherent in the entire species. Conditioned reflexes are acquired reflexes peculiar to an individual. They occur during the life of an individual under certain conditions and disappear in their absence. Conditioned reflexes are not fixed genetically (not inherited) and are formed on the basis of unconditioned reflexes with the participation of the higher parts of the brain.

4.2.1 Rooting Reflex of the Child

The rooting reflex is an innate reflex and belongs to the so-called primitive reflexes of the child. As the name of the reflex implies, its implementation is aimed at finding the part of the mother's breast where the nipple is located and capturing it with part of the areola into the oral cavity. According to the morphofunctional structure of the reflex, the rooting reflex must include afferent and efferent links, paths, and a central representation. Unfortunately, for methodological and ethical reasons, there is no morphofunctional information at the cellular level of the child's rooting reflex. There are only clinical psychophysical data of the rooting reflex. Numerous clinical studies show that the initial receptor link of the child's rooting reflex can be receptors of several modalities and primarily mechanoreceptors localized in the lips and skin of the perioral area. The rooting reflex formed by mechanoreceptors can be designated as a mechanosensory rooting reflex.

Mechanosensory Rooting Reflex Stimulation of mechanoreceptors in contact with the surface of the lips and adjacent skin areas of the newborn with the surface of the mother's gland or other parts of her body is accompanied by movement of the lips and tongue, as well as turns of the child's head. Special mechanical stimulation of the mechanoreceptors of the lips and skin around the mouth of the newborn also causes a number of reflex motor reactions. Thus, when stroking the skin in the area of the corner of the child's mouth, the upper lip is lowered, the tongue is deflected and the head is turned toward the stimulus. Touching the skin of the middle of the upper lip of the newborn causes the opening of the mouth and the extension of the head. When pressing on the skin of the middle of the lower lip of the child, there is a lowering of his lower jaw and bending of the head. Moreover, all these reflex reactions are most pronounced in a hungry child. Here it should be noted that the highest mechanical sensitivity of the human face is observed in the skin surface of the lips and the perioral area. This is evidenced by the data of clinical studies started in the 40 years of the last century by V. Penfield (Penfield and Boldrey 1937) on mapping the sensory sensitivity of the surface of the torso and face of a person. The map of the "sensory" surface of the human body and face, located on the post-central gyrus of the cortex, was called the "sensory homunculus." Images on this map of parts of the human body are represented according to the density of their sensory innervation. The highest density of sensory receptors is observed in the area around the mouth and lips of a person, as well as in the skin of the fingers and especially the thumb. Therefore, the "sensory homunculus" looks like a man with disproportionately large hands and fingers, and the largest size on the face are the lips. The emergence of a technique for registering the electrical impulse activity of single nerve fibers forming sensory receptors in the skin and, in particular, mechanoreceptors in a waking person (Johansson and Vallbo 1983; Vallbo and Johansson 1984) allowed us to study in more detail their functional characteristics, receptive fields, and distribution in various parts of the body and based on these data to classify several types of mechanoreceptors. In addition, it is very important to note that it was possible to identify the types of mechanoreceptors with a high degree of probability (Johansson and Vallbo 1983; Vallbo and Johansson 1984).

According to the conducted surveys, the skin mechanosensory sensitivity of the newborn is quite high but decreases with age (Lagercrantz and Changeux 2009; Cornelissen et al. 2013). It is believed that central structures are responsible for changes insensitivity, and the composition of skin mechanoreceptors and their functional characteristics do not change with age. In the study of mechanoreceptors localized in the skin of the face, lips, and oral mucosa adjacent to the skin of the lips of adults (Trulsson and Johansson 2002; Bukowska et al. 2010) several types of low-threshold mechanoreceptors were found, which are undoubtedly present in newborns. It is interesting to note that the type and density of low-threshold mechanoreceptors of the skin of the lips and the perioral area showed a great similarity with the low-threshold mechanoreceptors of the hairless skin of the fingers. For a better understanding of the mechanisms of the rooting reflex, it seems appropriate to make a digression from the presentation of information about the rooting reflex of the child and briefly focus on the properties of these types of mechanoreceptors. In addition, looking ahead, it should be noted that these types of mechanoreceptors were the initial link not only of the rooting reflex but also the milk ejection reflex of the child, as well as maternal milk reflexes.

Encoding of information in the nervous system of humans and animals is carried out using a frequency code. In this regard, one of the most frequently used physiological methods for studying the functional characteristics and classification of skin sensory receptors is the study of impulse activity in the nerve fibers that form these receptors under the action of trapezoidal mechanical stimuli. Use trapezoidal stimuli with different amplitude, duration, and rate of rise and reduction of the stimulus allows you to determine what is an adequate stimulus for receptor—amplitude of displacement or velocity of its change (Burgess and Perl 1973; Johansson and Vallbo 1983). Mechanoreceptors, the frequency of impulse activity, which depends on the rate of change of the mechanical stimulus and does not depend on the amplitude, are defined as velocity detectors. If the frequency does not depend on the rate of rise of the stimulus and is determined only by the amplitude of the stimulus, mechanoreceptors are classified as amplitude detectors. When the frequency of impulse activity changes with the change in the stimulus amplitude and the rate of its rise or fall, mechanoreceptors are classified as amplitude–velocity detectors (Burgess and Perl 1973; Johansson and Vallbo 1983).

Figure 4.5 shows the responses of various types of low-threshold mechanoreceptors of the human lip skin and near the oral region to trapezoidal stimuli, as well as their receptive fields (Bukowska et al. 2010). In particular, mechanoreceptors were found (Fig. 4.5a), which generated impulse activity only during the increase and in some cases during the decline of the amplitude of the trapezoidal tactile mechanical stimulus. The frequency of impulse activity increased with the increase in the rate of increase of mechanical stimulation. At the beginning of the action of the static component of the mechanical stimulus, the frequency of the impulse was rapidly reduced (adapted) and in the future, it was not generated for permanent mechanical stimulation. This type of receptor is called rapidly adapting type I receptors (RAI) and has been classified as velocity detectors (Johansson and Vallbo 1983; Vallbo and Johansson 1984). The area of receptive fields averaged 5.6 mm^2. The threshold value of the stimulus force inside the receptive field was 0.19 mN. Impulse responses of mechanoreceptors of hair follicles (HF) to trapezoidal deviations of the hair are similar to the responses of RAI. The receptive fields of the HF are dotted and almost coincide in size with the cross-sectional area of the hair follicle (Trulsson and Essick 2010).

In addition to RAI, two other types of mechanoreceptors were found in the skin of the lips and near the oral region of a person. The first distinctive feature of these mechanoreceptors was the presence of spontaneous impulse activity, the second the generation of impulse activity during the action of the static component of the stimulus (Fig. 4.5b, c). In both types of mechanoreceptors, during the action of the static component of the mechanical stimulus, the frequency of impulse activity slowly decreased over time—slow adaptation occurred, hence these types of mechanoreceptors were called slow-adapting (SA). According to their characteristics, SA receptors were divided into two groups—SAI and SAII (Trulsson and Essick 2010). In some cases, SAI and SAII had spontaneous impulse activity. The frequency of action potentials during the dynamic response of SA as RA changed with the change of the velocity of mechanical stimulation; however, SAI impulse response to changes in the rate of increase was more pronounced than SAII. In addition, the adaptation of the frequency of pulse activity during the static component in SAI occurred faster and the frequency of action potentials was less regular than in SAI (Fig. 4.2b, c). SAI and SAII increased the frequency of action potentials with an increase in the amplitude of the stimulus. These receptors were classified as amplitude–velocity detectors. It should be noted that the study of the functional characteristics of the mechanoreceptors of the lip skin and the adjacent skin surface was conducted using tactile mechanical stimuli. However, it turned out that SAII, unlike other types of mechanoreceptors, showed high sensitivity to skin stretching and responded by increasing the frequency of action potentials to lateral stretching relative to the zone of maximum sensitivity determined by a tactile stimulus (Trulsson and Johansson 2002). The area of receptive fields averaged 4.2 mm^2 for SAI and 5.65 mm^2 for SAII. The threshold value of the stimulus force inside the receptive field was on average 0.15 mN for SAI and 0.35 mN for SAII. Here are the data for the lower lip and adjacent skin surface (Bukowska et al. 2010; Trulsson and Essick 2010). Similar indicators were found earlier for

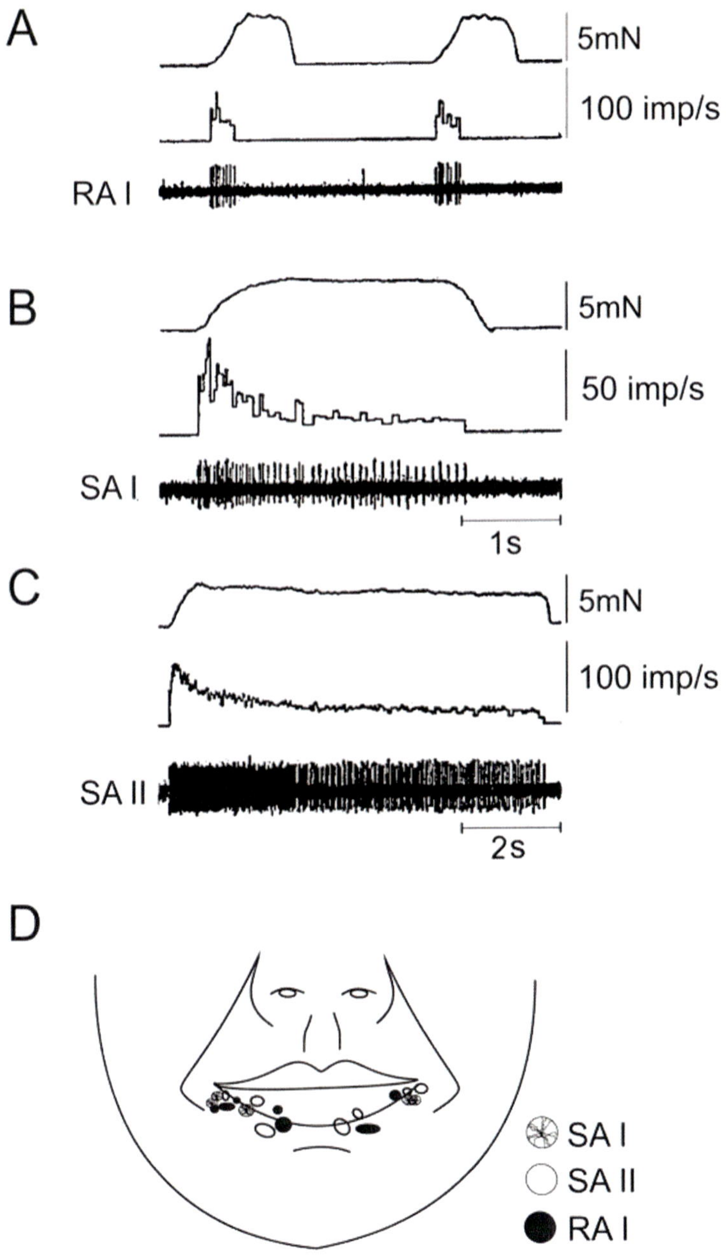

Fig. 4.5 Responses of low-threshold mechanoreceptors of the skin of the lower lip and near-mouth region (**a–c**) to trapezoidal stimuli and their receptive fields (**d**) (from Bukowska et al. 2010). (**a–c**) On all electrograms, the upper beam is the mark of mechanical irritation, the middle beam is the instantaneous frequency of action potentials, the lower beam is the impulse activity recorded from nerve fibers. Vertical calibration of force and frequency, respectively, horizontal—time. G. Receptive fields of various types of mechanoreceptors

low-threshold receptors located in the skin of the upper lip and in the areas of the face that lie adjacent to it (Johansson et al. 1988). The highest density of mechanoreceptors was found in the corners of the lips. Extremely interesting data were obtained in surveys on the occurrence of sensations in a waking person with electrical stimulation of single nerve fibers, from which the pulse activity of low-threshold mechanoreceptors of the skin of the oral region was recorded when they were adequately stimulated (Trulsson and Essick 2010). It turned out that electrical stimulation of single fibers forming RAI by a series of pulses at a frequency of 5 imp/s and 60 imp/s was accompanied by a sense of successive touches and vibrations in areas close to the skin surface, respectively. Similar stimulation of the nerve fibers of SAI caused in most cases a feeling of pressure on the skin in its deeper areas. At the same time, electrical stimulation of the nerve fibers that form SAII was not accompanied by any sense of mechanical action. These data coincide with the results of similar studies conducted earlier on mechanoreceptors of hairless human hand skin (Ochoa and Torebjörk 1983). Unfortunately, to date, there is no morphological data in the literature about the types of mechanoreceptors located in the skin of the lips and the human near the mouth region. We can only assume, based on a large experimental material on the study of the functional characteristics of low-threshold mechanoreceptors located in various areas of the human skin, including in the skin of the lips and skin around the mouth, that RAI correspond to Meissner's corpuscles or Meissner-like corpuscles, and SAI—the Merkel complex. SAII corresponds to skin stretching receptors Ruffini corpuscles or muscle spindles localized in the muscles of the mouth (Johansson and Vallbo 1983).

Now let us go back to the child's rooting reflex. It should be noted here that the study of the functional capabilities of mechanoreceptors of the fingers (Phillips et al. 1992; Johnson and Hsiao 1992) in determining the surface texture showed a high ability of SAI and RAI to distinguish the degree of roughness, i.e., the presence of small bumps and depressions in the examined area. Given the similarity in the types, density, and size of the receptive fields of mechanoreceptors located in the skin of the fingers and the skin of the lips and near the mouth area region, it can be assumed that the latter also has a high "scanning" ability. In the previous chapter, it was noted that the skin of the nipple and areola is markedly different in structure from the rest of the skin surface of the breast of a woman. In particular, on the surface of the nipple and areola, the epidermis forms numerous grooves. In addition, there are folds on the skin of the nipple and areola (see, for example, Figs. 3.6A and 3.12). During pregnancy, the diameter of the areolar area usually increases significantly. The rough surface of the nipple and areola, as well as high ability SA and RA lips and the skin near the mouth area in determining the smallest bumps and depressions on the surface of the areola and nipple will undoubtedly facilitate newborn search and seizure in the mouth of the nipple and of the areola.

Olfactory Rooting Reflex As already noted, in addition to mechanoreceptors (main receptors), receptors of other modalities can participate in the formation of the rooting reflex. Here, first of all, it is necessary to note the olfactory receptors of the child. In the literature, there are a number of works, the results of which indicate the direct involvement of the child's sense of smell in the search for the breast nipple. In particular, surveys conducted on newborns, who were placed at the same distance from the mother's mammary glands found that when one of the mammary glands was washed with water immediately after birth, newborns in 73% captured the nipple of the gland that was not washed with water (Varendi et al. 1994). It was assumed that in addition to milk and colostrum, volatile components of the secret of other glands: sebaceous, sweat, and Montgomery glands, located in the area of the nipple and areola, are used by the child to find the nipple of the mother's breast. The data available to date indicate that one of these glands may be the Montgomery glands (Fig. 3.6). Often before feeding the child, droplets of secretions are observed at the outlet of the ducts of the Montgomery glands (Fig. 3.6b). Due to the fact that the amount of secret produced in the Montgomery glands is insignificant and apparently not related to the baby's nutrition, it has been suggested that its volatile components contribute to the effective contact of the baby with the mother's gland before feeding (Doucet et al. 2009, 2012; Schaal 2010). The conducted surveys (Doucet et al. 2012) showed that bringing the tip of a glass rod, on which there was a drop of Montgomery's gland secret, to the nostrils of the child's nose caused characteristic oro-cephalic reactions: head movement, pursing of the lips, tongue protruding similar to the reactions of the muscles of the lips and tongue when stimulating the mechanoreceptors of the skin of the lips and the near-mouth area. Oro-cephalic reactions could be caused by odors of other biologically significant and neutral substances: women's milk, cow's milk, a product of sebaceous skin glands secretion, milk formula, water, and vanillin. A quantitative assessment of oro-cephalic reactions over the 10-s period of olfactory stimulus presentation revealed that the secret of Montgomery glands in 63% of the examined children increased the duration of oro-cephalic reactions by 20%. For other odorants, the number of newborns with a 20% reaction was significantly less and amounted to 26% for woman milk, 26% for cow's milk, 21% for sebaceous secretions, 21% for water, 16% for milk formula, and 5% for vanillin.

Visual Rooting Reflex Clinical studies indicate that the rooting reflex may occur on adequate stimulation of the child's photoreceptors—the visual rooting reflex (Minderaa et al. 1985). By now, it is well known that a newborn child is able to see and distinguish colors normally. A child of 1-hour old can fix his gaze for 10–15 s and move his eyes and head behind the object. Moreover, the newborn is able to analyze what he sees and prefer one object to another (Vorontsov et al. 1993; Lagercrantz and Changeux 2009). Here it is important to note that the visual sensory system of animals and humans is not tuned to the perception of diffuse highlights, but to the moving edges of objects or contrasting areas of images. Motor responses of the visual rooting reflex can be triggered by approaching and undulating an object in front of the child's face, for example, a hammer to trigger the tendon reflex, a hand with clenched fingers (Minderaa et al. 1985). The child's responses are a movement of the head in the direction of an approaching object, opening the mouth or pulling out the lips. The appearance of the areolar–nipple area of the gland and in particular the difference in color between the nipple and the areola, the areola and the rest of the gland, as well as the presence of folds and grooves on the nipple and areola will be an effective stimulus for the child's visual sensory system, and facilitate the occurrence of a rooting reflex.

In conclusion, we note that other features of the breast of a lactating woman can contribute to the formation of the rooting reflex of the child. In particular, studies of the temperature of various areas of the skin surface of the breast (Zanardo and Straface 2015; Zanardo et al. 2017) found that in lactating women, the skin surface of the areola has a temperature 0.6 °C higher than other areas of the skin surface of the gland. In psychophysical surveys on adults (Essick et al. 2004), it was found that the lowest threshold of thermal sensitivity of the skin surface of the human face is observed for the perioral area. Thus, the thermal threshold of the transition point of the intermediate part of the lips to the mucous membrane—the red border is 0.2 – 0.3 °C. The thermal threshold of the child's orofacial surface was not studied. However, if we assume that the same high thermal sensitivity is observed for the surface of the lips of newborns, the child can use the heat receptors of the lips to localize the location of the areola and the nipple of the mother's breast. The heat of the areola may be an additional stimulus for the emergence of a rooting reflex in a child (Zanardo et al. 2017).

Thus, it can be concluded that the joint participation of the mechanosensory, olfactory, visual, and thermal sensory systems of the newborn in the formation of the rooting reflex provides an effective search and capture of the nipple and part of the areola of the mother's breast for further milk excretion. It is important to note that normally the rooting reflex in a child is observed for 3–4 months, but in the future, as the central nervous system develops, it normally fades. However, clinical data show that in some cases, the rooting reflex can be observed in later life and even in adults, due to pathological phenomena in the central nervous system (Zafeiriou 2004).

4.2.2 Infant Milk Ejection Reflex

So, as a result of the rooting reflex, the nipple with part of the areola was in the child's mouth, and the newborn can begin to withdraw milk from the mother's gland while stimulating the sensory receptors of the mother's breast. This happens with the help of another primitive reflex of the child, for which the name "sucking reflex" is often used in the literature. Here it should be pointed out that in order to remove milk from the mother's breast capacitance system, it is necessary to create a pressure difference between the milk inside the alveolar-ductal system and the external environment. This can be done by increasing the pressure of milk in the milk ducts. This method is used by a nursing woman when manually pumping milk. The woman squeezes the areolar area in front of the nipple with her fingers, where there are wide parts of the milk ducts (Fig. 3.3), increasing the pressure in them and squeezing milk through the milk ducts in the nipple. The pressure difference can be created by reducing the pressure in the area surrounding the nipple and areola below atmospheric pressure (creating a vacuum). This method is used in vacuum milk-removing devices—breast pump. The third method consists of simultaneously

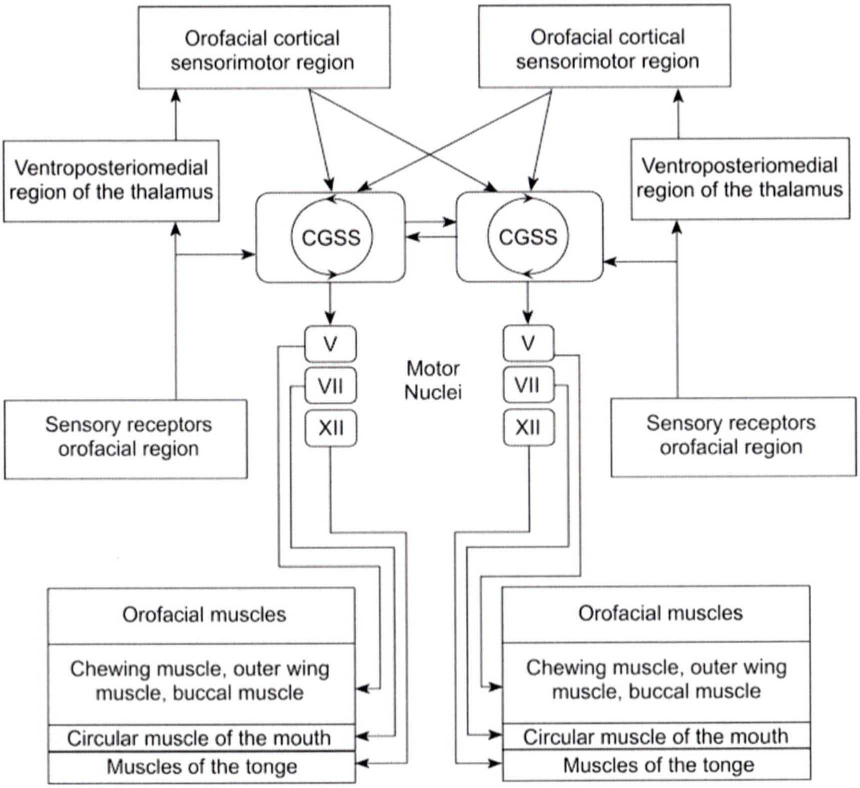

Fig. 4.6 Estimated neuronal scheme of milk excretion by a child (from Finan and Barlow 1998). CGSS is the central generator of the pattern of sucking stimuli

creating a positive pressure inside the alveolar-ductal system and a vacuum around the areolar–nipple area of the breast. Numerous surveys of the process of removing milk from the mother's breast (we will discuss this in more detail later) show that this method is used by the child when feeding it. Therefore, the term "sucking reflex" to refer to the process of milk excretion by a child is not quite correct (Akre 1989). In this regard, in the future, the text will use a more appropriate designation of this process—the milk ejection reflex.

Figure 4.6 shows the morphofunctional scheme of this reflex. It should be noted that to date, most of the morphological and functional characteristics of the infant's milk ejection reflex remain poorly understood. The initial receptor link of the baby's milk ejection reflex is mainly mechanoreceptors located in the skin and muscles of the tongue, in the walls of the oral cavity, and partially in the muscles and skin of the inner side of the lips. In addition, for the formation and modulating of the reflex of excretion of milk of the child, taste receptors located on the tongue and the walls of the oral cavity have a certain value. The final effector link is the muscles of the tongue, as well as the jaw and facial muscles, which directly remove milk from the alveolar-ductal system of the breast. Sensory and effector innervation is performed by branches of the trigeminal (V), facial (VII), lingual (IX), vagus (X), and sublingual (XII) nerves.

To date, the literature has data on the properties of mechanoreceptors of the skin and muscles of the tongue, the oral mucosa of an adult (Johansson et al. 1988; Trulsson and Essick 1997; Trulsson and Johansson 2002). It can be assumed, as in the case of the rooting reflex, that the mechanoreceptors of these sites in adults have functional and morphological characteristics similar to those of newborns. Responses from several types of low-threshold mechanoreceptors of the tongue and oral mucosa in a waking person were registered using the method of recording electrical impulse activity from single nerve fibers of the lingual and subglacial nerves that are part of the trigeminal nerve. Analysis of the patterns of impulse activity arising on mechanical trapezoidal stimuli showed that the skin and muscles of the tongue have the same functional types of mechanoreceptors as in the skin and muscles of the lips and near the oral region: RAI, SAI, and SAII (Fig. 4.5). However, the frequency of occurrence of various types of mechanoreceptors, the size of receptive fields, and values of the mechanical threshold of the mechanoreceptors of the tongue had noticeable differences from the mechanoreceptors of the skin of the lips.

So, in comparison with the skin of the lips, the skin of the tongue is dominated by RAI. Of the total number of language receptors studied, two-third of the receptors were RAI (Trulsson and Essick 1997). The RAI receptive fields were circular or oval, averaging 2.0 mm^2, which is about two times less than the RAI receptive fields in the skin of the lips. The mechanical threshold was equal to an average of 0.11 mN and was also twice less than that of the RAI skin of the lips. RAI of the skin of the tongue in all cases generated impulse activity at the beginning and end of the mechanical stimulus (Trulsson and Essick 1997; Bukowska et al. 2010).

SA in the skin of the tongue had round or oval receptive fields with an average diameter of about 1 mm^2, less than SA lips, but with a mechanical threshold is not different from the respective threshold for SA lips. SA localized in the skin of the tongue also had a round or oval receptive fields and the values of average diameter 5.3 mm^2, mechanical threshold 0.31mN. In addition to SA mechanoreceptors localized in the skin of the tongue of this type, mechanoreceptors were found in the deeper muscle layers of the tongue. According to the pattern of impulse activity generated by trapezoidal stimuli, the SA of the deep layers coincided with the SAI and SAII of the skin of the tongue. However, the receptive fields of these mechanoreceptors had a significant size (200–650 mm^2) and could exceed 4–10 times the receptive fields of SAI and SAII of the skin of the tongue. The mechanical threshold (4 mN) was also about 20–40 times higher than that of the SAI and SAII skin of the tongue (Trulsson and Essick 1997; Bukowska et al. 2010). Electrophysiological examinations have shown that the highest density of mechanoreceptors is observed for the tip of the tongue.

In the oral mucosa, in contrast to the skin of the tongue, only SAI and SAII mechanoreceptors were detected (Johansson et al. 1988), which did not differ in the size of receptive fields and mechanical thresholds from the SAI and SAII of the skin of the lips and near the oral area.

In morphological studies of sensory innervation of the adult tongue skin (Gairns 1953), as well as the human fetus at 28 weeks of pregnancy (Hewer 1935), no such variety of sensory receptors was found as in electrophysiological surveys (Trulsson and Essick 1997; Bukowska et al. 2010). Only receptor formations similar in structure to Meissner bodies were found, as well as nerve terminals of various degrees of branching that do not have auxiliary structures that would allow them to be attributed to known encapsulated skin mechanoreceptors. It is believed that one of the reasons for morphological "poverty" may not be a sufficiently detailed study of the mechanoreceptor structures of the tongue (Trulsson and Essick 1997). However, it is important to note that to date, there are numerous data on mechanoreceptors of internal organs of animals and humans, the patterns of impulse activity of which in response to trapezoidal stimuli do not differ from those of RAI, SAI, and SAII. However, they are formed by branching nerve fibers without any special auxiliary structures (Alekseev 2009; Zagorodnyuk et al. 2010). Therefore, it is possible that the mechanoreceptor function of the skin of the tongue and the mucous walls of the oral cavity of a child and an adult can also be performed by free nerve endings of varying complexity of branching.

High-threshold mechanoreceptors located in the deep layers of the tongue can be attributed with a high degree of probability to the receptors of the muscles of the tongue—muscle spindles. These mechanoreceptors had typical responses of primary and secondary endings of muscle spindles and, unlike the mechanoreceptors of the skin of the tongue, generated impulse activity when the tongue moved (Trulsson and Essick 1997).

When the nipple and areola of the mother's breast are inserted into the child's oral cavity, first of all, adequate stimulation of the mechanoreceptors of the tongue and the walls of the child's oral cavity occurs. The impulse activity that occurs in the nerve fibers that form these mechanoreceptors goes to the neurons of the reticular formation of the brain stem (Fig. 4.6), which, as it is assumed (Finan and Barlow 1998), form a special bilaterally located neural network. When it is activated by afferent impulses from the receptors of the orofacial region, the neural network automatically begins to send efferent action potentials in a certain sequence to the motor endings of the muscles of the child's tongue, lips, and cheeks. As a result of a coordinated rhythmic movement of the muscles, the baby outputs milk. This supposed neural structure is called the central generator of the sucking stimulus pattern (CGSS). It is interesting to note that fluctuations in the jaws and lips are already observed in 12-week-old fetuses (de Vries et al. 1984). However, stable rhythmic movements of the lips and jaws, which can occur spontaneously, are established by 34 weeks of pregnancy (Miller et al. 2003), i.e., it seems that the formation of the CGSS ends by this time. In addition to the reticular formation of the brain stem, impulse activity simultaneously enters the ventroposteriomedial part of the thalamus, and then into the orofacial section of the primary sensorimotor region of the cortex. However, given that the infant's milk ejection reflex, as well as other primitive reflexes, is most pronounced in the first months of life, the orofacial section of the cortex is not the main structure for milk ejection for a certain time after birth. The CGSS performs this function. However, surveys show that the work of the CGSS is modulated by afferent impulses coming from neurons of the orofacial section of the primary sensorimotor region of the child's cortex (Finan and Barlow 1998; Zimmerman and Barlow 2008; Oder et al. 2013) (Fig. 4.6).

There is no information in the literature about the structural and functional features of the child's CGSS at the present time. There is only a small amount of data on the localization and functional characteristics of CGSS in newborn guinea pigs (Iriki et al. 1988) and rat pups (Tanaka et al. 1999). In particular, it was found that the algorithm of movement of the muscles of the oral apparatus is provided by the impulse activity that occurs in the nerve cells of the segments of the varolian bridge located between the trigeminal and facial motor nuclei. In addition, this impulse activity can be modulated via descending pathways from the cortical structures of the brain (Iriki et al. 1988; Tanaka et al. 1999). In particular, in newborn animals, a site was found in the frontal region of the cortex, the electrical impulse stimulation of which was accompanied by movements of the jaws similar to those when removing milk from the gland. This area is called the cortical sucking area (CSA). CSA neurons were projected into the dorsal part of the paragigantocellular reticular nucleus on the contralateral side and further on to the CGSS nerve cells.

In adult guinea pigs, electrical impulse stimulation of a certain part of the frontal cortex also caused a rhythmic movement of the jaws, but similar to chewing movements. This area was called the cortical masticatory area (CMA) and was located caudal to the CSA. CMA neurons through the pyramidal tract gave projections to the neurons of the dorsal part of the paragigantocellular reticular nucleus on the contralateral side and then projected to the nerve cells of the neural structure located in the medulla oblongata and called the chewing movement generator (CMG). It is interesting to note that pulsed electrical stimulation in newborn animals of areas of the frontal cortex lying caudal to the CSA approximately at the level of the CMA of adult animals did not cause movement of the jaws. Accordingly, similar stimulation of the cortex of adult animals located rostral to the CMA at the CSA level of newborns was also not accompanied by movement of the jaws (Iriki et al. 1988). According to the authors, with the development of the animal, changes occur in the neural organization of areas of the frontal cortex so that the projections from the CSA to the paragigantocellular reticular nucleus are replaced by projections from the CMA, and the CGSS is apparently transformed into the CMG. Moreover, during the transition from food consumption by sucking to chewing, two types of generators function simultaneously: CGSS and CMG (Iriki et al. 1988; Alberts and Pickler 2012).

It can be assumed that a similar situation exists for the child, i.e., as the child grows and develops its central nervous system, the milk ejection reflex gradually "fades" and the CGSS is replaced by the CMG (Alberts and Pickler 2012).

The most studied part of the infant's milk ejection reflex was its effector part, which ensures the removal of milk from the mother's gland. Additionally, it is important to note that the impact of the tongue and jaws of the child on the areola–nipple area of the breast is needed not only for immediate removal of milk but are also adequate stimulus receptors that area of the breast in the formation of the second reflex is required for removing milk from the gland—maternal reflex excretion of milk.

Before we begin to describe the movement of the orofacial structures of the child during the removal of milk, it is advisable to briefly focus on the morphological features of the oral cavity of newborns. The oral cavity of a newborn differs in structure from the oral cavity of an adult, and, according to the general opinion of experts, is quite well adapted for the excretion of milk by a child. Since the child uses a vacuum to place and hold the nipple with the areola in the oral cavity, it is necessary that the structures of the orofacial area, first of all, provide isolation of the oral vacuum from the atmospheric pressure of the external environment. For this purpose, the newborn has a number of "devices" in the oral cavity. In particular, the lips of a newborn have a proboscis-shaped shape. The circular muscle of the mouth is well expressed. On the lips are present in the form of thickening pads located along the surface of the lips. These formations are called sucking pads and they occur on the lips of the fetus at 25 weeks of age (Hendrik 2013). Thus, the child's lips tightly cover the skin of the areola located at the base of the breast nipple and ensures that the vacuum is maintained and the nipple with a part of the areola is held inside the oral cavity. In addition, the gingival membrane can additionally enhance the isolation of the child's oral cavity. The gingival membrane is a comb-like fold of the mucous membrane located on the alveolar processes of the upper and lower jaws in the frontal region. It abounds in small papillary tubercles and is rich in blood vessels, as a result of which it has the ability to seal when in contact with the skin of the areola of a woman's breast. It is believed that the retention of the nipple with part of the areola in the child's oral cavity is facilitated by transverse palatine folds, which increase the adhesion of the skin of the nipple and areola to the palate. They are observed in newborns in a much more pronounced form than in adults. On average, there are 4–5 pairs of transverse folds, of which 2–3 pairs depart from the palatine sagittal suture.

The oral cavity of newborns and children of the first year of life is smaller in comparison with the oral cavity of older children. At the same time, the tongue of a newborn occupies a significant part of the oral cavity and its volume in relation to the volume of the oral cavity of a newborn is much larger than that of older children. Since the vacuum in the oral cavity is formed mainly by reducing the volume of the tongue when its muscles contract, to create, for example, the same vacuum as an adult, a newborn needs to reduce the volume of the tongue to a lesser extent than an adult. It should be noted that when creating a vacuum in the oral cavity due to the pressure difference, the skin of the cheeks should be drawn inside the oral cavity and make it difficult to remove milk. But this does not happen in a newborn, because there is a multilobed fat layer in the thickness of the cheeks, enclosed in a connective tissue capsule and preventing the cheeks from being drawn into the oral cavity.

We will briefly note another very important structural and functional feature of the oral cavity of newborns. In an infant, the topography of the larynx is different from that of an adult. The adult is unable to breathe while swallowing. An infant at the same time swallows and breathes. It is believed that this is due to the fact that the epiglottis is located in an adult deeper than in a child. In an infant, the entrance to the larynx lies high above the lower-posterior edge of the palatine curtain and is connected only to the nasal cavity. The path for food is located in the child on the sides of the protruding larynx, where there is a communication between the mouth and pharynx. These anatomical relationships in the area of the root of the tongue and larynx in the child allow him to simultaneously breathe and swallow food. This feature is extremely important for the child because he does not have to interrupt the excretion of milk during respiratory movements.

Patterns of efferent impulse activity generated by the child's CGSS in response to stimulation of the orofacial region receptors arrive in a certain sequence at the motor nuclei (Fig. 4.6), from whose neurons the impulse activity directly comes to the synaptic endings of various muscles in the orofacial region, causing them to contract and relax in a certain sequence.

Data on the dynamics of movements of the structures of the child's oral cavity during the removal of milk from the mother's gland is of great interest. However, for ethical reasons, this can only be done by noninvasive methods. A series of studies on the movement of structures of the child's oral cavity: the tongue, jaws, gums, palate, as well as the nipple and areola of the mother's breast during the removal of milk using a noninvasive method, cineradiographic shooting, was first conducted by G. M. Ardran with colleagues in the 1950s (Ardran et al. 1958a, b; Ardran and Kemp 1959). These works can be considered classic and they are among the most cited in articles, textbooks, and manuals on breastfeeding and the physiology of lactation of women. In the first work, the removal of milk from a bottle with a teat was investigated (Ardran et al. 1958a). To get a clearer image of milk excretion, barium sulfate was used, a suspension of which was added to the donor milk in a bottle. Examinations showed that when the milk was removed from the teat, the lower jaw of the child was alternately lowered and raised. When the mandible was moved down, the teat filled with milk was located between the hard palate and the tongue. The end of the teat was near the transition of the hard palate to the soft, and the base was in close contact with the baby's lips. When the lower jaw moved up, the base of the teat was compressed between the upper gum and the tip of the tongue, located on the lower gum. The movement of the milk contrasted with barium sulfate was quite noticeable when the teat was compressed. Part of the milk from the teat was squeezed into the baby's mouth, and a certain amount passed back into the bottle. Milk from the oral cavity then passed through the gap between the tongue and the soft palate into the throat. Then the tongue and lower jaw moved down from the upper palate, and the volume of the nipple ducts was filled with milk again. It is interesting to note that similar results were obtained later using this technique when studying the movement of orofacial structures in premature infants when they removed milk from the teat (Goldfield et al. 2010, 2013). It was already noted above that spontaneous rhythmic movements of the fetal lips and jaws in the womb were observed by 34 weeks of pregnancy (Miller et al. 2003), indicating the beginning of functioning in the child's central nervous system of the CGSS. At the same time, the data obtained on prematurely born children of 26–32 weeks

(Goldfield et al. 2010, 2013) are experimental confirmation that the formation of the structure and function of the CGSS in fetuses occurs much earlier.

When registering movements of the child's orofacial structures and the areolar–nipple area of the mother's breast, difficulties arose (Ardran et al. 1958b). In particular, to get images, the child had to be in an unnatural position during feeding: he lay on his back, and the mother had to bend over him to give the breast. For a clearer image, as in the case of teat feeding, the nipple and areola of the nursing mother were smeared with a paste consisting of barium sulfate and lanolin powder. The quality of images obtained during breastfeeding was lower than when feeding from a bottle with a pacifier. The structure of the nipple and the areolar–nipple area of the breast has significant differences. However, the analysis of the images revealed similarities in the dynamics of movements of the child's orofacial structures when removing milk from the breast and nipple. After the baby grasped the nipple of the breast in the mouth, it pressed the lips tightly against the skin of the areolar area of the mother's gland, thereby isolating the volume inside the oral cavity from the surrounding atmosphere. Then the lower jaw and the back of the tongue moved down. As a result of increasing the volume of the oral cavity, a vacuum was created in it. Under the influence of vacuum, the nipple with the areola was drawn inside the oral cavity. The tip of the nipple was close to the point where the hard palate met the soft part. The child formed a structure in the mouth resembling a teat, consisting of an elastic part of the areola and a dense, low-tension nipple. The length of the formed " teat," depending on the elasticity of the areola after a certain time of milk excretion, could be three times longer than the original length of the part of the areola and the nipple that the child grasped before milk excretion. Then the lower jaw, as well as the front of the tongue, rose toward the hard palate while flattening and narrowing the stretched part of the areola. Since the mother's milk was not contrasted in this case, its movement was not noticeable in the images. After the rise of the lower jaw and tongue occurred, the volume of the oral cavity decreased, the length of the "teat" formed from the areola and the nipple decreased. Then the mandible with the nipple began to descend again and the cycle was repeated. For comparison, it should be noted that in the case of removing milk from the rubber teat due to the lower extensibility of its walls than the areola section, the length of the teat changed little under the influence of vacuum, which cyclically varied in the child's oral cavity (Ardran et al. 1958a). Unfortunately, the technique did not allow us to identify exact time relationships between the sequence of movements of the jaws, tongue, and areolar–nipple area of the breast. Based on the data obtained in surveys when removing milk from the baby's nipple and breast, G. M. Ardran concluded that the baby's milk during breastfeeding is removed by squeezing it from the expanded ducts located under the areolar area in front of the nipple. The vacuum created in the oral cavity serves to hold the "teat" and transport the milk to the stomach (Ardran and Kemp 1959).

After the work of G. M. Ardran and co., there was a 30-year break in the study of the dynamics of movements of the orofacial structures of the child and the areolar-nipple area of the mother's breast during breastfeeding. Possible reasons could be the inconvenience of the method, as well as the use of X-rays, which could be dangerous

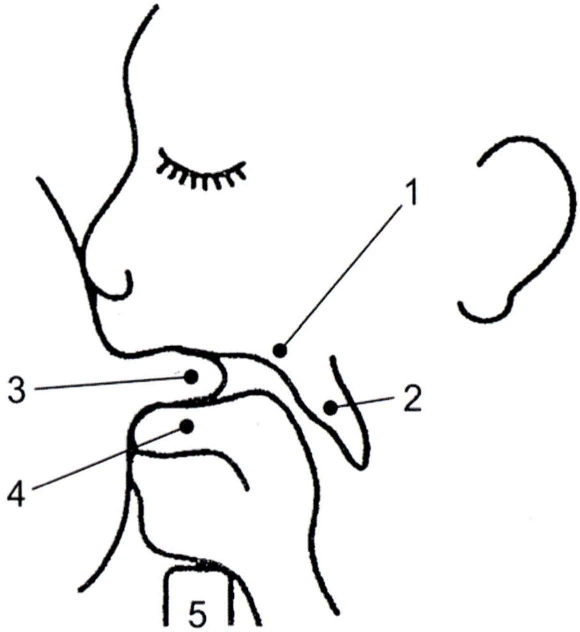

Fig. 4.7 Scheme for recording movements of orofacial structures of the child and mother's nipple using an ultrasound apparatus (from Jacobs et al. 2007)

for the mother and child when studying the excretion of milk during breastfeeding. At the same time, as already mentioned, recently there were again works using the X-ray method to study the process of removing milk from a bottle with a teat in premature babies (Goldfield et al. 2010, 2013).

Research resumed using a safe noninvasive ultrasound method (Smith et al. 1985). The scheme of this technique is shown in Fig. 4.7. When removing milk from the breast, the child was in a natural position, and the remote head of the ultrasound device was placed under his chin. However, records of the movement of the orofacial structures of the child and the nipple–areolar complex of the mother's breast, obtained using ultrasound in the first works (Weber et al. 1986; Bu'Lock et al. 1990; Nowak et al. 1995) as noted by the authors themselves, were less clear than when using the cinoradiographic method. Based on the data obtained using cinoradiographic and ultrasound methods by M. V. Woolridge (1986), a scheme of movement of the structures of the child's oral apparatus, as well as changes in the shape and size of the areola and nipple of the mother's breast during feeding was proposed. The scheme basically corresponded to the experimental data, but there were additions made to it that were not registered using the methods mentioned above in the process of milk excretion by the child. According to the author's scheme, during the excretion of milk, a peristaltic movement occurs on the surface of the child's tongue, which is in contact with the " teat" breast formed from a part of the areola and nipple. The wave starts at the tip of the tongue and ends at the back of the tongue. In this case, the "teat" is subjected to peristaltic compression from the base to the tip of the teat. Such movements on the images were not presented in any of the experimental works available at that time. In comparison with the scheme of G. M. Ardran (Ardran et al. 1958b), the nipple disappeared from the tissue "teat," which in a woman is a fairly dense, low-tension skin and muscle formation. Inside it are the milk ducts, narrowed in comparison with the milk ducts of the under areolar region by 3–7 times (see Chap. 3). As the clinic shows, squeezing the nipples with the fingers, gums, or tongue of a child is accompanied by pain in a woman. In addition, the volume of milk that can be found in the milk ducts of the nipple is hundredths of a milliliter, which is about two orders of magnitude less than the volume of the milk ducts of the stretched area of the areola of the teat. Thus, the compression of the nipple is painful and practically no amount of milk can be squeezed out of its milk ducts. However, the Woolridge scheme has become widespread and has been used in popular articles and scientific reviews, as well as breastfeeding manuals.

As a result of the improvement of ultrasonic devices, it was possible to increase their resolution. Improving the accuracy of recording the movement of the orophacial structures of the child and the areolar-nipple region of the breast made it possible to observe the exit of milk from the nipple, as well as to investigate the change in the diameter and length of the milk ducts supplying the nipple (Ramsay et al. 2004; Jacobs et al. 2007; Geddes et al. 2008; Geddes 2009a, b; McClellan et al. 2010; Sakalidis et al. 2013; Elad et al. 2014). The analysis of images obtained with the help of ultrasound devices with higher resolution supplemented and clarified the data of the X-ray survey of G. M. Ardran with co. (1958b), as well as the results of early ultrasound studies. However, it should be noted that all ultrasound examinations were performed when placing the remote ultrasound head under the child's chin (Fig. 4.7). In this position, the tip of the tongue and the base of the nipple were not clearly visible in the images. While using X-ray photography (Ardran et al. 1958b), this area was represented on radiographs.

Experimental data from X-ray and ultrasound studies allowed us to present an algorithm for the movement of the orofacial structures of the child and the areolar–nipple complex of the mother's breast when removing milk. After the nipple with part of the areola was in the mouth of the child, the cycle of milk excretion began with the lowering of the back of the tongue and lower jaw of the child. At the same time, there was an increase in the volume of the oral cavity, and the atmospheric pressure in it began to decrease. The nipple and areola were stretched, and a "teat" began to form in the child's mouth. The tip of the " teat" moved in the direction of the transition of the hard palate to the soft one. This was clearly visible on both X-ray and ultrasound images. Due to the fact that the area of the transition of the nipple to the areola and the anterior part of the nipple are not clearly visible on ultrasound images, D. Elad et al. (2014) reanalyzed the X-ray images previously recorded by G. M. Ardran et al. (1958b). As a result of processing radiographs and ultrasonograms, it turned out that the front and back of the child's tongue make periodic oscillations in a perpendicular direction relative to the hard palate and is guided by movements of the lower jaw (Ardran et al. 1958b; Elad et al. 2014). Figure 4.8b shows an image of the contours of the tongue, taken from 150 consecutive images taken during the removal of milk by an ultrasound device. Since the

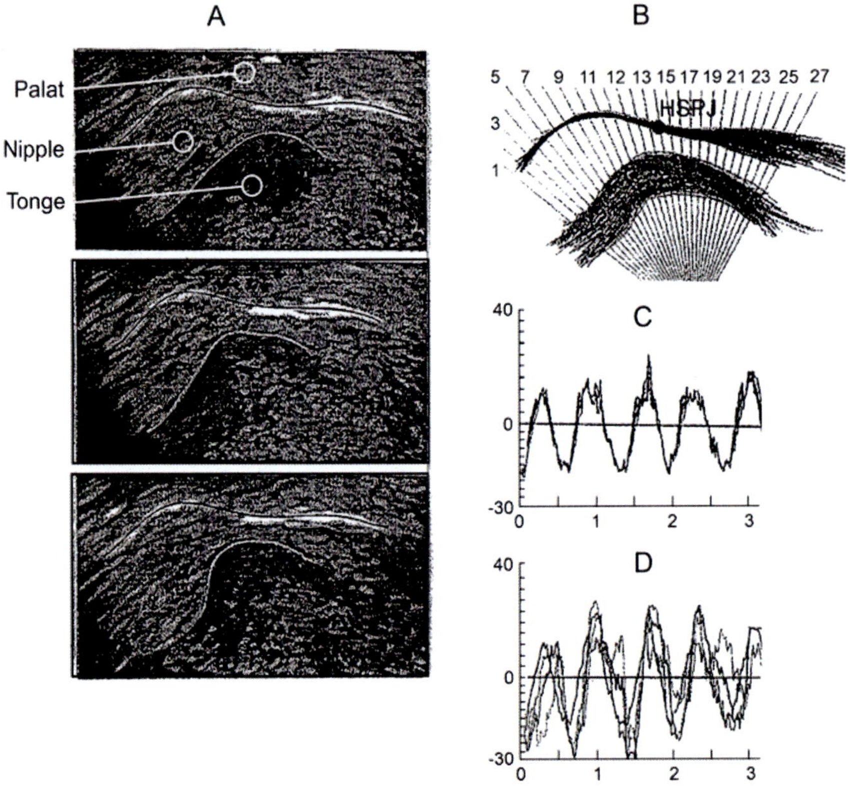

Fig. 4.8 Child tongue movements recorded by ultrasound during breastfeeding (from Elad et al. 2014). (**a**) Image of the mother's breast nipple and the upper part of the child's tongue when placing an ultrasound head under the child's chin. The contours of the tongue and palate on the frames are highlighted by special lines. On the right, the images show the frame numbers of the ultrasonic survey. (**b**) Changing the position of the contours of the palate and tongue, taken from 150 consecutive ultrasonic frames in polar coordinates, when milk is taken out by the same child. (**c**) Graphic representation of the movement of the front of the tongue in polar coordinates 1–8. (**d**) Graphic image of the motion of the rear part of the tongue in polar coordinates 8, 13, 17, 22. For **c**, **d** on the abscissa axis—time in s, on the ordinate axis—movement in pixels

amplitude of the tongue oscillation varies for each cycle, the sequence of tongue contours represents a series of lines offset relative to each other and because of the selected scale looks like a ribbon (Fig. 4.8b). The displacement of the hard palate during milk withdrawal is minimal, so its image is represented as a "thick" line in the drawing. At the same time, the position of the soft palate in the process of milk excretion varies. For Fig. 4.8b the transition point of the hard palate to the soft sky is marked with a dot and the letters HSPJ. After the transition, the thick line expands and takes on the appearance of a ribbon.

Data processing found that the next fluctuations in the back of the tongue occur with a time shift in the direction of movement to the esophagus. This is clearly seen in the graphs of Fig. 4.8d. The peaks of the contours of successive movements of the back of the tongue are shifted along the lines of 8, 13, 17, 22 polar coordinates with a shift of about 0.14 s, resembling peristalsis. The front part of the tongue, moving to the hard palate, "wedges" between the areolar–nipple complex and the lower lip and, as the authors note, presses the "nipple" to the hard palate (Elad et al. 2014). In contrast to the back of the tongue, there was no peristaltic displacement of the front part of the tongue along the "teat" (Fig. 4.8c). A phase shift can be observed between the movement of the back and front of the tongue. In particular, the front part of the tongue begins to move toward the hard palate, squeezing the areolar part at the base of the nipple at the moment when the displacement of the back part of the tongue downward is greatest (Fig. 4.8c, d).

The compression of the teat "can be quite strong." This is evidenced by the data of milk excretion by a child with ankyloglossia. Ultrasonograms show that due to the short frenulum, the front part of the baby's tongue acts directly on the nipple (Geddes et al. 2008; Garbin et al. 2013), causing pain in the mother. As already indicated the nipple has a higher pain sensitivity than the area of the areola adjacent to the nipple. After frenotomy during feeding, the tip of the child's tongue already affected the areolar part bordering the base of the nipple (Garbin et al. 2013). Pain after phrenotomy did not occur in a woman.

As the lower part of the tongue is lowered and the vacuum in the baby's mouth increases, milk begins to be released from the nipple. In addition, at this time, the milk ducts become noticeable when they enter the nipple (Geddes et al. 2008). When the back of the tongue was lowered, along with the movement of the nipple to the transition point of the hard palate to the soft palate, an increase in the diameter of the " teat" was observed (McClellan et al. 2010; Sakalidis et al. 2013). The increase in diameter was not the same in the length of the "teat" and varied significantly in different women. In some women, the diameter of the "teat" could increase twice, in others, the diameter practically did not change. Here it is important to note that in the mentioned works there was no peristaltic, i.e., undulating influence of the front part of the tongue, extending along the length of "teat."

In the second part of the milk withdrawal cycle, the back of the tongue began to rise, and the front part moved away from the " teat." At the same time, the tip of the nipple moved away from the transition point of the hard palate to the soft one, and the diameter of the "teat" decreased along its entire length. As a result of the lift, the back of the tongue occupied a position in the child's oral cavity higher than the original one (Fig. 4.8). Then the cycle was repeated. Fluctuations of the tip of the nipple during the cycle relative to the transition point of the hard palate to the soft palate were within 3.7 mm, and the front of the tongue was within 4.4 mm. The duration of the oscillation cycle of various parts of the child's tongue was about 0.64 s (Elad et al. 2014).

It should be noted that when milk was withdrawn, changes in the length and diameter of the "nipple" were observed not only during the cycle but over time there was a gradual increase in the length of the " teat," as well as its diameter. This was clearly visible immediately after the end of feeding, when the nipple with part of the areola was removed from the baby's mouth. The reason is the special viscoelastic

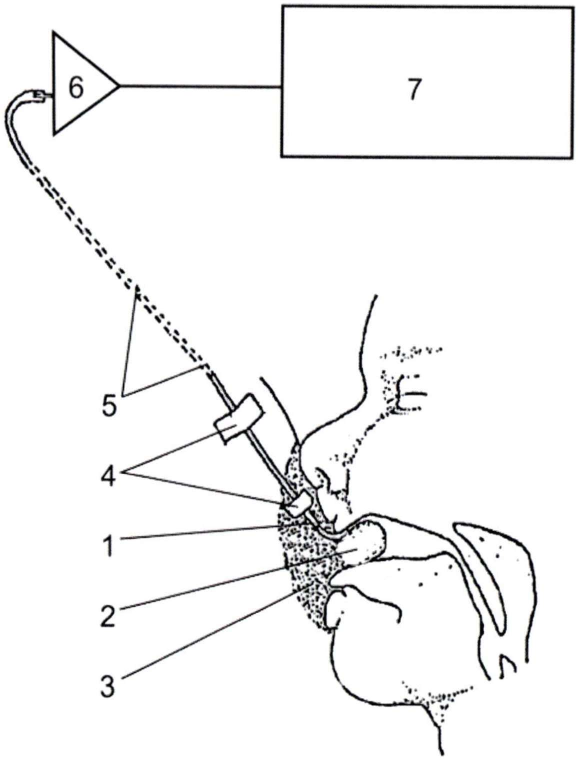

Fig. 4.9 A scheme for recording vacuum or compression affecting the areolar–nipple complex of the breast when milk is excreted by a child (from Alekseev et al. 2003). 1-plastic tube with rigid walls and an open end filled with distilled water for detecting vacuum stimuli or a plastic tube with elastic walls with a closed end for detecting compression stimuli; 2-nipple; 3-areola; 4-strips of leukoplast; 5-tube supplying pressure change to the sensor

properties of the nipple and areola, from which the child forms a "teat" in the process of milk excretion.

Thus, as a result of the movement of the child's orofacial structures, the areola and the mother's nipple are stimulated by vacuum and compression. With their help, a pressure difference is created between the milk located in the ductal system of the breast and the environment surrounding the nipple and areola, and due to this, milk is removed from the gland. Information about the dynamics of vacuum and compression stimuli and their amplitude–frequency characteristics during milk excretion is of great interest, since they are a significant addition to the cineradiographic and ultrasound methods for elucidating the process of milk excretion.

It should be noted that the study of the dynamics of pressures affecting the nipple and areola during milk excretion was started before the cineradiographic and ultrasound examinations of the movement of orofacial structures of the child (Gunter 1945). These examinations are performed using sensors that are placed on the areolar-nipple area of the breast. Therefore, these examinations can be called invasive. In particular, registration of the vacuum in the child's oral cavity was performed using a plastic tube with rigid walls filled with sterile water, which was placed on the nipple-areolar part of the gland (Fig. 4.9). One end of the tube with an internal diameter of 1–1.5 mm, located on the areolar-nipple part was open, and the other was connected to vacuum sensors. To register compression (positive pressure), a tube with elastic walls was used, also filled with sterile water. The end of the tube, located on the areolar-nipple part was closed, the other was connected to the pressure sensor. Signals from the sensors were fed to the recording device via an amplifier (Fig. 4.9). Despite the simplicity of the method for registering pressures, and the fact that surveys using this method began to appear since the 1940s (Gunter 1945), experimental work conducted on mother–child dyads is relatively small to date.

The most interesting are examinations in which it was possible to register a sequence of vacuum stimuli or compression stimuli that affect the nipple-areolar part of the gland during the entire period of feeding the child. For Fig. 4.10A as an example, records of patterns of vacuum stimuli during milk withdrawal by four 6-month-old children are presented (Prieto et al. 1996). In order for the entire record to be present in the drawing, registration was performed when the recorder tape was moving slowly. The records clearly show that the vacuum stimuli act periodically. At the same time, the periods have different durations and the frequency of vacuum stimuli in the periods is also different. Periods with a high frequency of vacuum pulses were accompanied by a decrease in their frequency down to zero. It can be assumed that the time of intensive milk excretion was replaced by rest. However, all records show a decrease in the frequency of vacuum stimuli toward the end of feeding.

The vacuum amplitude of the stimulus is changed in periods. This is consistent with data on ultrasonic registration of movement of the back of the tongue. Displacement of the back of the tongue down, which creates a vacuum in the child's mouth, as seen in Fig. 4.5b varies markedly. Accordingly, a small downward movement will reduce the pressure in the child's mouth to a lesser extent than a deeper lowering of the back of the tongue. The minimum value of the amplitude of the vacuum stimuli was -50 ± 5.7 mm Hg, and the maximum value was -197 ± 10 mm Hg. It should be noted that the shape and duration of the vacuum stimuli also coincide with the shape and duration of the movement of the back of the child's tongue when removing milk. In Fig. 4.10B, vacuum stimuli were recorded at a higher speed of the recorder tape (Segami et al. 2013). Vacuum stimuli had the form of triangles or trapezoids. Lowering and lifting of the tongue also occurred along a trajectory resembling a triangle (Fig. 4.8c) (Elad et al. 2014). The duration of the vacuum stimuli at the base could vary from 0.5 s to 1.2 s. In trapezoidal stimuli, the duration of the plateau could reach 0.65 s (Segami et al. 2013). The period of lowering and raising the back of the tongue was 0.64 s on average (Elad et al. 2014). When measuring the interval between the peaks of vacuum stimuli, the average interval was 0.7 ± 0.1 s (Prieto et al. 1996) (Fig. 4.12b).

In our work (Alekseev et al. 2003), we were able to register a sequence of separate compression stimuli that affect the areola of the mother's breast during the entire time of feeding—sessions of children 4–6 days of age. Eleven feeding sessions were studied. The entire feeding process in the examined mother–child dyads took 7–12 min. Since vacuum stimuli also act on the nipple and areola during feeding, the elastic tube's stiffness was selected so that the vacuum stimuli did not change its geometric dimensions, and this system would only register compression stimuli. Analysis of recordings on a tape recorder showed that the child acts with compression stimuli on the areola and nipple of the breast, as well as with vacuum stimuli

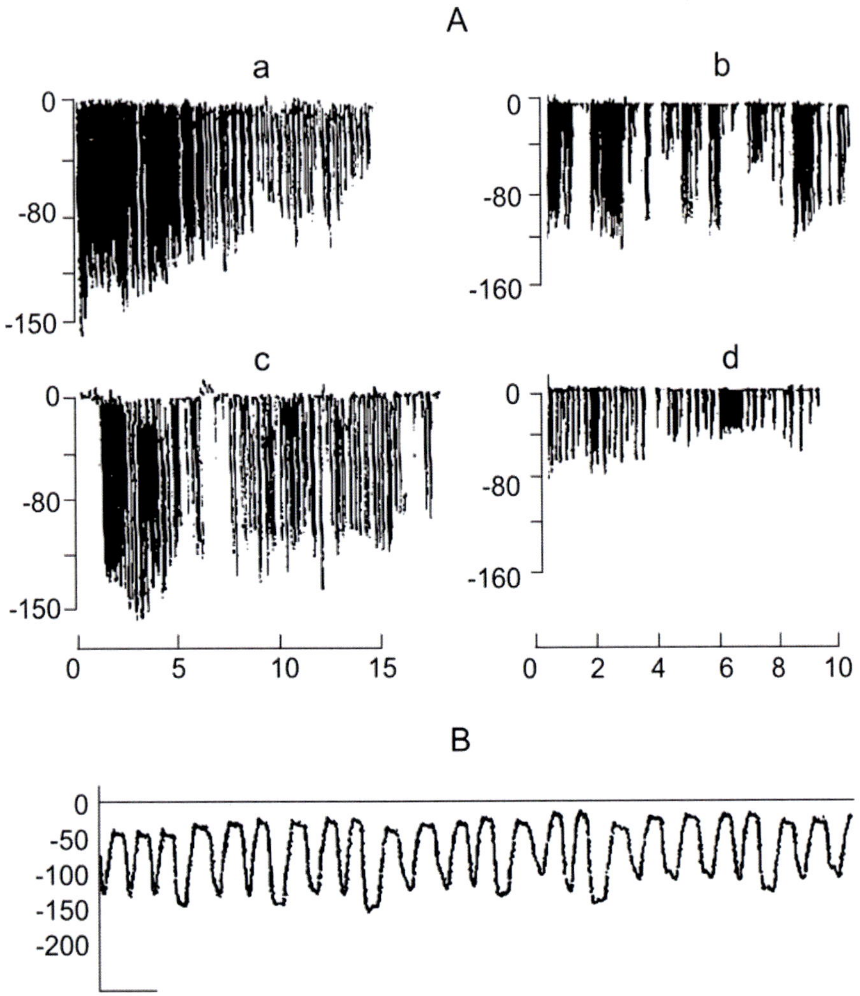

Fig. 4.10 The sequence of vacuum stimuli affecting the nipple-areolar part of the mammary gland during infant feeding. (**A**) is a sequence of vacuum stimuli affecting the nipple-areolar part of the mammary gland throughout the feeding period in four children (from Prieto et al. 1996). On all records (a, b, c, d) ordinate axis: vacuum in mm Hg, abscissa axis: time in min. (**B**) is a fragment of the recording of vacuum stimuli at a higher speed of the tape of the recorder during the withdrawal of milk by the child from the mother's breast (from Segami et al. 2013). Ordinate axis: vacuum in mm Hg, abscissa axis: horizontal line time—1 s

periodically. The number of periods per session varied in different children within the range of 17–24, and there was no direct relationship between the number of periods and the time of the feeding session. For example, for 7 min of feeding, the number of periods could be 24, and 17 for 12 min. The duration of periods per session also varied significantly and ranged from 4 s to 1 min. Compressive forces in the period are represented by triangles that vary in amplitude and duration (Fig. 4.11). In contrast to vacuum stimuli, trapezoidal compression stimuli were

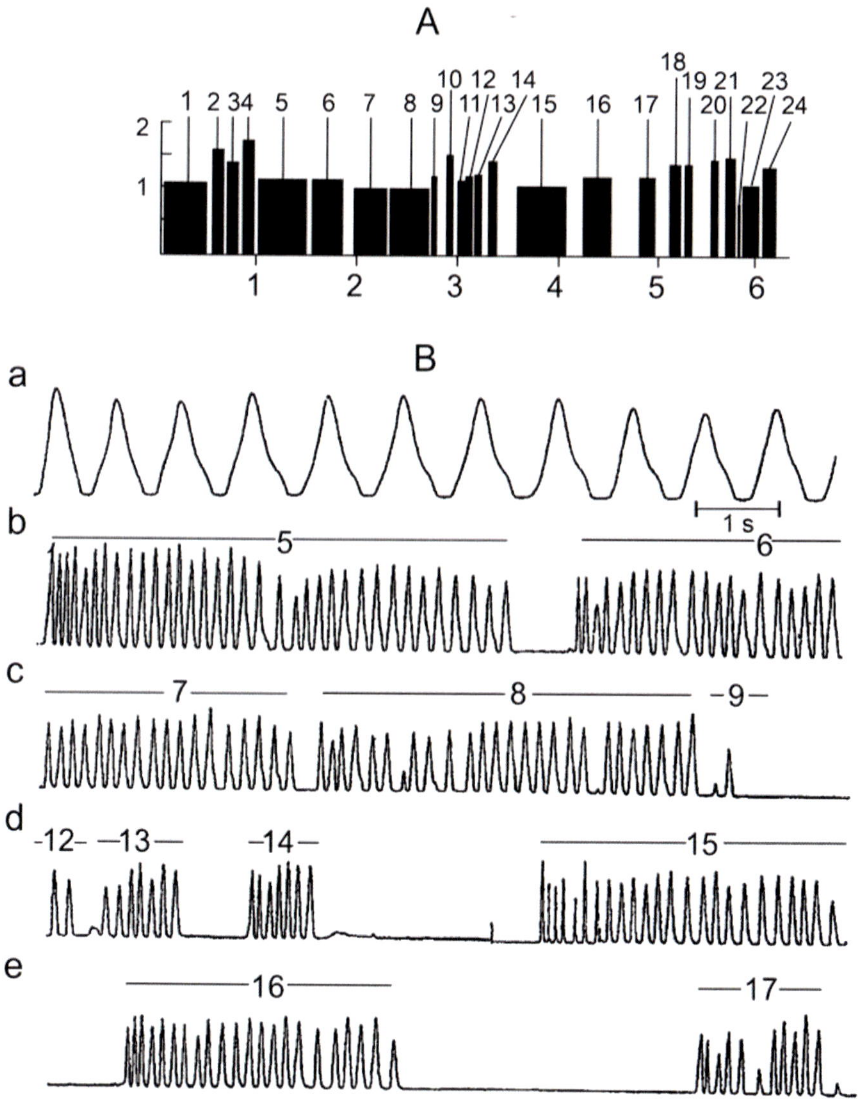

Fig. 4.11 Sequence of compression stimuli affecting the nipple-areolar part of the mammary gland when feeding the child (from Alekseev et al. 2003). **A**—is a graphical image of the dynamics of the duration of periods and the average frequency in the period of compression stimuli affecting the areola of the mother for the entire feeding time of the child. Lines indicate periods of exposure to compression stimuli on the mammary areola. On the abscissa axis: period duration, min; ordinate axis: average frequency in the period, pulse/s. The numbers on top denote period numbers. **B**—records of compression stimuli at different speed of tape movement of recorder. a—recording speed 25 mm/s, b–e—recording speed 5 mm/s. Figures with horizontal lines denote the duration and period number corresponding to the period on **A**. Time calibration: 1 s for a, 10 s to d for b–e

not observed. Figure 4.11A shows a graphical representation of a sequence of 24 periods of compression stimuli during the entire feeding session of one of the children on a reduced time scale. The length of the shaded rectangles horizontally corresponds to the time of action of compression stimuli (period) on the areola of the breast, and vertically—to the average frequency. Note that in all cases, the duration of the compression stimulus periods was 3–6 times longer than the pause time in the first 4–5 min of feeding. In the future, the duration of the compression stimulus periods became shorter and the pauses longer. The pause was taken into account if it exceeded the time between two tactile actions equal to 1.3 s (Drewett and Woolridge 1979).

When calculating the frequency of peaks in periods, it was found that for this "mother–child" dyad, it is unstable and varies from 0.9 imp/s to 1.7 imp/s. As a rule, the average frequency of compression peaks in longer periods was less than in short periods (Fig. 4.11A). Figure 4.11B shows a partial or complete sequence of compression stimuli at a fast speed of the recording device tape of various periods of milk excretion, represented graphically in Fig. 4.11A. Compression effects in the period are also like vacuum stimuli triangles, varying in amplitude and duration. You can see that there is a tendency to decrease the amplitude of compression stimuli in later periods. In addition, compression stimuli as well as vacuum pulses vary in amplitude. This is consistent with the data on ultrasonic registration of the movement of the front part of the tongue (Fig. 4.8b). With a slight movement of the front part of the tongue, there is a slight compression of the "teat." Increasing the offset presses the "teat" with greater force against the hard palate.

The maximum amplitude of compression peaks could reach 70 mmHg. The duration of compression stimuli at the base varied from 0.3 to 0.8 s (Fig. 4.12a). The most common stimuli are 0.45–0.75 s. The duration of stimuli for moving the front part of the tongue at the base was also within these limits (Fig. 4.8c). At the same time, the histogram of the duration distribution of intervals between compression stimuli (Fig. 4.12b) is shifted to the left relative to the histogram of stimulus durations. The area of the most common values is in the range of 0.1–0.25 s.

It should be noted that variations in the amplitude, duration, and frequency of vacuum and compressive effects exerted by the child on the nipple and areola of the breast maybe for different reasons. Thus, single fluctuations in the amplitude and duration of vacuum and compression can be the result of natural fluctuations in the movement of the tongue. However, the cause of longer oscillation parameters orofacial structures of the child, apparently, mainly is the modulation operation CGSS flow of milk, which is excreted by the baby during feeding.

Of particular interest is the information about the time relations between the vacuum and compression stimuli that affect the areolar-teat area when the child is excreting milk. These data can be obtained by simultaneously registering the vacuum and compression when feeding a child, i.e., by placing two sensors on the areolar-nipple area: pressure and vacuum, according to the method presented in

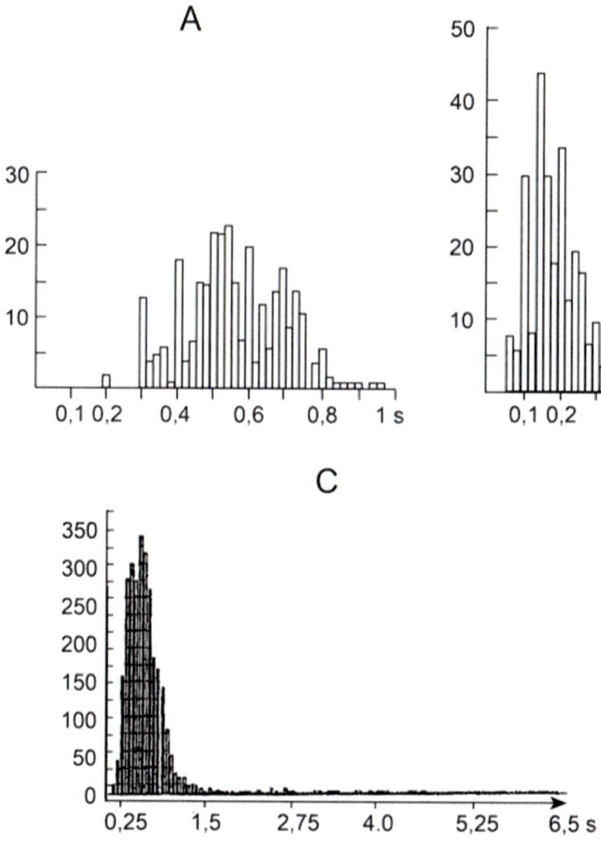

Fig. 4.12 Frequency of distribution of duration (**a**), inter-stimulus intervals (**b**) of compression stimuli for the whole feeding period (from Alekseev et al. 2003) and inter-stimulus intervals of vacuum stimuli (**b**) for the whole feeding period of three children (from Prieto et al. 1999). Abscissa axis—time for all histograms in s, ordinate axis—the number of stimuli (**a**) and inter-stimulus intervals (**b**, **c**)

Fig. 4.7. In the literature, one survey was found (Luther et al. 1974), in which it was possible to simultaneously record the vacuum and compression stimuli acting on the areolar-nipple area of the breast when the child ejects milk. Unfortunately, details of this method are not provided in the article. Despite the low time resolution of the records in the figure (Fig. 4.13a), it can be seen that the vacuum and compression pulses with a shift relative to each other act on the areolar-nipple area of the gland for 1–1.5 s. A vacuum is initially created in the oral cavity. When approximately the maximum vacuum value is reached, the compression action begins. The amplitude of the compression pulse reaches a maximum and then decreases to zero along with the vacuum pulse. Here it should be noted that in parallel with the study of the removal of milk from the breast by a child using the above-described method, the

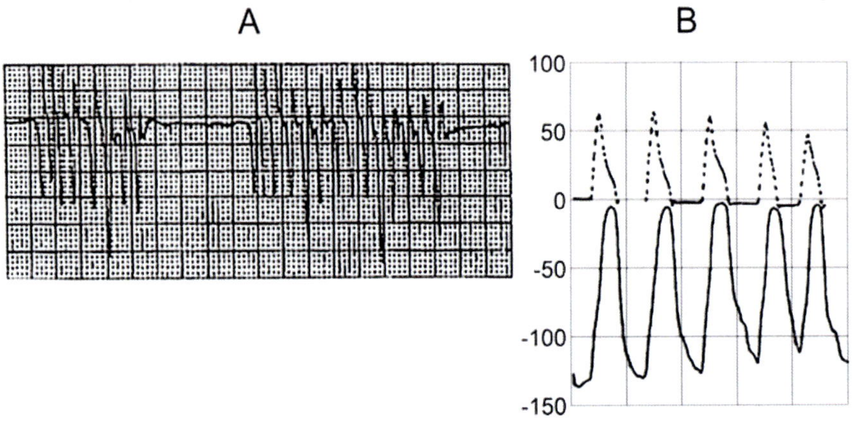

Fig. 4.13 Sequence of vacuum and compression stimuli acting during milk excretion on the areolar-nipple region of the mammary gland (**a**) (from Luther et al. 1974) and nipple (**b**) (from Mizuno and Ueda 2001). (**a**) Upward deviation of the signal from the horizontal (zero) line corresponds to positive pressure (compression), downward—to negative pressure (vacuum). Calibration: vertical one cage 25 mm Hg; horizontally, three cells 5 s. (**b**) Ordinate axis: pressure in mm Hg; abscissa axis: time: one cell—1 s

removal of milk from a bottle with a teat was studied. As already noted, the structure of the artificial teat differs from the "teat," which is formed by the child during the removal of milk from the mother's breast. At the same time, it can be assumed that the patterns of impulse activity that will occur in the nerve fibers that form the mechanoreceptors of the orofacial region when an artificial teat is inserted into the child's mouth will have a similarity to afferent impulses to the "natural teat." Therefore, the CGSS in response to afferent impulse activity will send appropriate efferent signals that will cause contractions or relaxation of the muscles of the orofacial area similar to those during the removal of milk from the breast. Figure 4.13b shows recordings of compression and vacuum stimuli when removing milk from a teat bottle. The figure clearly shows that, just as in the case of removing milk from the breast (Fig. 4.13a), the vacuum stimulus initially begins to act on the nipple. Then, when the vacuum approaches the maximum value, the baby begins to compress the nipple. The compression amplitude reaches its maximum value and then decreases to zero along with the vacuum stimulus. It should be noted here that the phase relations between the action of the vacuum and compression stimuli coincide with those between the movement of the back and front of the child's tongue (Fig. 4.8c, d), as a result of which there are vacuum and compression stimuli. Thus, one cycle of exposure to the areolar-nipple area of the breast during the removal of milk by the child includes both the stimulus of vacuum (sucking milk) and compression (squeezing milks). For brevity, this effect is designated as SSE.

Unique studies on measuring the amount of milk produced by a child during a single SSE cycle were conducted by M. V. Woolridge and colleagues (Woolridge et al. 1982). A thin rubber funnel was placed on the nipple and part of the areola,

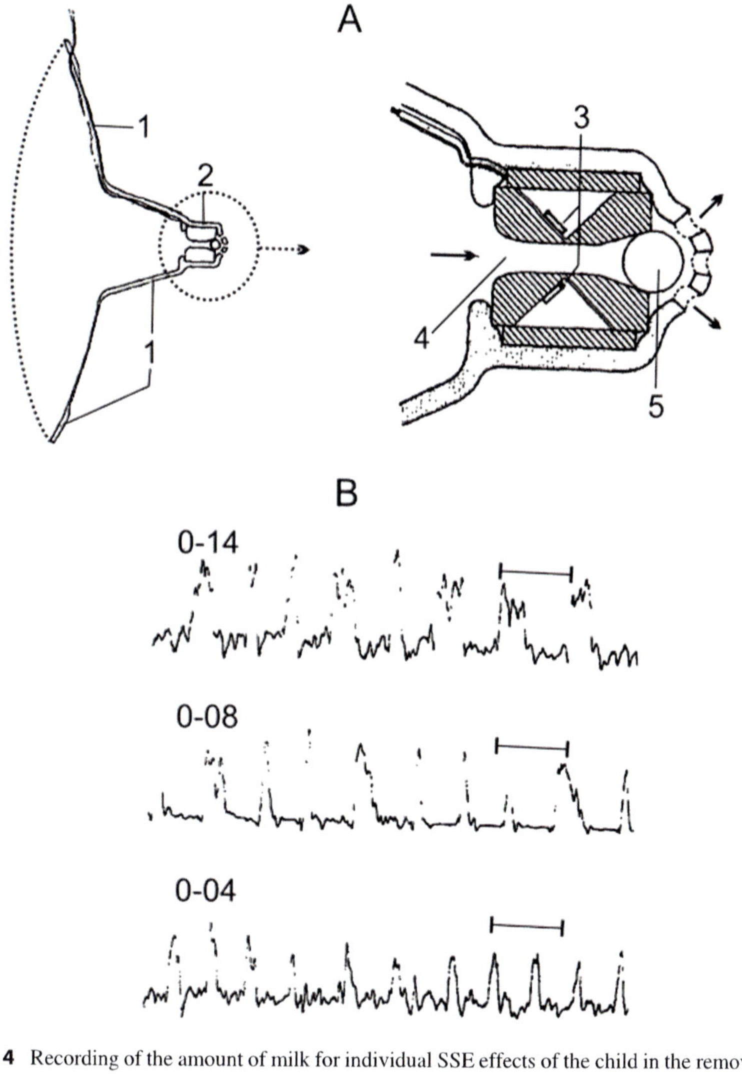

Fig. 4.14 Recording of the amount of milk for individual SSE effects of the child in the removal of milk from the mother's breast (from Woolridge et al. 1982). (**a**) Scheme of the experimental installation. 1—thin rubber sheath covering the areola and the breast nipple; 2—a milk flow sensor; 3—piezo crystals recording milk flow; 4—hole for milk movement; 5—valve. The arrows show the direction of the movement of milk. (**b**) Milk flow for individual SSE at different periods of infant feeding. The numbers denote the flow of milk for individual squeezing effects in ml. Horizontal line—time s

inside which a miniature sensor was placed in the area of the tip of the nipple, which allowed registering the flow of milk during feeding of the child (Fig. 4.14a). As the authors note, the funnel with the sensor did not affect the behavior of the child when feeding, so it was possible to register the amount of milk for each SSE during the entire time the child removed milk from the mother's gland. It turned out that the child excretes milk by periods alternating with rest periods. As an example, Fig. 4.14b shows records of milk excretion for each SSE at different intervals of one of the periods. The change in the speed of the milk flow had the shape of a triangle and its duration at the base was in the range of 0.25–0.7 s on the presented records. The amount of milk withdrawn per SSE at the beginning of the period is the largest and amounted to approximately 0.14 ml. By the end of the period, it decreased to 0.04–0.02 ml.

Figure 4.15 shows the patterns of milk withdrawal pulses during the entire feeding session at a slow speed of the recording, so that all periods of milk withdrawal are represented on the record. The figure clearly shows that the patterns of milk excretion show a great similarity with the patterns of action on the nipple by the stimuli of vacuum and compression (Figs. 4.10A and 4.11A). It is of considerable interest to simultaneously record the stimuli of vacuum, compression, and the rate of milk excretion during the entire feeding session in the same child. However, for methodological reasons, this is quite difficult to do.

Here it should be noted that when feeding a baby from a bottle with a pacifier, two types of effects were found that the child had on the pacifier. The first type was registered when removing donor milk or formula and had a similarity to the patterns of effects of vacuum stimuli (Fig. 4.10) and compression (Fig. 4.11) on the areolar-nipple area of the breast when removing milk from it. This type of exposure is called "nutritive sucking." However, given that the child during feeding uses compression stimuli along with vacuum, it is more correct to call this effect as " nutritive SSE." The second type was first described initially by K. V. Shuleikina (1967), and then in more detail by P. Wolff (1968). This type of exposure occurs when the donor milk or formula is blocked, or a baby is placed in the mouth with a pacifier. Stopping the flow of milk to the teat causes a decrease in the frequency of SSE, and then there are characteristic periodic reactions consisting of short series of SSE, interrupted by pauses (Shuleikina 1967) (Fig. 4.16a). These periodic reactions are called "nonnutritive sucking" or more correctly, nonnutritive SSE. By its structure, each SSE for nonnutritive sucking does not differ from SSE for nutritive sucking. The series consists of an average of 7 ± 1 individual SSE with a frequency in the series of 2.1 ± 0.2 imp/s and with intervals between series of 7 ± 2 s (Wolff 1968; Hafström et al. 1997) (Fig. 4.16b). Here it is interesting to note that the ultrasonic registration of the movement of the orofacial structures of the child and the areolar–nipple complex of the mother in nonnutritive SSE also did not show any fundamental differences from those nutritive SSE (Sakalidis et al. 2013). The presented data indicate that the functioning of the CGSS is modulated by afferent impulses from the child's taste receptors. In particular, in the case of milk entering the oral cavity of a child, along with stimulation of mechanoreceptors, its taste receptors are stimulated. The child begins to affect the areolar–nipple complex for longer periods of time, especially at the beginning of feeding (Figs. 4.10 and 4.11). At the same time, the frequency of SSE in periods decreases and becomes less than 1 imp/s, and the amplitude of SSE increases. Probably, the cessation of milk intake makes unnecessary long-term exposure to SSE. In this regard, the patterns of efferent activity of the CGSS coming to the muscles of the orofacial region become shorter and begin to

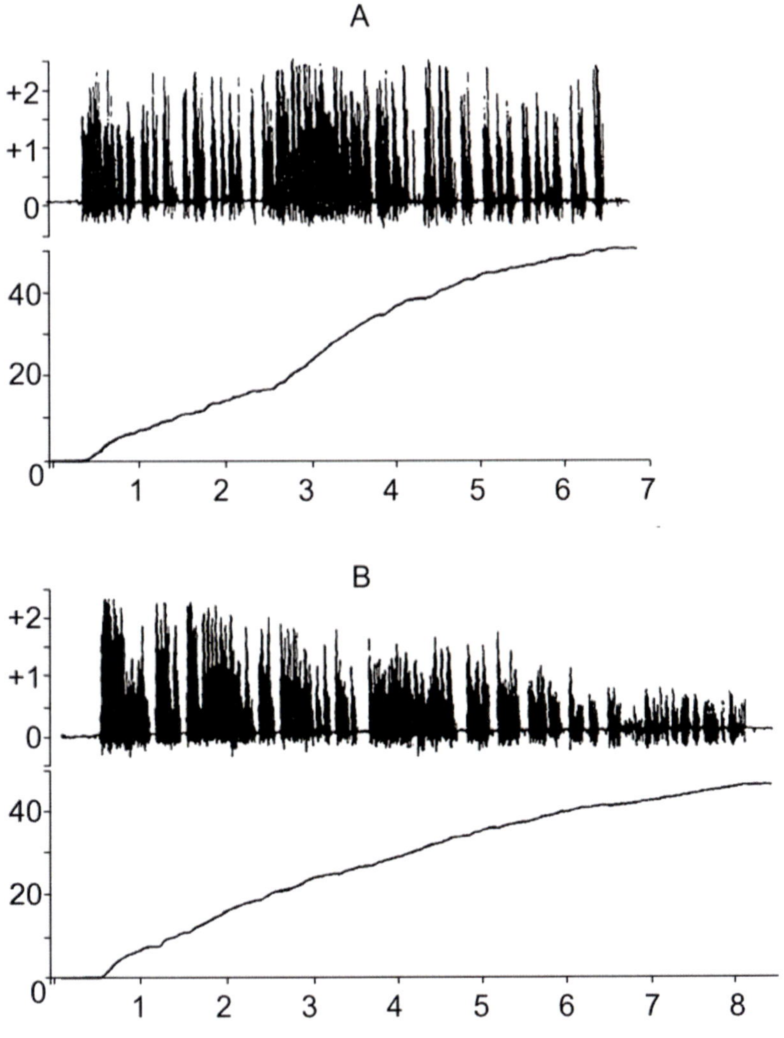

Fig. 4.15 Recording the amount of milk excreted by the child during the whole feeding session from the mother's right (**a**) and left (**b**) breast (from Woolridge et al. 1982). (**a**, **b**) is the upper graph recording the amount of milk for individual SSE of the child, the lower graph is the general integral curve of the dependence on the amount of milk withdrawn during the entire session. Abscissa axis—time in min, ordinate axis for the upper plots, the flow of milk in ml for SSE, for lower plots—the amount of milk in ml

perform mainly a stimulating function (we will focus on this in more detail when describing the maternal reflex of milk excretion). The first studies of parameters of nonnutritive sucking SSE in children found their change depending on various abnormalities in the child's development (Wolff 1968). For example, in children with Down syndrome, the frequency of SSE in the series (1.6/s) is significantly lower than in normal children (Wolff 1968). Registration of nonnutritive SSE is

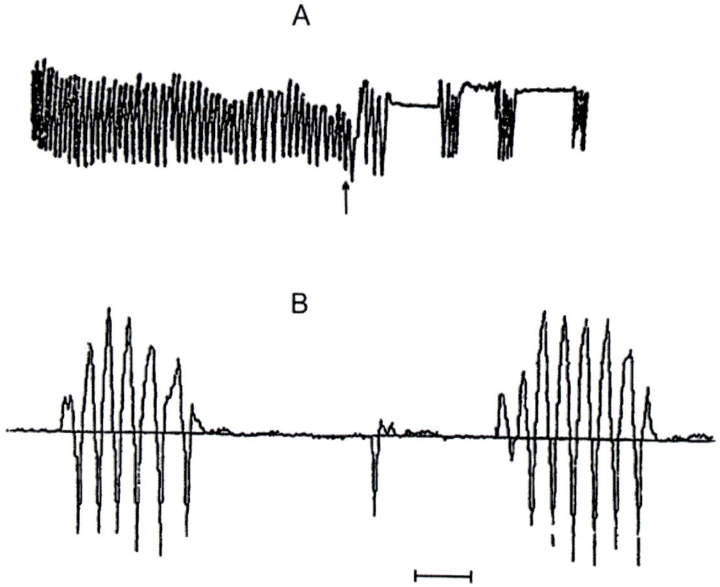

Fig. 4.16 Change in pressure exerted by the child on the teat during nonnutritive SSE. (**a**) Stimuli acting on the teat when feeding a baby from a bottle with maternal milk and a pacifier (the time of switching to a pacifier is indicated by the arrow). Only sucking (vacuum) stimuli are registered (by: Shuleikina 1967). (**b**) Patterns of squeezing influences exerted by a child on a pacifier. Vacuum (deflection of the line down) and positive pressure (deflection of the line up) were recorded by special sensors located in the pacifier (from Hafström et al. 1997). Time calibration for (**b**)—1 s for (**b**) and 4 s for (**a**)

quite easy to implement in a clinic, so this method has been used additionally for the diagnosis of disorders in the development of the nervous system in a child (Pinelli and Symington 2005).

At the end of the section, based on the comparison of data obtained in radiographic and ultrasound examinations of the movement of orofacial structures, as well as the dynamics of pressures affecting the nipple and areola of a woman when feeding a child, an updated and corrected scheme of milk excretion by a child is presented (Fig. 4.17). As a result of the rooting reflex at the beginning of the cycle of milk excretion, the nipple with part of the areola is captured in the oral cavity by the child (Fig. 4.17a). There is a stimulation of the mechanoreceptors of the orofacial region of the child, the afferent impulse from which enters the central nervous system of the child (Fig. 4.6). In turn, efferent impulse activity is transmitted from the CGSS to the muscles of the orofacial region. The child tightly presses the lips to the skin surface of the areola, thus isolating the volume of the oral cavity from the environment. Then there is a contraction of the muscles of the tongue and lowering of its back, as well as a downward shift of the mandible. As a result, a vacuum is created in the child's oral cavity. The nipple and the adjacent part of the areola are drawn into the oral cavity and the tip of the nipple approaches the transition point of

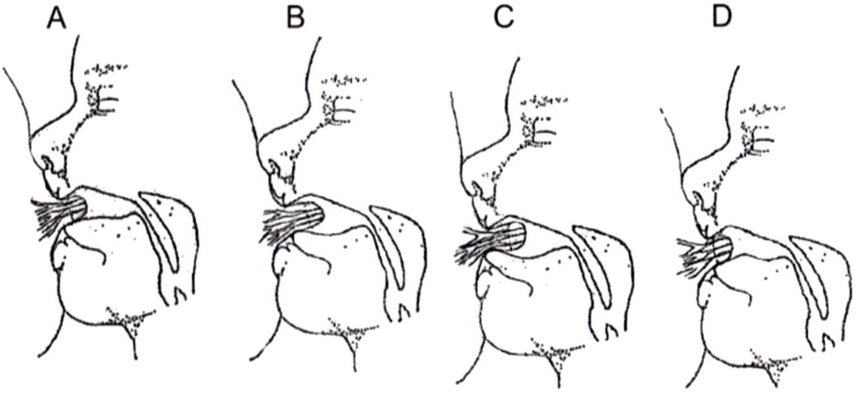

Fig. 4.17 Scheme of the cycle of milk withdrawal by the child from the mother's breast. A dashed line in all figures indicates the border of the nipple

the hard palate to the soft palate (Fig. 4.17b). There is a pressure difference between the pressure of milk in the alveolar-ductal system and the pressure in the oral cavity. Milk begins to flow into the baby's mouth. With a time delay against the background of vacuum, the lower jaw and the front part of the tongue begin to move up, pressing the "nipple" to the hard palate (Fig. 4.17b). Here it is important to note that the front part of the tongue should not affect the entire "nipple" but the area of the areola adjacent to the nipple. In this area of the areola, the milk ducts have the largest diameter (Figs. 3.3b and 3.4a). Compression of this part of the ducts (Fig. 4.17c) will further increase the pressure difference between the intraductal volume and the volume of the oral cavity, which will increase the rate of milk excretion by the child. As already noted, if the front of the tongue squeezes the nipple directly, the woman feels pain. This phenomenon is often observed at the beginning of feeding the child if the areolar part is not sufficiently stretched and during the first cycles of milk excretion, the action of the front part of the tongue is mainly on the nipple. In the future, as a result of residual deformation, the length of the areolar part before the nipple, as well as the diameter of the entire "teat" increases with each cycle. The most painful part of the "teat"—the nipple is in the child's mouth, and the front part of the tongue already affects the areolar area in front of the nipple (Fig. 4.17c), "wedging" between the areolar–nipple complex and the lower lip (Elad et al. 2014). In the second part of the milk withdrawal cycle, the back of the tongue began to rise, and the front part moved away from the "teat." In this case, the tip of the nipple was removed from the transition point of the hard palate to the soft one, and the diameter of the" teat" decreases along the entire length (Fig. 4.17d). Then the cycle repeats. It is interesting to note that the next fluctuations of the back of the tongue occurred with a shift in the direction of movement to the esophagus (Elad et al. 2014) (Fig. 4.8d), resembling peristalsis and probably contributing to the movement of milk from the oral cavity to the esophagus. It should be noted that immediately after the extraction of the nipple and part of the areola—"teat" from the child's mouth at the end of feeding, the " teat" due to residual deformation has an increased size. However, after a few minutes, the size of the nipple and areola decreases to the original level. Thus, as a result of creating a periodic pressure difference between the milk located in the alveolar-ductal system of the breast and the oral cavity, the child excretes milk. However, the child can withdraw relatively easily milk localized in the thick ducts of the breast (Fig. 3.3b). The volume of this milk is several milliliters. The main part of milk in women, as well as in all mammals, is found in the alveoli, thin and medium ducts that have a high resistance to the flow of milk (Mortazavi et al. 2015). Therefore, it is necessary that in the thick ducts there is a" pumping out" of milk from the upper parts of the alveolar-ductal system of the mother's breast. In addition, after the baby's milk is removed from the gland, it is necessary to resume the volume of milk for the next feeding. This is done, as already mentioned, at the expense of two maternal reflexes: the reflex of excretion and the reflex of milk secretion.

4.2.3 Maternal Milk Excretion Reflex

Maternal reflexes of milk excretion and secretion are often called neuroendocrine reflexes. Their reflex arc includes the same links as infant milk ejection reflex. However, the efferent pathway consists of nervous and humoral links. In particular, the final part of the efferent link for the milk excretion reflex is carried out with the help of the hormone oxytocin, which enters the alveolar-ductal section of the breast through the bloodstream of the breast from the endocrine gland-pituitary. Figure 4.18 shows the scheme of the maternal reflex of milk excretion. When the areolar-teat area of the mother's breast is exposed to vacuum and compression stimuli during milk excretion by the child, the mechanoreceptors localized in the skin of this part of the breast are stimulated. The impulse activity that occurs in the afferent nerve fibers that form these receptors enters the somatosensory cortex, where a woman has a sense of mechanical action on the areolar-nipple area of the gland. Experimental data indicate that the cerebral cortex has a significant effect on the formation of the milk withdrawal reflex, but this effect is modulating. The main central link of the maternal milk excretion reflex is the hypothalamus, where oxytocinergic neurosecretory cells are localized, which under the influence of afferent impulse activity that occurs in the child's SSE release oxytocin into the bloodstream. Here it should be noted that in a nursing woman, stimulation of the mechanoreceptors of the areolar–nipple complex can occur in the intervals between feeding the child. For example, when the skin of the nipple and areola rubs against clothing, or when the mammary glands accidentally touch objects surrounding the woman. The impulse activity that occurs in the afferent nerve fibers innervating the mammary gland of a woman should not cause the release of oxytocin and, accordingly, the release of milk. Therefore, it is believed that the impulse activity from the areolar–nipple complex receptors passes not "directly" to the neurosecretory cells, but through a special neuronal network localized in the hypothalamus and surrounding brain structures. It is assumed that this neuronal formation selects afferent impulse activity coming from breast receptors, and forms periodic "bursts" of action potentials that provide, respectively, a periodic release of oxytocin into the bloodstream. In the diagram, this structure is

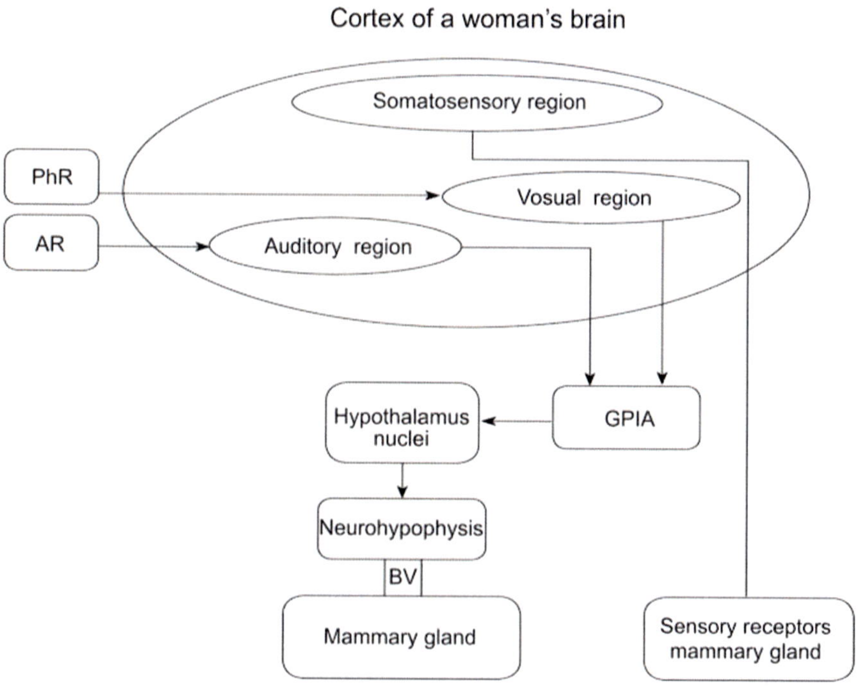

Fig. 4.18 Scheme of maternal reflex of milk excretion. *FR* photoreceptors, *AP* auditory receptors, *GPIA* generator of periodic impulse activity, *bv* blood vessels

designated as a generator periodic impulse activity (GPIA) (Fig. 4.18). Pulse activity from the GPIA is transmitted to the neural secretory cells located mainly in the paraventricular and supraoptic nuclei of the hypothalamus. They are the beginning of the effector link of maternal the milk excretion reflex. Axons of neurosecretory cells pass into the posterior lobe of the pituitary gland (neurohypophysis). The hormone oxytocin is released from the terminals of these fibers into the bloodstream in accordance with the pattern of nerve impulse activity. With the blood flow, oxytocin reaches the alveolar-ductal section of the breast. Effectors of the maternal reflex of milk excretion are myoepithelial cells of the alveoli and ducts of the gland, which directly " squeeze" milk from the alveoli. Here it should be noted that the excretion of milk in the mother is mainly due to an unconditional innate reflex. However, observations in the clinic of breastfeeding women found that milk can be released from the nipple breast at the sight of the child or when it cries (McNeilly et al. 1983). This indicates that the visual and auditory sensory systems are involved in the excretion of milk in a woman through conditionally reflex connections. In addition, we can cite experimental data obtained using the method of nuclear magnetic resonance on rats when feeding their young. It turned out that when milk is excreted by cubs of rats, there is an increase in blood flow in the auditory, olfactory, and gustatory cortex in nursing rats (Febo 2011). Thus, nerve impulses from the visual and auditory cortex enter the GPIA, modulating its functioning

(Fig. 4.18). Now I will focus in more detail on the morphofunctional characteristics of the links of the maternal reflex of milk excretion.

The skin surface of the nipple and areola of the breast is part of the overall skin surface of the body and, as already noted in the previous chapter, with the help of sensory receptors, it receives information about the characteristics of external stimuli of various modalities: mechanical, cold, heat, and pain. However, during the lactation period, sensory receptors in this area additionally provide lactation function of the breast. Taking into account that when milk is excreted by a child, there is mainly a mechanical effect on the areolar-nipple area of the gland, the main receptors for the occurrence of maternal reflexes are mechanoreceptors. The previous chapter presented morphological data on the distribution of mechanoreceptors in the skin of the nipple and areola of the breast of a woman. As follows from the description, these data are not numerous and are presented only at the light-microscopic level. There is no information about the functional characteristics of mechanoreceptors of a woman's breast. However, there is currently data on the properties of mechanoreceptors in other areas of the human skin surface (Johansson and Vallbo 1983; Macefield 2005; McGlone and Reilly 2010; Roudaut et al. 2012; Abraira and Ginty 2013). Given the similarity in the structure of mechanoreceptors located in the skin surface of the breast and other areas of human skin, it is highly likely that the functional characteristics of the corresponding receptors will be the same.

When milk is excreted by a child, the areolar-nipple area of the gland is subjected to stretching and compression (Fig. 4.17). Stretching occurs under the action of vacuum and stretches mainly the areolar part of the gland (Ilyin and Alekseev 2014). The stimuli of the vacuum and, accordingly, the stretching of the areolar-nipple area are in the form of triangles or trapezoids (Fig. 4.10B). Tactile effects (compression) normally also occur on the areolar area adjacent to the nipple and resemble triangles in shape (Fig. 4.11). The duration of the increase in stretching and compression of the areola varies between 0.3 and 0.5 s. The time of constant stretching of the areola in the case of trapezoidal vacuum stimuli could reach 0.6 s. In accordance with the functional characteristics of the studied human skin mechanoreceptors, the areolar–nipple complex skin stretching reception can be adequately performed by Ruffini corpuscles (SAII). These mechanoreceptors are called skin stretching receptors by analogy with muscle stretching receptors (Chambers et al. 1972; Johansson and Vallbo 1983). Ruffini's corpuscles (SAII) also respond to tactile stimuli (Johansson and Vallbo 1983). Tactile mechanoreception of the areolar-nipple area of the breast can be performed by two more types of encapsulated mechanoreceptors: Meissner's corpuscles (RAI) and Merkel's corpuscles (SAI). At the same time, unlike Ruffini bodies, the reaction of these mechanoreceptors to skin stretching is insignificant or absent. Detailed functional characteristics of the mentioned mechanoreceptors, as well as examples of their impulse activity, are presented in Sect. 4.2.1 (Fig. 4.5), dedicated to the child's mechanosensory rooting reflex. According to the functional properties of the mentioned mechanoreceptors (Burgess and Perl 1973; Johansson and Vallbo 1983), they can adequately encode information about the time and amplitude characteristics of stretching and compression of the areolar–nipple complex of the breast of a woman during the excretion of milk by a child. In addition to

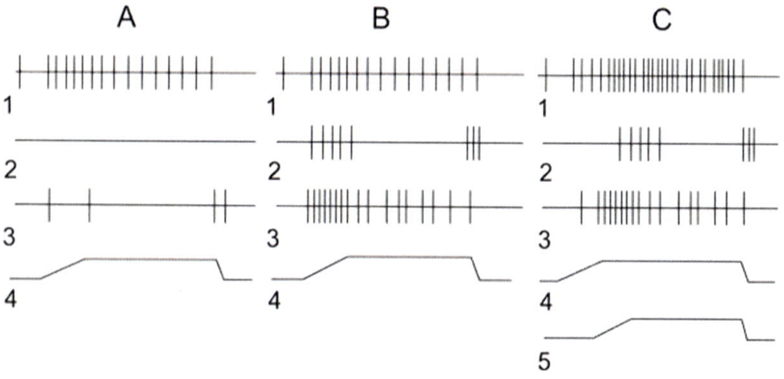

Fig. 4.19 Schematic representation of patterns of impulse activity occurring in nerve fibers forming bodies of Ruffini (1), Meissner (2) and Merkel (3) in response to stretching (**a**), compression (**b**) and joint action of stimuli of stretching and compression (**b**) of the areolar-nipple region of the breast

encapsulated bodies, mechanical stimulation in the skin of the areolar–nipple complex can probably be perceived by low-threshold mechanoreceptors, which are free nerve endings (Fig. 3.12) (Roudaut et al. 2012; Abraira and Ginty 2013). However, there is no information about their functional characteristics.

To elucidate the mechanisms of formation of the maternal reflex of milk excretion, it is interesting to study the patterns of afferent impulse activity that occur in the nerve fibers that form the mechanoreceptors of the areolar–nipple complex when they are adequately stimulated, and then enter the central link of the reflex of milk excretion.

When studying the impulse activity of the Ruffini (MAII) and Merkel (MAI) bodies, which occurred on trapezoidal mechanical stimuli with different rates of rise and fall, it turned out that the generation of impulse activity occurred mainly only on the increase in the amplitude of the mechanical stimulus and the stationary component of the stimulus. On the decline of the mechanical stimulus, impulse activity was not observed. In Meissner's bodies, impulse activity was generated on the growth of tactile stimulus. There was no impulse activity in response to the static part of the mechanical stimuli. Only 1–2 action potentials could be generated in response to a decrease in mechanical stimulation (Burgess and Perl 1973; Johansson and Vallbo 1983; Roudaut et al. 2012; Abraira and Ginty 2013). Figure 4.19 shows schematically the patterns of afferent impulse activity that can be generated in the nerve fibers that form encapsulated mechanoreceptors of the areolar-nipple region in response to stimuli of stretching and compression of the areolar-nipple region. Ruffini corpuscles (SAII) will react mainly to the stretching of the areolar-nipple area. With trapezoidal stretching, in addition to the impulse activity that occurs during the period of increasing tension, afferent action potentials will be generated for the static part of the stretch. On the decline of the stretching amplitude, the action potentials will not be generated (Fig. 4.19a1). Merkel's (SAI) and Meissner's bodies do not generate impulses for stretching (Fig. 4.19a2,3). In the case of compression of the areolar area of the breast, impulse activity will be generated in all mechanoreceptors (Fig. 4.19b). As the compression amplitude increases, the current frequency of action potentials increases. On the decline of compression, only the Meissner bodies (RAI) have 1–2 action potentials (Fig. 4.19b2). Figure 4.19c shows the expected dynamics of the impulse activity of mechanoreceptors when the areolar-nipple area of the breast is stimulated by stretching and compression, as it normally occurs when feeding a child (Fig. 4.17). The beginning of the compression pulse, as well as when feeding a child, is represented with a delay relative to the stretching pulse, which is generated as a result of creating a vacuum in the child's mouth. In accordance with the functional characteristics of mechanoreceptors, the most intense impulse will be observed in Ruffini bodies (SAII), since they respond to both stretching and compression of the skin surface.

Here it should be noted that we have found in the skin of the nipple and areola of laboratory animals (rats, guinea pigs) mechanoreceptors that have similarities in functional characteristics with the SAI, SAI, RAI of human skin (Fig. 4.5). These mechanoreceptors, respectively, can detect the amplitude, duration, and rate of change in the amplitude of a mechanical stimulus (Alekseev et al. 1975, 1976; Grachev et al. 1976b, 1977; Grachev and Alekseev 1980). However, morphological studies indicate that the most common receptors in the nipple and areola are free nerve endings. Therefore, it can be assumed that in the areolar–nipple complex of the breast of a woman, in addition to encapsulated receptors, the mechanosensory function can additionally, or perhaps mainly, be performed by low-threshold receptors formed by free nerve endings. Thus, according to the simulation of the pulse activity of the mechanoreceptors of the nipple and areola of the breast of a woman, in response to each SSE, "bursts" of pulse activity will occur in the nerve fibers that form the mechanoreceptors, which will be transmitted to the CNS of the woman. In this regard, we are interested in data on registration of pulse activity in afferent fibers of the nipple and areola, which we managed to obtain in experiments on lactating rats when feeding pups (Grachev and Alekseev 1980). In some prepared nerve branches innervating the nipple and areola of the rat gland, responses of different types of mechanoreceptors were simultaneously registered from the nerve fibers that make up the branch (Fig. 4.20A) (Grachev and Alekseev 1980). Note that in accordance with the concentration gradients of ions across the membrane of nerve fibers that form receptors in the nipple and areola, the amplitude of afferent action potentials in different nerve fibers is the same. At the same time, since the conditions for registering electrical activity for different nerve fibers differ in the experiment, the amplitude of afferent action potentials has a different value (Fig. 4.20A) and the responses of several types of mechanoreceptors can be easily identified on oscillograms. It should be pointed out that the cub rat also affects the areola and the nipple simultaneously with vacuum and compression stimuli, which can be distinguished by dynamic and static components (Brake et al. 1986; Karyakin and Alekseev 1992). Comparison of visual observation of the cub, or rather the movement of its jaws, with the oscillograms of afferent impulse activity revealed a characteristic rhythmic pattern of afferent nerve impulses over time. At first, the cub retracted the nipple with the areola into the oral cavity, while the compressing

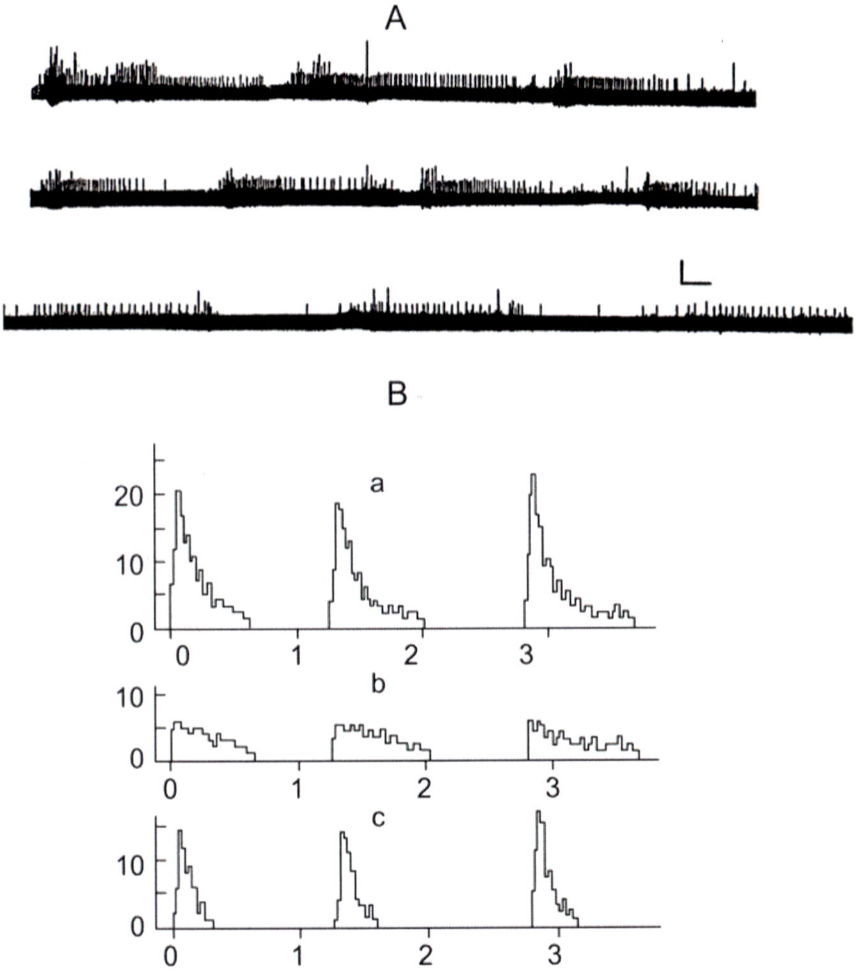

Fig. 4.20 Afferent impulse activity (**A**) and histogram (**B**) of impulse activity of receptor units of the areolar–nipple complex of lactating rat during milk excretion by cubs (from Grachev and Alekseev 1980). (**A**) Horizontal lines under oscillograms indicate the time of mechanical action on the nipple and areola of the cub. Calibration: vertical 0.2 mV, horizontal 250 ms. (**B**) a—total histogram of impulse activity; b—histograms of impulse activity respectively of slow and fast adapting mechanoreceptors. Abscissa axis—time, s; ordinate axis is the number of action potentials

effects of the jaws were observed. Accordingly, during this period (0.1–0.2 s), there was a sharp spike in impulse activity, and there was an increase in the frequency of all receptor units. Then the cub's jaws were clenched for 0.4–0.6 s. The frequency of medium and low-amplitude afferent action potentials gradually decreased over time. Further, the jaw compression decreased and during 0.4–1 s, the pulse activity was practically absent. This rhythm was repeated throughout the entire milk withdrawal session. The cub's SSE frequency was in the range of 30–60/min. Figure 4.20B shows histograms of afferent impulse activity for the milk withdrawal session. It can be seen that the maximum frequency of the action potentials at the initial burst of impulse activity exceeds its final value by 15–20 times (Figure 4.20Ba). Each SSE of the cub (Fig. 4.20A) can be represented as a superposition of slow- (Figure 4.20Bb) and fast-adapting (Figure 4.20Bc) mechanoreceptors functioning. It can be assumed that a similar type will have an impulse activity in response to the child's SSE when removing milk from the mother's breast.

As already noted, the sensory receptors of the areolar-nipple area of the breast during lactation perform a double function: (1) perform the reception of external stimulus and (2) participate in the formation and maintenance of lactation. Therefore, we can assume that there are two ways to transmit sensory information. Along the first path, the most accurate information is quickly transmitted to the brain, differentiating the amplitude, gradient, duration, and location of the impact on the mammary gland of extrasensory stimuli. The final link of the first pathway is the somatosensory region of the cortex, where a woman is believed to have a sense of the mechanical impact of the baby's mouth on the areolar-nipple area of the breast when feeding it. For the second pathway, the final structures are the neural secretory cells of the hypothalamus. In morphological experiments conducted on various animals, it was found that neurons forming receptors in the areolar-nipple region of the breast are localized in the spinal ganglia. They are the common starting point for both paths. The study of afferent pathways from the skin surface receptors of animals and humans to the somatosensory cortex showed their great similarity. We can assume that conducting afferent pathways to the somatosensory cortex of the brain from the mammary gland are carried out in the same way as from other parts of the external surface of the human body. In particular, the impulse activity that occurs in the afferent nerve fibers that form the mechanoreceptors of the areolar–nipple complex of the woman breast and are mainly part of the deep branch 4 of the intercostal nerve and sometimes partially in the branch 5 of the intercostal nerve (Fig. 3.11) enters the posterior horns of the spinal cord. Accordingly, afferent impulse activity passes ipsilaterally along the nerve fibers in the posterior columns of the spinal cord. Then synaptically switches in the medulla oblongata to second-order neurons, whose axons in the medial lemniscus go counterlateral, reaching the nuclei of the thalamus, and switch to third-order neurons. The nerve fibers of the thalamus cells pass to the somatosensory cortex, or rather the postcentral gyrus of the cerebral cortex (primary somatosensory cortex). As already mentioned, W. Penfield (Penfield and Boldrey 1937) was made in accordance with sensory sensitivity mapping of the surface of the body and face of a person. Images on this map of parts of the human body are represented according to the density of their sensory innervation and is called the sensory "homunculus". A study of breast localization in non-lactating women in the primary somatosensory cortex using electrophysiological and magnetic resonance methods (Rothemund et al. 2005) showed that the projection of the areolar-nipple area of the breast of a woman is mapped contralateral on the trunk of the "homunculus" close to the inguinal area.

The afferent pathways of the woman milk excretion reflex remain unexplored. The question of afferent pathways of the milk excretion reflex was studied most intensively in laboratory animals, and to a lesser extent in farm animals (rev.: Grachev and Alekseev 1980; Tsingotjidou and Papadopoulos 2003, 2008; Crowley 2015). In morphophysiological studies, it was found that the afferent pathway to the neurosecretory cells of the hypothalamus differs from the lemniscal system of pathways to the somatosensory cortex and is carried out along the spinocervical pathway. Nerve fibers pass ipsilateral to the lateral cervical nucleus. The nerve cells of this nucleus are activated when milk is excreted by the cubs. Then a part of the nerve fibers makes a cross, passes through the ventrolateral part of the medulla oblongata, and reaches the neuronal network—the Generator of Periodic Impulse Activity (GPIA), localized in the lateral covering of the midbrain (rev. Crowley 2015). However, accurate information about the course of neural pathways to the neural secretory cells of the hypothalamus is currently not available even for laboratory animals (Crowley 2015). Undoubtedly, because of the species differences, the pathways of the milk excretion reflex of women have their own characteristics. However, it can be assumed that in general they coincide with the pathways of other mammals, and the woman GPIA is located in the lateral covering of the midbrain.

The initial part of the efferent pathway of the woman milk excretion reflex is the neural secretory oxytocinergic cells located in the hypothalamus. Oxytocinergic neurons are quite large. In humans and animals, the diameter of their cell bodies reaches 25–30 microns (Fig. 4.21c). This type of cell is called magnocellular neurosecretory cells. At the same time, another type of magnocellular neurosecretory cells was found in the hypothalamus—vasopressinergic cells, which provide osmotic homeostasis of the body. Magnocellular cells are unevenly distributed in the hypothalamus. Their greatest density is observed near the third cerebral ventricle and optical chiasm (Fig. 4.21a, b). This is especially noticeable on histological preparations with small magnifications of the microscope. The clusters were designated as magnocellular nuclei and were named paraventricular (PVN) and supraoptic (SON) nuclei, respectively. Small clusters of both types of neurosecretory cells are located between SON and PVN, which are called additional (accessory) magnocellular nuclei. It should be noted that morphological studies in the nuclei did not reveal any noticeable isolation of oxytocin and vasopressinergic cells (Koutcherov et al. 2000; Brown et al. 2013). It should be noted that individual magnocellular cells are also observed in the areas between the nuclei. All types of magnocellular nuclei are permeated by a dense capillary network (Morton 1969; Goudsmit et al. 1990; Hofman et al. 1990; Swaab et al. 1993; Brown et al. 2013). In the human hypothalamus, SON has three sections: dorsolateral (about 7.2 mm long), dorsomedial, and ventromedial (together 5.7 mm long). PVN cells in the form of a spindle structure with a length of about 6.4 mm are relatively evenly distributed along the walls of the three ventricles (Fig. 4.21b) (Morton 1969). However, it also identified five sub-cores (Koutcherov et al. 2000). An approximate count of magnocellular neurons in human SON and PVN was made (Morton 1969; Fliers et al. 1985; Goudsmit et al. 1990; Swaab et al. 1993; Manaye et al. 2005). Numerical data from different authors show noticeable variations, which is apparently due to the different methods used. The largest number of works is devoted to SON

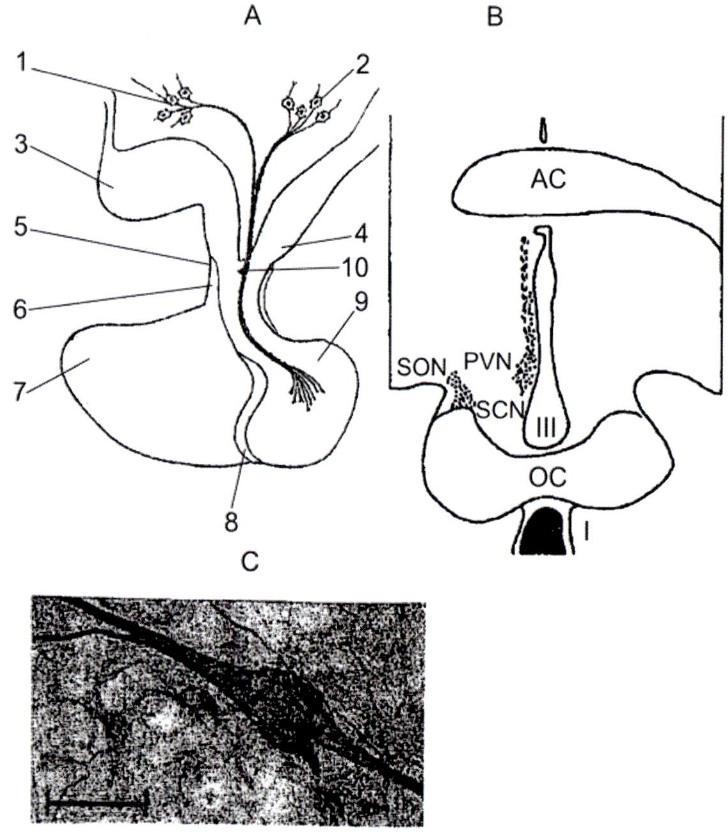

Fig. 4.21 Human pituitary structures involved in the formation of milk excretion reflex. (**a**) Schematic representation of the human pituitary gland showing the course of axons of magnocellular neurons from the paraventricular and supraoptic nuclei to the posterior lobe of the pituitary gland. 1—Supraoptic nucleus neurons, 2—paraventricular nucleus neurons, 3—optic chiasm, 4—median eminence, 5—pituitary stalk, 6—tuberal part of pituitary gland, 7—anterior part of the pituitary gland, 8—intermediate part of pituitary gland, 9—posterior part of pituitary gland, 10—pituitary tract. (**b**) Topography scheme of the supraoptic and paraventricular nucleus in the human hypothalamus (from Hofman et al. 1990). *PVN* paraventricular nucleus, *SON* supraoptic nucleus, *OC* optic chiasm, *SCN* suprachiasmatic nucleus, *III* third ventricle, *I* pituitary funnel, *AC* anterior commissure. (**b**) Magnocellular neuron of the paraventricular nucleus of the human hypothalamus (from Koutcherov et al. 2000). Horizontal calibration: 50 μm

morphometry. Thus, the number of magnocellular neurons for this nucleus according to morphometric calculations ranges from 40,000 to 94,000 (Manaye et al. 2005). One paper (Morton 1969) is devoted to estimating the number of magnocellular PVN cells. The total number of magnocellular neurons in the PVN was estimated at 56,000, 25,000 of which are oxytocinergic. In SON, about 7600 neurons were identified as oxytocinergic (Morton 1969; Fliers et al. 1985; Goudsmit et al. 1990; Swaab et al. 1993; Manaye et al. 2005). Given that oxytocinergic cells are also localized in the accessory nuclei, and are also located in the "extra-nuclear" space, their number in the human hypothalamus is estimated at about 40,000. For comparison, there are about 9000 oxytocinergic cells in the rat hypothalamus (Lincoln 1984). In addition to large oxytocin and vasopressinergic cells, small parvocellular oxytocin and vasopressinergic cells were found in the structures of the human hypothalamus. In this regard, it should be noted that oxytocin is a multifunctional hormone and in the woman's body, in addition to reproductive and lactation functions, it is involved in the regulation of other physiological and mental functions (Meyer-Lindenberg et al. 2011). Morphological examination showed that the axons and dendrites magno- and parvocellular oxytocinergic cells are in different parts of the brain. At the same time, it is believed that the efferent nervous part of the reflex of milk excretion in women are magnocellular oxytocinergic cells.

The soma of human oxytocinergic neurons has a fusiform shape (Fig. 4.21c) (Koutcherov et al. 2000). From it departs 2–3 dendrites and one axon. Despite the large size of the soma of magnocellular cells, the diameter of their dendrites and axons in the widest part does not exceed 1–0.5 microns, i.e., at the limit of the resolution of the light microscope. Therefore, it is extremely difficult to trace the course of axons and dendrites in the structures of the hypothalamus. It is important to note here that the axons cross the hemato-encephalic barrier on their way so that the posterior pituitary lobe, median eminence, and some brain structures surrounding the third ventricle are located behind the blood–brain barrier (Lincoln 1984) (Fig. 4.21a). The axons of magnocellular cells approach the median eminence of the hypothalamus and then pass through the pituitary stalk to the posterior lobe of the pituitary as a well-marked bundle of nerve fibers (Green 1948; Seyama et al. 1980a). (Fig. 4.21a). The length of the human pituitary stalk varies between 3.26 and 8.66 mm (Satogami et al. 2010). Detailed electron microscopic studies of the human pituitary gland found two types of nerve fibers in the pituitary stalk: A and B (Bergland and Torack 1969a, b; Seyama et al. 1980a; Dudas et al. 2010). Both types of nerve fibers are surrounded by processes of cells located in the neurohypophysis, called pituitary cells. Type A of axons, the most common, contains inside the neurotubules, neurofilaments, mitochondria, and neurosecretory granules with a diameter of 100–300 nm. Type B axons are much less common and contain the same intracellular organelles, but the neural secretory granules in these axons have a smaller diameter: 50–100 nm. It should be noted here that the results of studies of the human neurohypophysis (Bergland and Torack 1969a, b; Seyama et al. 1980a; Takei et al. 1980; Scheithauer et al. 1992) indicate that oxytocinergic and vasopressinergic fibers are axons of type A. Nerve fibers of type B are aminoergic and their function is not known. Therefore, in the future, the main attention will be directed to the presentation of morpho-functional characteristics of cells and their axons of type A.

Secretory granules are found along the entire length of the axons. However, their greatest concentration is observed in axon extensions, which are in most cases not uniform local axon increases along the entire diameter, but are located along the length of the nerve fiber in the form of asymmetric outgrowths (Bergland and Torack

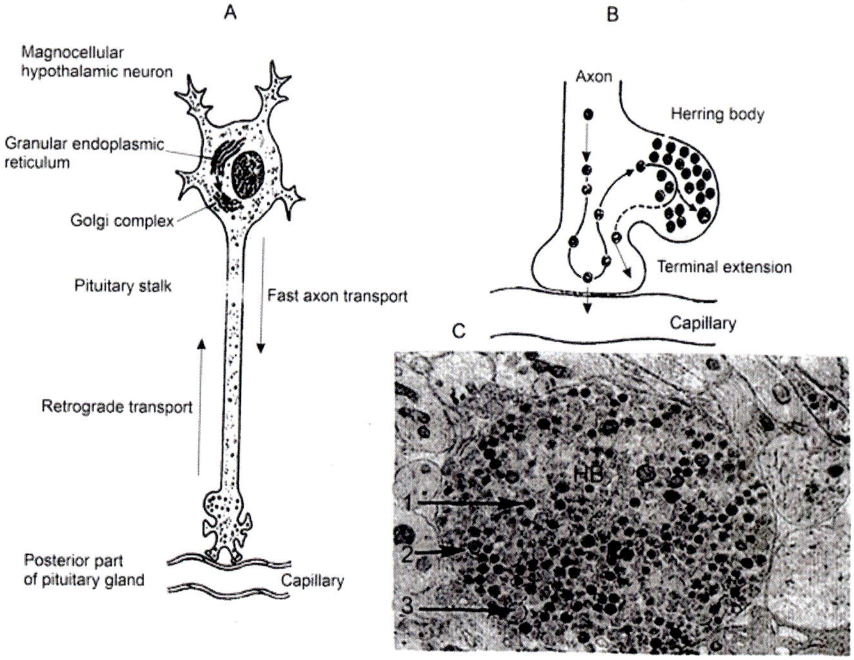

Fig. 4.22 Schematic representation of the structure and function of a magnocellular hypothalamic neuron. (**a**) On the right, the stages of hormone synthesis are noted, starting from the uptake of progenitors by the hypothalamic neuron to the accumulation of final products in the nerve terminal and posterior pituitary lobe. Axonal transport of peptides is carried out in both directions; shown on the left is a possible retrograde transport trapped after exocytosis of the granule membrane into the cell body (from Lincoln 1984). (**b**) Movement and accumulation of neurosecretory granules in axonal extensions and capillary-associated terminals of large-cell neurons in the posterior pituitary lobe (by: Lincoln 1984). (**b**) Micrograph of Goering's body. HB- Goering bodies, with arrows 1, 2, 3 different types of vesicles are indicated × 18,000 (from Bergland and Torack 1969a, b)

1969a, b; Seyama et al. 1980a) (Fig. 4.22a). The diameter of the extensions varies between 1 and 50 microns. The largest extensions were named Goering's body (Fig. 4.22b, c). In addition to Goering's bodies, several types of axon extensions have been identified (Seyama et al. 1980a). The difference was determined by the content of intracellular structures in axon extensions. The functional purpose of axon extensions is not defined in most cases. Most interesting is the first type of extensions (Goering's body), which is characterized by a high density of neural secretory granules. In addition, few mitochondria and microvesicles are found in these extensions (Seyama et al. 1980a). It is believed that most of the hormones are concentrated in the secretory granules of Goering's bodies.

The terminal parts of the axons of large-cell neurons that have penetrated the posterior lobe of the pituitary gland are characterized by numerous extensions and active branching to form terminals near the walls of fenestrated blood vessels (Seyama et al. 1980b). (Fig. 4.22a). Despite the fact that the morphological

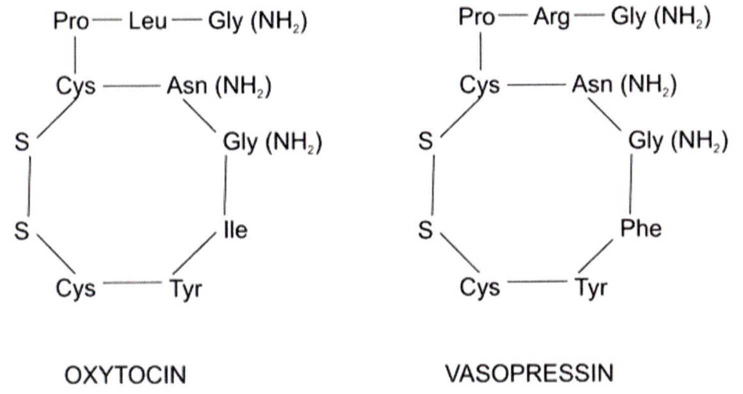

Fig. 4.23 The structural formula of oxytocin and vasopressin (from Bergland and Torack 1969a, b)

characteristics of human neurosecretory oxytocinergic neurons have been relatively well studied, their functional properties remain unexplored for methodological reasons. Therefore, to discuss the function of human neurosecretory cells, data obtained in animals will be used, and in particular, the rat's neurohypophysial system will be used as a model.

It should be noted that oxytocin and vasopressin were the first synthesized neuropeptides of animals and humans. This was done by Vincent du Vigneaud in 1951, for which he was awarded the 1955 Nobel prize in chemistry (Du Vigneaud 1955). Both hormones contain nine amino acids with cysteine residues in position 1 and 6 linked by disulfide bridges, resulting in a ring structure (Fig. 4.23). Seven amino acids are common to both peptides, which apparently provides them with interaction at the level of specific receptors. Thus, vasopressin can cause milk production in rats, but its lactogenic activity is only 25% of the activity of oxytocin (Lincoln 1984).

Oxytocin in the magnocellular neurons of the hypothalamus, as in other neuroendocrine cells, is formed as a result of the cleavage of a high-molecular-weight precursor, or prohormone. A glycosylated peptide consisting of 166 amino acids is synthesized in the granular endoplasmic network of the cell body. This peptide is then transferred to the Golgi apparatus, where it is packed in membrane-covered granules 100–300 nm in diameter. Neurosecretory granules are quickly transported to the posterior pituitary lobe via the axons of large-cell neurons (Fig. 4.22a). In a rat, the labeled peptide appears in the posterior pituitary lobe within 3 h of the introduction of the labeled precursor (^{35}S-cysteine) into the ventricles of the brain. Thus, the transport velocity is 1–3 mm/h, which is 50 times more than the normal axoplasm velocity along the axon. The granules move along the axon along the microtubules. This is indicated by data on the cessation of peptide transport during the treatment of axons with colchicine, which destroys microtubules. At the same time, the axoplasm current is not disturbed. During transport from the hypothalamus to the posterior pituitary, the precursor undergoes further cleavage, probably under the action of enzymes located inside the secretory granule, resulting in the formation of an active hormone. The main attention in the study of axon transport in large-cell neurons was paid to the movement of substances from the soma of the cell to the nerve terminal. However, it was found that the transport of substances in magnocellular neurons can occur in the opposite, retrograde direction (Fig. 4.22a). The physiological significance of retrograde transport is not clear enough. It is assumed, for example, that it can serve as a return of granule membranes to the cell body after the granule contents have been released (Lincoln 1984).

The posterior lobe of the mammalian pituitary gland contains a fairly large supply of oxytocin compared to the amounts needed for normal physiological functions (Lederis 1961; Lincoln and Paisley 1982). In women, the amount of oxytocin in the neurohypophysis is approximately 6–18 micrograms (Lederis 1961). At the same time, no more than 0.1–0.2 μg are required for milk excretion reflex (Cobo et al. 1967). In the rat, the ratio is even higher. She has about 1 μg of oxytocin in the posterior pituitary lobe, and the release of milk can only cause 0.001 μg of oxytocin. It is believed that high concentrations of oxytocin in the pituitary gland can play an important role in emergency situations, such as during childbirth (Lincoln 1984).

According to Morris (1980), the number of neural secretory granules in the posterior lobe of the rat pituitary gland is 2×10^{10}, so each granule contains about 60,000 oxytocin molecules. Only 30% of these granules are detected in terminals located near the capillaries. Most of them are located in axon extensions—Goering's bodies (Fig. 4.22b, c). These data served as an anatomical basis for the assumption of the existence of an easily isolated peptide fund. The easily released peptide fund can make up 10% of the hormone contained in the neurohypophysis. However, this is quite a large amount, if you consider that the release of milk in a rat requires only 0.1% of the supply of oxytocin in the neurohypophysis. It is interesting to note that the newly synthesized hormone is subjected to the preferred release into the bloodstream. New neural secretory granules entering the posterior lobe of the pituitary gland must first be directed to the terminals, and then move to less accessible areas of the axon where they accumulate (Fig. 4.22b) (Lincoln 1984).

Just like nerve cells in other parts of the nervous system, large-cell neurons in the hypothalamus generate action potentials. On the soma of rat magnocellular neurons, the processes of nerve cells located in various parts of the brain form about 5000 synaptic endings (Lincoln 1984). According to the mediators that are released from these synapses, the membrane potential of neurosecretory cells can change toward depolarization or hyperpolarization. In the case of depolarization, when the threshold level of the membrane potential is reached, the soma of the large-cell neuron generates action potentials that propagate along the axon, reaching the terminals of the posterior pituitary and causing the release of oxytocin. The frequency of action potentials is determined by the cell's membrane potential, which in turn is a function of the intensity and balance of numerous excitatory (depolarizing) and inhibitory (hyperpolarizing) synaptic signals. Since the soma of the neurosecretory cell has a large size (about 30 microns) (Fig. 4.21c), its electrical activity can be relatively easily studied using extra- and intracellular electrodes. Such experiments can only be performed on animals, of course. Currently, two main approaches are used to measure the electrical activity of magnocellular cells. The first method involves an acute experiment on an anesthetized animal. The advantage of this method is that additional physiological parameters can be recorded in parallel with the recording of electrical activity. Another method is to use sections of the hypothalamus 100 microns thick, which are placed in a culture medium (Jourdain et al. 1998). Slices can remain viable for a day, which makes it easier to study the response of neurons to various pharmacological substances (neurotransmitters, regulatory peptides, etc.). In this case, the size of the animal does not create difficulties, whether it is slices obtained from the hypothalamus of a rat or a cow (Lincoln 1984). It should be noted that experiments were carried out to register impulse activity in an unmobilized, conscious animal (rat) (Summerlee and Lincoln 1981). This approach is possible, but as practice has shown, it is quite difficult.

When registering impulse activity from rat magnocellular neurons, the described methods in all cases registered impulse activity in the absence of any sensory stimulation. This impulse is called spontaneous impulse activity. The frequency of spontaneous activity is in the range of 1–5 imp/s. When the soma membrane potential was reduced (depolarized), the cells generated a series of action potentials. Conversely, an increase in the membrane potential (hyperpolarization) was accompanied by the disappearance of spontaneous impulses. Here it is interesting to note that the axon terminals of neurosecretory cells are also able to generate action potentials. In experiments on isolated rat neurohypophyses (Bourque 1990), it was possible to insert the tip of the electrode inside the neural terminal. The amplitude and duration of the action potentials generated by the nerve terminal were analogical to those of the soma neuron. However, in contrast to the soma of the cell in response to prolonged depolarization of the membrane terminals generate only one action potential.

When generating the action potential of the axon terminal, sodium ions enter the terminal, which in turn leads to the opening of calcium channels through which calcium ions enter the terminal (Fig. 4.24). An increase in the concentration of calcium ions causes the movement of neural secretory granules to the outer membrane of the terminal. The granule membranes and terminals merge, and the granule contents are released from the cell into the extracellular space. As already mentioned, this method of removing substances from the cell is called exocytosis. Next, oxytocin from the extracellular space diffuses through fenestrated capillaries (Seyama et al. 1980b) into the general bloodstream. Then the status quo in the terminal is restored. To do this, it is necessary to reduce the intracellular concentration of calcium and restore the membrane potential. Data from studies using a ^{3}H-choline-labeled granule membrane show that the granule membrane is recaptured, which appears in the pituitary terminals of rats in the form of large bubbles directed to the soma of the cell. The excess amount of calcium ions is sequestered in the micro-bubbles and removed from the terminal. The membrane potential, as in other parts of the cell, is restored by active ion transport (Lincoln 1984). Here it should be noted that initially significant disagreements arose on the mechanism for removing oxytocin from terminals using exocytosis. The reason was that the electron microscopic study of the animal neurohypophysis (rev. Theodosis

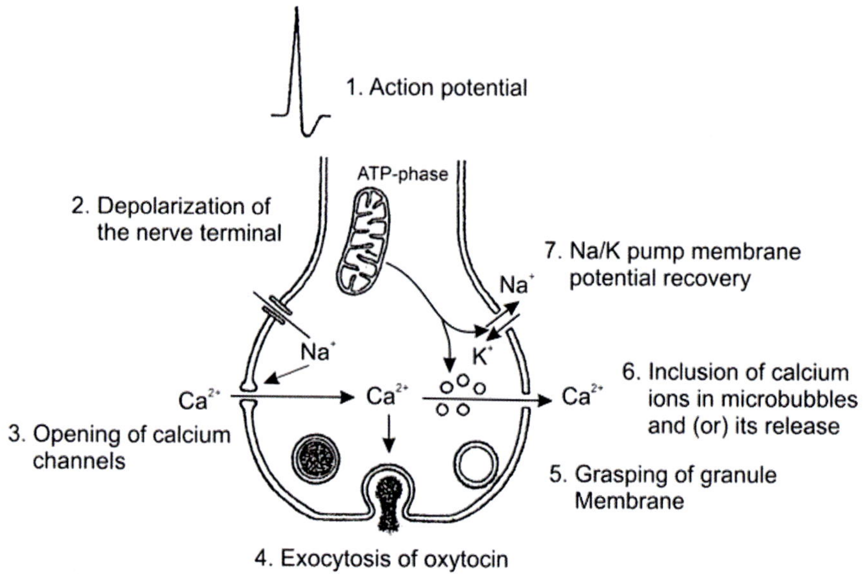

Fig. 4.24 Steps of the process linking stimulation to secretion in the axonal terminal of the posterior pituitary lobe (by: Lincoln 1984)

et al. 1978) and human (Bergland and Torack 1969a, b; Seyama et al. 1980a, b) data confirming the movement of granules from the terminals to the extracellular space were extremely rare or in most cases absent. The reason for the disagreement according to D. Lincoln (1984) was that the process of exocytosis is extremely fast, probably taking less than 10 ms. Therefore, even if we artificially increase the level of oxytocin release to 1% of its content in the neurohypophysis per minute, we can expect that only 1–2 granules per million will be involved in the exocytosis process! In accordance with the usual technique of electron microscopy, small sections of the terminal membrane can be viewed. In this case, the probability of detecting the exocytosis of granules is insignificant. More effective is the use of the electron microscopic freezing–chipping technique. Thanks to this method, it is possible to view large areas of the terminal membrane compared to those that are detected in sections with conventional electron microscopy. D. Theodosis and coworkers (Theodosis et al. 1978) conducted studies on the rat neurohypophyses using the freeze–fracture method and obtained numerous illustrations of exocytosis.

Thus, for each action potential received from the soma to the nerve terminal of the oxytocinergic cell, a certain amount of oxytocin is released into the bloodstream. Here it should be noted that most of the rat's neural secretory cells, in which oxytocin is synthesized, generate spontaneous (background) action potentials continuously (day and night) with an average frequency of 2–4 action potentials per second. It is possible that background impulses are generated with this frequency in the neural secretory oxytocinergic cells of the human hypothalamus. This frequency is small, but it is constant and therefore in rats, 9000 cells that synthesize oxytocin will generate about 2 billion action potentials per day, and accordingly, a woman has 40,000 cells that will generate about 9 billion action potentials. The presence of background impulse activity in neurosecretory oxytocinergic cells suggests the presence of a background (base) concentration of oxytocin in the blood. Measurement of the baseline blood concentration in lactating rats (Higuchi et al. 1986) and women (Nissen et al. 1996) showed that it does not exceed 15–20 pcmol/l. Recalculation of the amount of oxytocin released per spontaneous action potential (Lincoln 1984) shows that this is an extremely small amount and is measured by attograms (10^{-18} g). Therefore, despite the constant generation of background action potentials according to biochemical studies, the loss of oxytocin from the posterior pituitary in rats and probably in women per day is only 4% of the total content of it in the neurohypophysis (Lincoln 1984).

The most interesting is the impulse activity that is generated by magnocellular oxytocinergic cells of the hypothalamus in response to sensory impulses that occur in the nerve fibers of the receptors of the areolar-nipple area of the breast when adequately stimulated, i.e., when feeding a baby (Fig. 4.20). In this regard, the research methodology is much more complicated. Since along with the registration of pulse activity from magnocellular oxytocinergic cells of the hypothalamus nuclei, it is necessary to ensure simultaneous excretion of milk by the cubs. A study of the milk excretion reflex in various laboratory animals (Lincoln 1984) showed that the best model for these experiments is the lactating rat. It turned out that milk from lactating rats can be withdrawn only when the rats are asleep. Milk was also secreted in anesthetized animals. The reason was that most anesthetics can cause a rat brain condition similar to slow-wave sleep.

It turned out that when the mother's nipples were strongly stimulated by the young for a few minutes after they were applied to the nipples, the impulse activity recorded from large-cell neurons did not differ from the spontaneous activity of the neurons of rats that were not fed (Dreifuss et al. 1981). However, after 10–15 min of stimulation of the gland teats of the mother who was under anesthesia, the impulse activity increased for 2–4 s in an explosive manner from 0–5 imp/s to 30–80 imp/s. This was followed by a period of 20–40 s during which there was no spontaneous impulse (Fig. 4.25a) (Lincoln and Wakerley 1975; Lincoln 1984). After the first burst of impulse activity, a sequence of 2–4 s bursts of action potentials was observed on average every 6 min (Fig. 4.25a). One of the main conditions for the occurrence of high-frequency bursts of impulse activity in magnocellular neurons of anesthetized rats was the presence of a certain number of cubs rats attached to the nipples. Or in other words, the total intensity of adequate stimulation of the areolar-nipple areas of the mammary glands. Thus, the number of neurons in which a sharp increase in impulse activity was detected, as well as the interval between bursts of activity did not change when the number of rats attached to the teats increased from 7 to 11. When the number of cubs rats decreased to 6 or less, high-frequency bursts were not observed (Lincoln and Wakerley 1975; Lincoln 1984). Thus, the patterns of impulse activity (Fig. 4.20) that occurred in the nerve fibers that form receptors in the areola of the rat mammary glands during the excretion of milk by rat cubs did not reach "directly" to the magnocellular

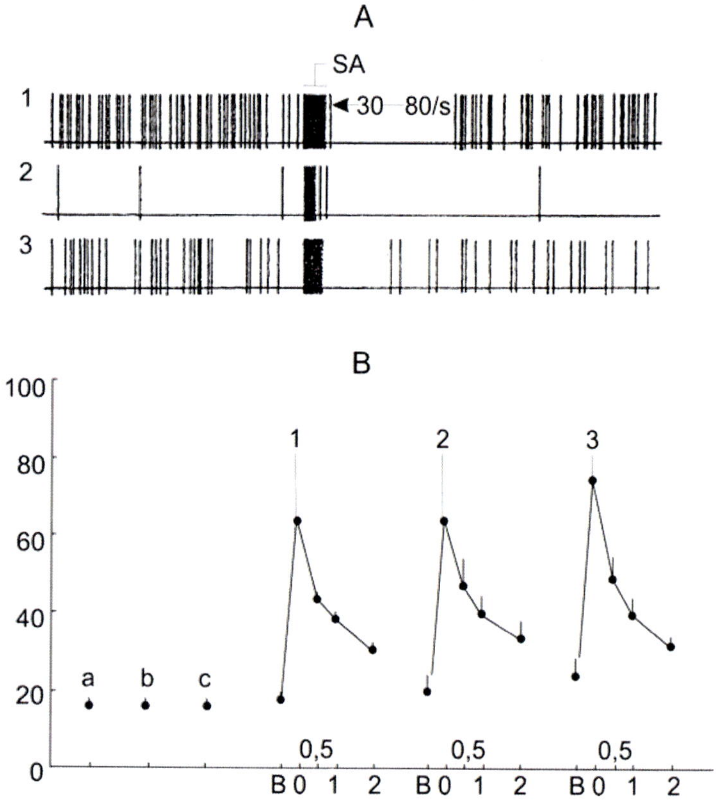

Fig. 4.25 Impulse activity of oxytocinergic neurons of the hypothalamus (**a**) and increase of concentration of oxytocin (**B**) in blood in rats during feeding of cubs. (**A**) 1,2,3—impulse activity of three oxytocin-producing neurons of the rat hypothalamus (by: Lincoln 1984). (**B**) Value of oxytocin concentrations in lactating rats (by Higuchi et al. 1986). (a, b, c is the basal concentration of oxytocin in the blood, respectively, before applying cubs to nipples, at the time of applying 4–5 cubs to nipples and 9–10 cubs to applications. 1,2,3—peaks of impulse increase of oxytocin concentration in blood in nursing rats in case 9–10 rats are applied to nipples. On the ordinate axis: concentration of oxytocin in pmol/l, on the abscissa axis for all peaks: **B**—the moment of beginning of reflex increase of oxytocin, 0, 0.5, 1, 2—the concentration of oxytocin, respectively, at 30 s intervals (0, 0.5, 1) and 2 min after the beginning of the reflex peak

oxytocinergic cells of SON and PVN. It can be assumed that the impulse activity that occurs in the nerve fibers that form receptors in the aeola and nipple, coming along the spinocervical pathway to the neuronal structures of the GPIA, localized in the lateral covering of the midbrain (Crowley 2015), is summed up for some time in the GPIA neurons. When a certain level of summation is reached (probably a certain level of the membrane potential of neurons), GPIA neurons are synchronously discharged to the magnocellular cells of SON and PVN, generating bursts of impulse activity (Fig. 4.25a). Since each action potential causes the release of oxytocin from the terminals of the posterior pituitary, respectively, during bursts in the frequency of action potentials, the amount of oxytocin released into the bloodstream will also increase impulsively. Here it is important to note that according to an approximate estimate (Lincoln and Paisley 1982; Lincoln 1984), the release of the amount of oxytocin per action potential during high-frequency generation of pulse activity will significantly increase and will be measured not by attograms (10^{-18} g), but by femtograms (10^{-15} g). One explanation for this phenomenon is an increase in the duration of action potentials that are generated in the nerve terminals of the posterior pituitary, with an increase in the frequency of generation of action potentials (Bourque 1990). An increase in the duration and frequency of action potentials will be accompanied by an increase in the accumulation of calcium ions inside the terminals and, accordingly, an increase in the output of granules with oxytocin into the bloodstream (Lincoln and Paisley 1982; Lincoln 1984). It should be added that in later studies on various types of secretory cells of the rat adenohypophysis (lactotrophs, somatotrophs), it was demonstrated that with patterns of impulse activity in the form of bursts, exocytosis of hormones significantly increases. In this case, the impulse output of hormones was caused by an impulse increase in the intracellular concentration of calcium ions (Stojilkovic et al. 2005). Mathematical modeling of the exocytosis process (Tagliavini et al. 2016) in secretory and neuro-endocrine cells also indicates the advantage of pulse activity in the form of frequency bursts for hormone exocytosis.

Experiments with simultaneous recording of the impulse activity of magnocellular neurons during feeding of young and measuring the concentration of oxytocin in the blood of nursing rats have not been possible to date. However, there is data in the literature on the determination of oxytocin in the blood of rats using the radioimmunological method during the feeding of cubs (Higuchi et al. 1986) (Fig. 4.25b). When determining the concentration of oxytocin in blood volumes that were taken after 30 s during feeding of cubs, the content of oxytocin in the blood increased impulsively from 15–20 pmol/l to 60–70 pmol/l. The duration of the peaks was about 2 min. Moreover, oxytocin peaks as well as peaks of impulse activity were registered on average after 6 min

As already noted, it is impossible to register electrical impulse activity from the oxytocinergic neurons of the hypothalamus of a woman for methodological and ethical reasons at the present time. However, it was not difficult to determine the concentration of oxytocin in the blood of women while feeding a child. The measurements showed that the release of oxytocin in the blood is of a pulsed nature. Here it is important to note that the pulse release of oxytocin is most clearly detected when blood samples are collected after 20–30 s (Lucas et al. 1980: McNeilly et al. 1983; Nissen et al. 1996; Jonas et al. 2009) (Fig. 4.26b). At intervals of determination of oxytocin in the blood after 3 min or more, only a smoothed increased level of oxytocin concentration during the feeding period is recorded (Weitzman et al. 1980; Dawood et al. 1981; Johnston and Amico 1986) (Fig. 4.26a). In the case of determination of oxytocin in the blood after 20–30 s intervals, the duration of oxytocin peaks was about 1–1.5 min, i.e., the same order as the time of oxytocin peaks in rats. In this regard, it can be assumed that oxytocin peaks in women when feeding a child are also preceded by 2–4 s periods of high-frequency pulsation of

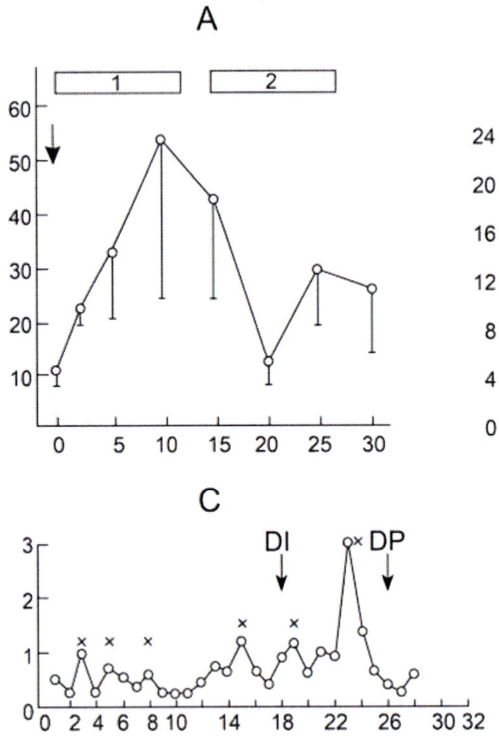

Fig. 4.26 Change in blood oxytocin concentration when a woman feeds a baby (**a**, **b**) and during childbirth (**b**). (**a**) Change in oxytocin concentration at blood sampling intervals after 3 min. 1—left breast feeding, 2—right breast feeding. The ordinate axis is the concentration of oxytocin pg/ml, the abscissa axis is the time in minutes (from Dawood et al. 1981). (**b**) The change in the concentration of oxytocin at blood sampling intervals after 25–30 min. The arrow indicates the moment of the first cry of the child. Hatched rectangles on top—the time when the baby takes out milk. The ordinate axis is the concentration of oxytocin ng/l, the abscissa axis is the time in minutes (from McNeilly et al. 1983). (**c**) Change in blood concentration during the second and third stages of childbirth. x—oxytocin peaks. DI—birth of baby, DP—removal of placenta. Ordinate–oxytocin concentration μIU/ml, abscissa—time in min (from Fuchs et al. 1991)

oxytocinergic cells of the SON and PVN hypothalamus. However, the interval between oxytocin peaks in a nursing woman was 2–3 times shorter than in a lactating rat. Regarding the determination of peak oxytocin concentrations during infant feeding, it should be noted that the results of measurements largely depend on the location of blood sampling. In lactating rats, blood was drawn from the large external jugular vein, which contains oxytocin-rich venous blood coming from the posterior pituitary lobe. In women, blood from the posterior pituitary also mainly enters the jugular vein, which is much less accessible, so most often blood is taken from the ulnar vein. But this means that the oxytocin contained in such blood has already passed through the lungs and the general circulatory system, and therefore must be diluted and its concentration reduced. Therefore, the measured average peak oxytocin concentrations in women (Nissen et al. 1996) are less than in rats (Higuchi et al. 1986).

Here it should be noted that oxytocin has long been known for its ability to stimulate (accelerate) the process of childbirth, hence its name (Greek: oxys—fast, tokos—childbirth). Oxytocin is released from the nerve endings of the magnocellular neurons of the SON and PVN hypothalamus of a woman in response to stimulation mainly of the uterine mechanoreceptors during childbirth. Unique surveys were conducted to measure the dynamics of oxytocin concentration in the blood of a woman during childbirth (Fuchs et al. 1991). It turned out that the patterns of oxytocin peaks in the second and third stages of labor (Fig. 4.26c) had a great similarity with those of women during breastfeeding. This suggests that GPIA is a common neuronal system for both impulse activity coming from breast and uterine receptors.

Through the bloodstream, "oxytocin waves" reach the alveolar-ductal area of the breast. A necessary condition for the action of oxytocin on the structures of the alveolar-ductal department of the breast is the presence of OR. As already noted, the effects of oxytocin are provided by one type of OR. This type of receptor is encoded by only one type of gene, which has been cloned and sequenced in various mammals (Lollivier et al. 2006) and in humans (Kimura et al. 1992). In a woman's lactating breast, OR was found in the membrane of myoepithelial and secretory cells (Kimura et al. 1998). In addition, they were found in endothelial cells of large blood vessels and capillaries of the human myometrium (Thibonnier et al. 1999; Weston et al. 2003).

It is believed that myoepithelial cells are the main effector of the maternal reflex of milk excretion. Externally, the effect of oxytocin on myoepithelial cells is manifested in the shortening of their processes (Fig. 4.27), as a result of the reduction of myofilaments located in them. Data on the intracellular mechanisms that reduce the processes of myoepithelial cells of a woman's breast are currently unavailable (Raymond et al. 2011). However, detailed studies of the functioning of myoepithelial cells were conducted on the mammary glands of lactating mice and rats. However, there were also great methodological difficulties, the main of which was that myoepithelial cells consist mainly of thin processes with a diameter of 1–2 microns. The soma of cells has the largest diameter of 5–6 microns and the same length (Fig. 3.1). The way out of this situation was to develop a method for studying myoepithelial cells isolated from lactating mammary glands of mice and rats in tissue culture (Soloff et al. 1980; Nakano et al. 1997, 2001; Olins and Bremel 1982, 1984).

It should be noted that isolated myoepithelial cells in tissue culture have a different appearance than in the composition of the alveoli, because in the free state of the cells tend to take a spherical shape. In particular, myoepithelial cells significantly shorten the processes, and the diameter of the soma increases (Soloff et al. 1980). However, all intracellular structures, including myofibrils, are preserved (Soloff et al. 1980; Nakano et al. 1997). Reduced in length, myoepithelial cell processes retain the ability to contract under the action of oxytocin (Soloff et al. 1980). Reduction of myoepithelial cell processes occurs due to an increase in the

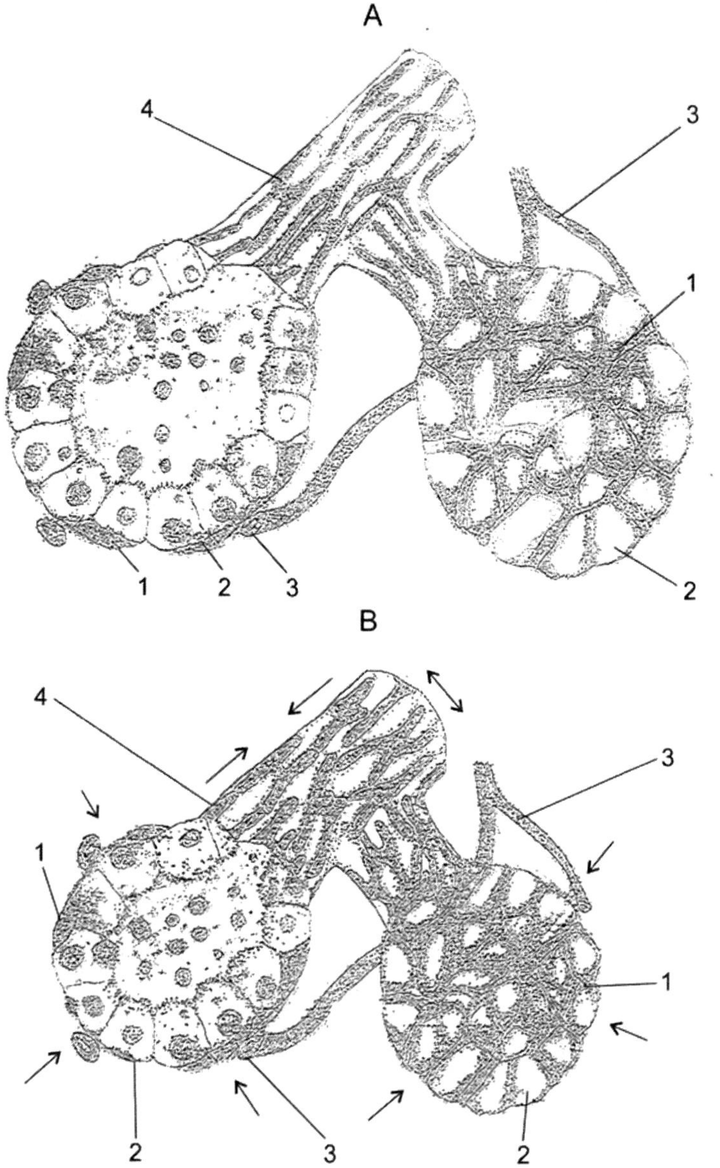

Fig. 4.27 Action of oxytocin on alveola and breast duct. (**a**) Schematic representation of the external (right) and internal structure (left) of the alveoli and ducts. (**b**) Too, but after the action of oxytocin. Arrows show compression of alveoli and ducts during the contraction of myoepithelial cells. An arrow with two heads shows an increase in the diameter of the duct with contraction of myoepithelial cells. For all figures: 1—myoepithelial cell of alveoli, 2—secretory cell of alveola, 3—blood capillary, 4—myoepithelial cell of duct

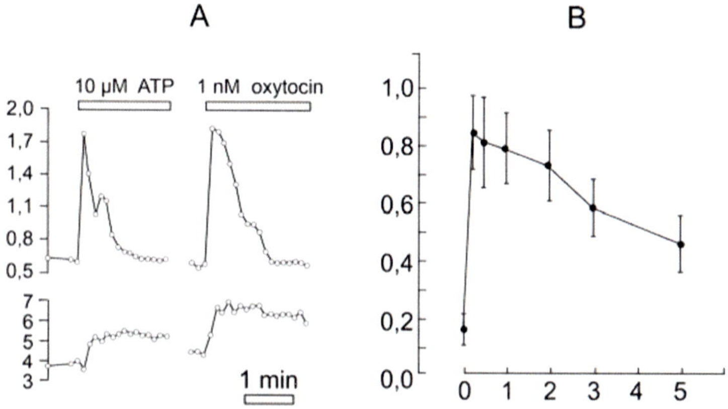

Fig. 4.28 Effects of oxytocin and ATP on breast myoepithelial cells. (**a**) Effects of oxytocin and ATP on intracellular calcium content (upper graphs) and contraction (lower graphs) of mouse breast myoepithelial cell (from Nakano et al. 2001). The intracellular calcium concentration is represented in relative units calculated as the radiation ratio caused Fura-2 the wavelength of 340 nm and 360 nm. During cell contraction, its thickness changed and evaluation was carried out at a wavelength of 360 nm. (**b**) Myosin light chain phosphorylation in rat breast myoepithelial cells from Olins and Bremel 1984). On the ordinate axis: the ratio of the number of phosphate molecules to the number of molecules of the light chain of the myosin, on the abscissa axis, time in minutes

intracellular concentration of calcium ions (Fig. 4.28a) as a result of a cascade of intracellular reactions caused by oxytocin. The dynamics of intracellular calcium concentration was determined photometrically using fura-2 fluorescent dye (Grynkiewicz et al. 1985). The cascade of intracellular reactions in myoepithelial cells caused by oxytocin is similar to the cascade of reactions in cells of uterine smooth muscles in response to oxytocin. In particular, oxytocin activates phospholipase C via OR G-linked proteins, resulting in the formation of a secondary intracellular mediator—Inositol triphosphate. In turn, Inositol triphosphate stimulates the release of calcium ions from the intracellular depot. Smooth muscle and non-muscle myosin contain a light chain with a molecular weight of 20,000, which can be phosphorylated by a special calcium-dependent kinase. An increase in the concentration of calcium inside the myoepithelial cell activates kinase, which causes rapid (within 30s) phosphorylation of the light chain and reduction of myofilaments in myoepithelial cells (Fig. 4.28b). Subsequent dephosphorylation of the light chain of myosin with a special phosphatase causes relaxation of myofilaments (Olins and Bremel 1982, 1984; Hartshorne et al. 1989). In microelectrode studies of electrical reactions of myoepithelial cells in tissue culture, it was found that oxytocin causes hyperpolarization of the cell membrane associated with increased intracellular concentration of calcium ions and the opening of Ca2 + – activated potassium channels (Nakano et al. 2001). These data are consistent with the above results of photometric studies (Nakano et al. 1997, 2001).

Epithelial (secretory) cells of the alveoli also have OR in the membrane. This suggests that secretory cells during the" oxytocin wave" will respond to the action of

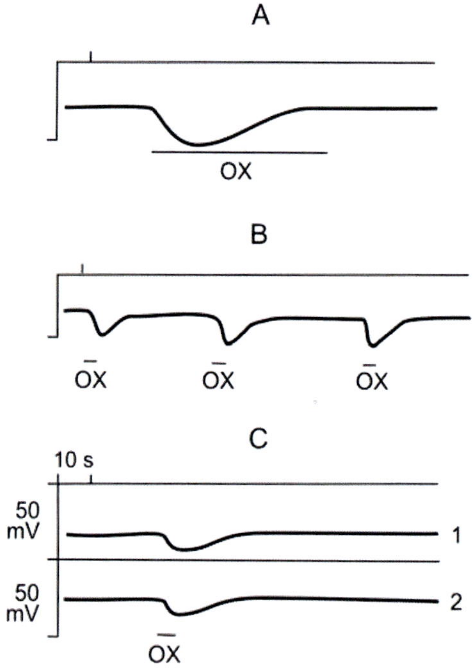

Fig. 4.29 Action of oxytocin on the membrane potential of secretory cells (**a**, **b**) and simultaneously on the membrane potential of secretory cell and transepithelial potential difference (**c**) of mouse mammary alveola (from Tolkunov Yu and Markov 2005). Abscissa axis, the time mark (10 s) on all electrograms, ordinate axis is the mark of the membrane potential (50 mV) of secretory cells (**a**, **b**, **c**1). On **c**2—change of transepithelial potential difference. For all electrograms, the lower line is the timestamp of the action of oxytocin

oxytocin. The size of secretory cells and their location in the alveoli make it possible to perform experiments in vivo. However, such experiments were conducted only on secretory cells of mammary glands of laboratory animals (Tolkunov Yu and Markov 2005). Observations of single alveoli under the action of oxytocin did not reveal any changes in the form of secretory cells that make up the alveolus (Grachev et al. 1976a). A possible cause was the masking effect of myoepithelial cells located in the form of a mesh layer over the secretory cells forming the alveolus (Fig. 4.27). In experiments on isolated secretory cells of the mammary glands of rats (Soloff et al. 1980) using a scanning electron microscope, it was found that isolated secretory cells, like myoepithelial cells, tend to take a spherical shape, but unlike myoepithelial cells, the surface of secretory cells is not smooth but contains numerous microvilli. Under the action of oxytocin, there was no change in the size of cells, as well as the number, size, and shape of microvilli. At the same time, in experiments with microelectrode recording of the membrane potential of secretory cells, it was shown that oxytocin, as in myoepithelial cells, causes hyperpolarization of the membrane of secretory cells (Fig. 4.29). Moreover, the ion mechanisms of hyperpolarization of secretory cells under the action of oxytocin are the same as in myoepithelial cells: an increase in intracellular calcium concentration and the opening of Ca2+- activated potassium channels (Furuya and Enomoto 1990; Tolkunov Yu and Markov 2005).

During the experiments, it was found that between the alveoli milk and the extracellular fluid, there is a potential difference, called transepithelial potential difference (TPD), which in relation to the extracellular environment as well as the inner part of the secretory cells is negative. Under the influence of oxytocin, there was a change in the polarity (hyperpolarization) and the amplitude of the TPD, which largely repeated the change (hyperpolarization) of the membrane potential of a separate secretory cell that is part of the alveolus (Berga 1984; Tolkunov Yu and Markov 2005) (Fig. 4.29). Given that epithelial (secretory) cells using dense contacts, in contrast to the mesh structure of myoepithelial cells create a closed internal volume of the alveolus isolated from the intercellular fluid, the change in TPD when oxytocin acts on the alveolus will be an integral response of the secretory cells that form the alveolus.

The action of oxytocin on secretory cells, in addition to the electrical reaction, is accompanied by changes in the localization of certain intracellular structures (Lollivier et al. 2006). In particular, on isolated fragments of the lactating mammary gland of a rabbit, it was found that secretory vesicles located in the cytoplasmic region between the nucleus and the apical membrane shifted when oxytocin entered the solution washing a fragment of glandular tissue and came into close contact with the apical membrane of the secretory cell. These data suggest that oxytocin during reflex peak promotes the release of milk components from secretory cells into the alveolar cavity (Lollivier et al. 2006). Moreover, the movement of secretory vesicles in the secretory cells of the mammary glands as well as in the axon terminals of oxytocinergic neurosecretory cells in the neurohypophysis occurs due to an increase in the intracellular concentration of calcium ions.

Another structure of the internal part of the breast that contains OR is the endothelial cells of blood vessels and capillaries. Histochemical experiments to determine OR in breast blood vessels of lactating women similar to those on endothelial cells of large blood vessels and capillaries of the woman's myometrium (Thibonnier et al. 1999; Weston et al. 2003) were not conducted. However, to date, there are physiological data (Janbu et al. 1985) that can be considered evidence of the presence of OR in the endothelium of the woman's breast vessels. Let us recall (Chap. 2) that when studying the average blood flow rate weekly in the main branch of the thoracic lateral artery of the breast of a woman from the 12th week of pregnancy (Thoresen and Wesche 1988), it was found that during lactation the blood flow rate in the glandular tissue was increased (Fig. 2.14). However, as already mentioned (Naccarato et al. 2003) microvascularization of the ducts and lobules differs. Thus, the ducts are surrounded by a large number of "typical capillaries." In lobules, the number of capillaries is less, but they are wider in size and have a sinuous shape. It is assumed that the shape and size of the capillaries in the lobules are of great functional significance. In particular, an increase in the diameter and "sinuosity" of capillaries will slow down the blood flow near the alveoli, contributing to a longer contact of hormones with the cells forming the alveola (Naccarato et al. 2003). It is well known that oxytocin has a vasodilating effect. Therefore, we can assume that when the reflex flow of the "oxytocin wave" into the capillaries of the lobule, the blood flow rate in it will further decrease. Indeed in the study of blood flow velocity in the distal branches of the lateral thoracic artery in lactating women during milk ejection reflex (Janbu et al. 1985) was found

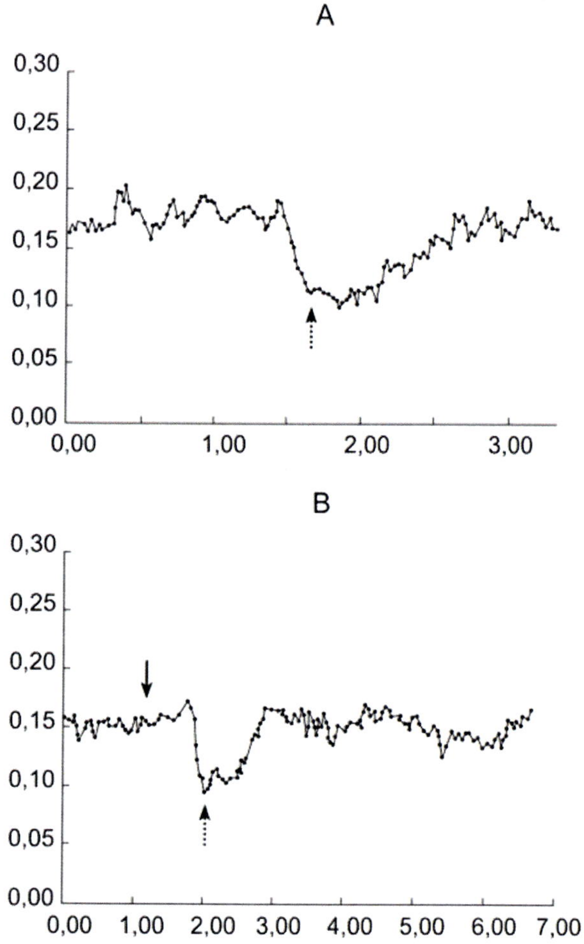

Fig. 4.30 Change in blood flow rate in lateral thoracic artery of lactating woman during milk excretion reflex (by Janbu et al. 1985). (**a**) Change in mean blood flow rate during infant milk withdrawal. The beginning of the baby's feeding coincided with the origin feeding. At the bottom, a dashed arrow indicates the moment when a woman began to feel mechanical changes in the gland caused by the reflex entry of oxytocin into the gland. (**b**) Change in mean blood flow rate when 25 mIE synthetic oxytocin is injected into a woman's blood. At the top of the arrow is the time of administration of oxytocin. At the bottom, a dashed arrow indicates the moment when a woman began to feel mechanical changes in the gland caused by the introduction of synthetic oxytocin. For all plots: ordinate axis—average blood flow rate m/s, abscissa axis—time in min

undulating for about 1–1.5 min reduction of flow velocity (Fig. 4.30a) (Janbu et al. 1985), i.e. time-coincident with the "oxytocin wave" recorded in the blood of a woman nursing infant (Nissen et al. 1996). Moreover, it is important to note that in both cases, women felt a characteristic mechanical reaction (filling of the gland, tingling) to the intravascular action of oxytocin.

Thus, the reduction of myoepithelial cell processes is crucial for milk excretion. Since the location of the processes of myoepithelial cells on the alveoli and milk ducts of the alveoli is different (Figs. 3.1 and 4.27), the reduction of the processes on the alveoli will compress the alveola mechanically affecting the secretory (epithelial) cells, reducing the volume of the alveoli and pushing the contents of the alveolar cavity into the ducts (Fig. 4.27b). In experiments on the mammary gland of a lactating mouse, it was possible to register a change in the size of a single areola in response to the action of oxytocin under conditions of vital microscopy, in which innervation and vascularization of the mammary gland were preserved (Grachev et al. 1976a). A photosensor was placed under a separate alveola. The alveola was illuminated by a narrow light beam. Under the action of oxytocin, the alveola shrank. In this case, the surface of the light sensor, which could be affected by light, was increased and, accordingly, the signal from the sensor was increased. The alveolar contraction was also impulsive, lasting 1–1.5 min in most cases.

The processes of myoepithelial cells on the ducts are located along the longitudinal axis of the ducts (Fig. 4.27b). Reducing the processes on the ducts will also mechanically affect the epithelial cells of the ducts, while on the one hand shortening the length of the ducts, and on the other hand increasing their cross-section, reducing the resistance to movement of milk. All this will accelerate the movement of milk in the ductal system toward its terminal part, located in the areolar region of the breast (Fig. 3.3).

In connection with the mechanical movement of the alveoli and ducts during the contraction of myoepithelial cell processes, data on the mechanosensitivity of secretory (epithelial) cells, which were obtained in experiments on the epithelial cells of the mammary glands of mice and rats, is of great interest. It was found that when mechanical tactile stimulation of rat epithelial cells in tissue culture with a glass capillary, the cells generated a hyperpolarization response (Enomoto et al. 1987). Detailed studies of mechano-activated hyperpolarization responses of epithelial cells have shown that they are generated by CA^{+2}-activated K+ channels with high conductivity (Furuya et al. 1989). In addition to changes in the membrane potential of secretory (epithelial) cells in response to mechanical stimulation, adenosine triphosphate (ATP) was released to the external environment (Enomoto et al. 1994). Special experiments on the culture of mammary gland tissues in mice found that ATP in micron concentrations, as well as oxytocin, impulsively increased the intracellular concentration of Ca^{+2}, due to which there was a reduction in the processes of myoepithelial cells (Fig. 4.28a). These data indicate that there is probably a positive feedback between the functioning of secretory and myoepithelial cells. In particular, given that myoepithelial cells on the alveoli and ducts are closely adjacent to secretory (epithelial) cells, the release of ATP from secretory cells when mechanically acting on them during the reduction of myoepithelial cell processes with a minimum diffusion time will further enhance the reduction of myoepithelial cells through purinergic receptors (P2U) (Nakano et al. 1997). As a result, the degree of emptying of the alveoli and the rate of milk excretion will increase. Perhaps this connection exists in the alveoli of the breast of women. If it is present, this is the physiological basis for the use of mechanical massage of the breast of a woman to more effectively remove milk. It should be noted here that in addition to the coordinated excretion of milk through the ducts in the system of myoepithelial cells, special cilia may be used (Stirling and Chandler 1976a, b, 1977) (Fig. 3.2). The structure of cilia of myoepithelial cells suggests that they have mechanical sensitivity. In addition, as noted (Stirling and Chandler 1976b), their location between the surfaces of myoepithelial and epithelial cells is ideal for the reception of mechanical stimulus. In this regard, the movement of the duct walls will cause stimulation of the cilia, followed by the transformation of the energy of the mechanical stimulus into electrical or chemical signals inside the myoepithelial cells. Accordingly, this can cause or increase the contraction of myoepithelial cells. As an assumption, it can also be assumed that in addition, the transmission of successive contractions of myoepithelial cells during the movement of milk in the ducts can occur due to their slotted contacts (Monaghan and Moss 1996). Unfortunately, there are no experimental data to confirm the assumptions made to date for the mammary glands of women, as well as for other mammals.

As a result of compression of the alveoli and the flow of milk into the ductal system, there is an increase in pressure (Fig. 4.31a). The pressure in the ductal system of the breast lobe is quite easy to measure using a catheter inserted through the milk channel of the nipple into the common duct of the lobe, located in the areolar area of the breast. However, this procedure is painful for a woman. It should be noted here that among mammals, intravascular pressure was first recorded in the mammary gland of a woman (Sica-Blanco et al. 1959), and then in the mammary glands of other mammals. Changes in the pressure in the mammary gland of a woman occur in waves. Moreover, the duration of the "pressure wave" coincides with the duration of the "oxytocin wave" and is within 1–1.5 minutes. The pressure amplitude can reach quite large values—30 mm Hg (Sandholm 1968; Cobo 1993). In this regard, when a woman feeds a child with one gland, as a result of increased pressure, milk is often released from the other gland in the form of drops or trickles.

A good illustration of this phenomenon is the picture of the famous Italian artist Jacopo Tintoretto (1570) "the Origin of the milky way." Tintoretto took the plot for his work from Greek mythology. Zeus wanted to make his son Hercules, born of an earth woman, immortal. To do this, he put his wife, the goddess Hera, into a deep sleep and put the baby to her breast so that it could drink the divine milk that grants immortality. The infant Hercules began to effectively stimulate and withdraw milk from one breast so that as a result of a reflex increase in pressure in the ductal system, milk began to be released from the other in strong jets. At the same time, the drops of milk that spilled into the sky turned into stars, which formed the Milky Way. The drops of milk that fell to the ground became snow-white lilies.

When the ducts are filled with milk, especially before feeding the child, a reflex increase in pressure as a result of stretching the walls of the ducts can cause stimulation of pain receptors located in the walls of the ducts (Fig. 3.12). It is known from the literature (Gebhart 2000; Alekseev 2009) that when the pressure reaches 25–30 mm Hg and higher, pain mechanoreceptors located in the walls of hollow organs begin to function. Perhaps for this reason, some women feel periodic "tingling" (the reaction of pain receptors to the stretching of the walls of the ducts) in the gland while feeding a child.

An increase in intramammary pressure will facilitate the removal of milk to the child, as well as increase the velocity of pumping milk with a breast pump device or hands. We conducted examinations on lactating women who had a catheter inserted into the duct of one gland to measure intravascular pressure, and milk was removed from the other gland by a device. At the same time, the amount of milk excretion was measured every 30 s using a graded milk collector (Alekseev et al. 1994, 1998). It

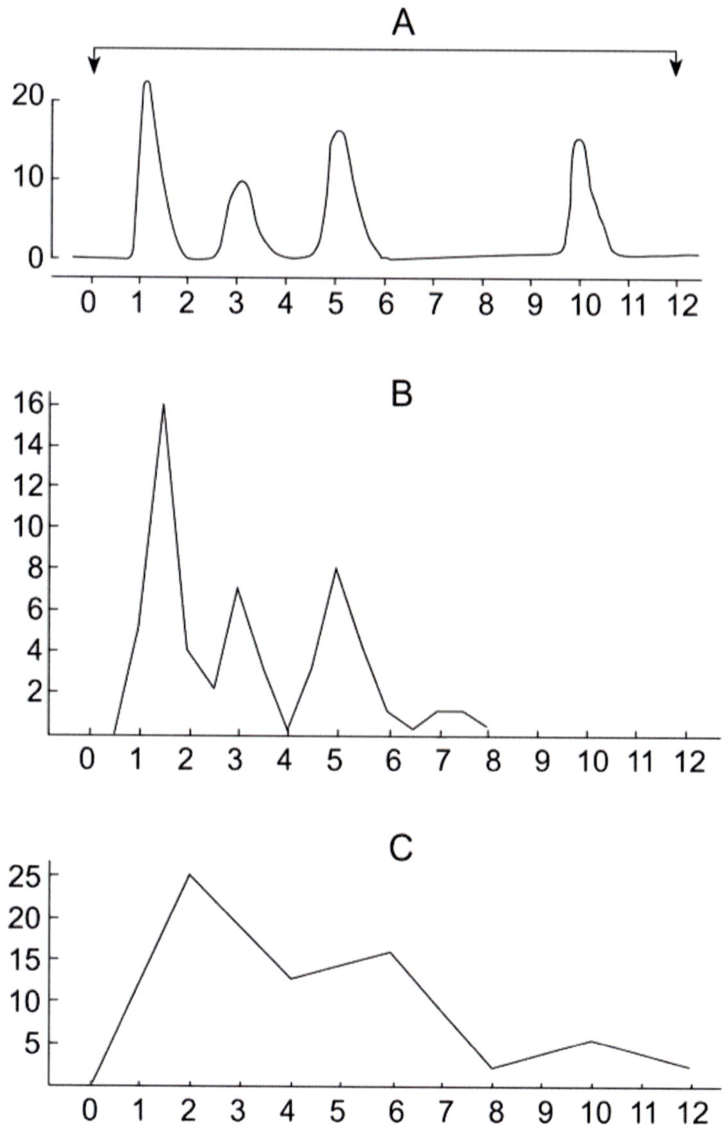

Fig. 4.31 Measurement of intramammary gland pressure (**a**), and milk excretion rate (**b**, **c**) during milk excretion by breast pump with tactile and vacuum component (from Alekseev et al. 1994). Ordinate: **a**—pressure in mm Hg, **b**—milk excretion rate in ml/0.5 min, **c**—ml/2 min. Abscissa for all graphs time min. The horizontal line on (**a**) indicates the time of milk excretion by the breast pump

turned out that the change in the rate of milk excretion was also undulating. The peaks of increasing the rate of milk excretion had a duration of 1.5–2 min and coincided with the time of appearance with the peaks of increasing intramammary

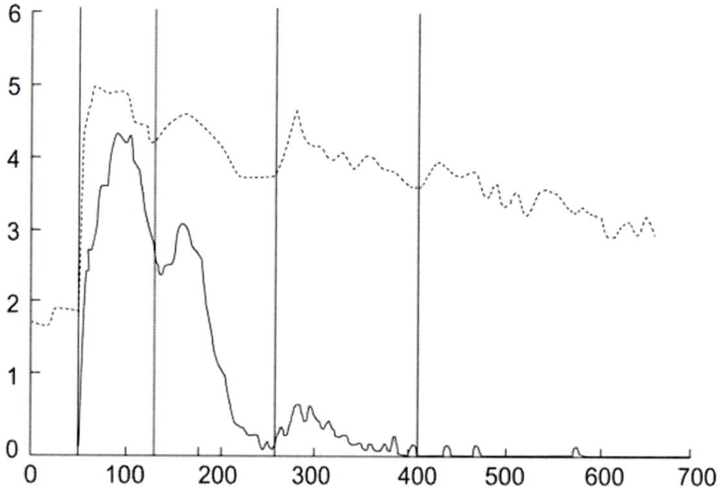

Fig. 4.32 Change in milk flow (continuous line) and milk duct diameter in the areola region (dashed line) when milk is taken out by breast pump (from Ramsay et al. 2006). Vertical lines indicate the time of milk flow peaks and milk duct diameter increase peaks. Ordinate: duct diameter in mm and the weight of excreted milk in g, abscissa: time s

pressure (Fig. 4.31). It is important to note that in the surveys we used a breast pump developed by us (Alekseev and Ilyin 2014, 2016), in which both vacuum and compression stimuli were applied to the gland for the removal of milk, as well as for the removal of milk by a child. Thus, the rate of change in milk excretion, along with the peaks of oxytocin and the peaks of intramammary pressure, can be used to study the reflexes of milk excretion in women when removing milk by a child or using a breast pump. This technique is noninvasive, does not require complex equipment, and is easy to use in comparison with methods for determining oxytocin in blood and pressure in glandular tissue. In our work, a milk collector with a division price of 1 ml was used to measure the rate of milk excretion, in the studies of other authors (Ramsay et al. 2006), a scale with a resolution of 0.1 g. As already noted, during reflex contraction of myoepithelial cells of the ducts, the ducts shorten and expand (Fig. 4.27b). Additionally, the expansion of the ducts will occur due to an increase in intramammary pressure. The change in the diameter of the ducts located in the areolar region of the gland during the reflex peak was detected by a noninvasive method using a high-resolution ultrasound device (Ramsay et al. 2004; Geddes 2009b). It turned out that the diameter of the duct in waves has changed during removal of milk by breast pump from the adjacent gland (Fig. 4.32), and the peaks of increase in the diameter of the ducts are coincident with peaks an increase of the rate of excretion of milk (Ramsay et al. 2006). However, the peaks of increasing the diameter of the ducts were less pronounced than the peaks of the rate of milk excretion. In addition, the authors noted that the ultrasound method is more time-consuming, since along with the availability of special equipment, it requires a highly qualified specialist to register changes in the milk ducts (Ramsay et al. 2006).

Extremely interesting data were obtained when studying the dynamics of the rate of milk excretion using various types of milk-removing pumps (for more information about the design of pumps and the features of their functioning, see Chap. 5, p. 5.2.4). It was found that in different women, the sequence of peaks of increasing the rate of milk excretion (lactogram) differs and is individual (Alekseev et al. 2000; Prime et al. 2011). Based on the analysis of 265 lactograms in 35 lactating women, we classified three types of lactograms (Alekseev et al. 2000) (Fig. 4.33). A similar classification of lactograms was presented by Prime and colleagues (2011) (Fig. 4.34). In lactograms of the first type (Fig. 4.33A), the rate of milk excretion increased periodically to the maximum, and then decreased, but not to zero, but to a certain level, and the number of peaks for 10 min was the maximum possible: 6–8. In the second type of lactograms (Fig. 4.34B), in most cases, the rate of milk excretion was reduced to almost zero, and the number of peaks for 10 min was 4–5. The third type of lactograms had a minimum number of peaks: 2–3 in 10 min, the amplitude of which also decreased to zero (Fig. 4.33C). Accordingly, the types of lactogram patients were divided into three groups. The second group was the largest with 24 patients (68.8%) (Fig. 4.33B), the first group included 4 patients (11.4%) (Fig. 4.33A), and the third group included 7 patients (20%) (Fig. 4.33B). It should be noted here that the surveys included women with different clinical characteristics. It was found that regardless of the characteristics of the course of pregnancy and childbirth, the parity of pregnancy and childbirth, the timing of applying newborns to the breast and evaluating newborns on the Apgar scale, the largest number of maternity hospitals was in group 2 and less in groups 1 and 3. Statistical analysis of lactograms of the same woman registered at different times of the day and day of lactation revealed a great similarity in the time distribution of reflex peaks. It turned out that for each patient there are individual type lactogram. Figure 4.33 shows the distribution of peaks in the rate of milk excretion after statistical analysis of lactograms of the right and left mammary glands under each type of lactogram, given as a single example for this patient. Vertical lines of the same height show the time of the maximum value of the milk withdrawal rate from the beginning of pumping, and horizontal lines indicate the standard deviation of the peak time. For all patients in their lactograms, the difference between the time of reaching the maximum value for each neighboring peak according to the student's criterion is significant ($p < 0.05$). In our surveys, lactograms in the same patient were registered within 3 days. Prime D. and colleagues (Prime et al. 2011; Gardner et al. 2015, 2017) were able to register lactograms in the same patients at intervals between pumping sessions of several weeks, months, and years (5 years). It turned out that the dynamics of reflex peaks in the rate of milk excretion in the same woman remained almost unchanged (Fig. 4.34B). In addition to these examinations, ultrasound recording of changes in the diameter of the ducts in the areolar part of the breast was performed when milk was withdrawn from a neighboring gland by a child and a breast pump (Gardner et al. 2015). A comparison of reflex peaks of increasing the diameter of the ducts during milk excretion by the child and the breast pumps revealed similarities in their time sequence and duration. These data, as well as the results of recording the amplitude–time parameters of reflex peaks of milk excretion

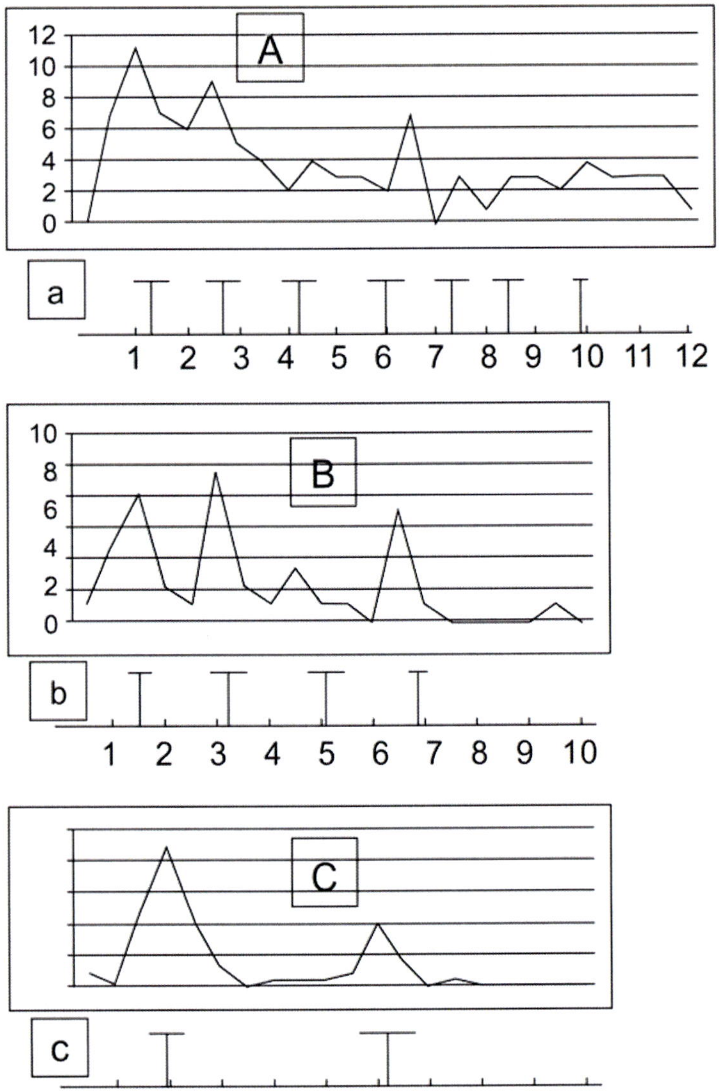

Fig. 4.33 Milk breeding rate in various breastfeeding women (from Alekseev et al. 2000). **A**, **B**, **C**—results of single experiments of three various patients. Abscissa: time—min, ordinate: the rate of milk excretion by the brast pump—ml/30s. a, b, c—vertical lines show the time of the maximum value of reflex peaks for increasing the rate of milk withdrawal for the same patients (according to 9 experiments for a and 7 experiments for b, c) horizontal lines at the top of vertical—mean square deviation

rate, suggest that each lactating woman has an internal genetically determined mechanism that provides a special sequence of reflex peaks of milk excretion for her. That is, formed in response to the impulse activity of afferent (sensory) receptors

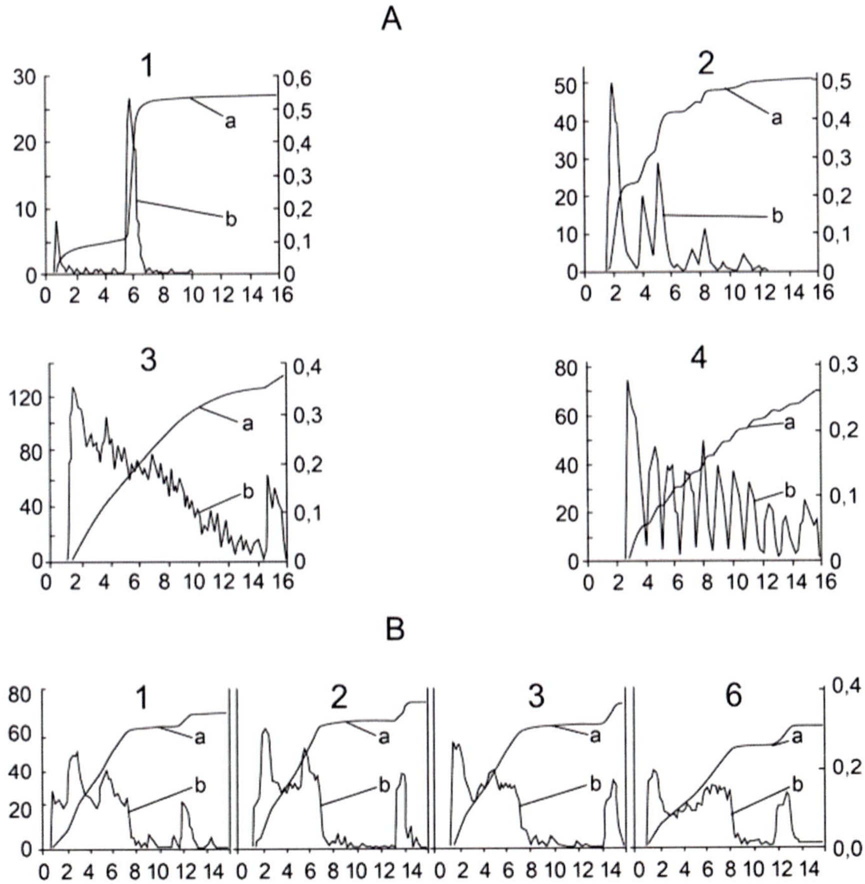

Fig. 4.34 Rate of milk excretion in different women (**a**) during one session of milk excretion and the same woman in different sessions (from Prime et al. 2011). (**A**) Dynamics of milk excretion of four different women per session. (**B**) Dynamics of milk excretion in the same woman (1, 2, 3) at weekly intervals and (4) at a 6-month interval. On all plots the ordinate axis: the total weight of milk taken out (1 on the right) in g and the weight of milk in g/s (4, on the left). On the abcissa axis: time, min

of the nipple and areola of the breast, a series of action potentials of a certain duration and a certain sequence characteristic of each woman from the GPIA go to the neural secretory cells of the SON and PVN. After synaptic switching, the action potentials that arise in the neural secretory oxytocinergic neurons of SON and PVN, also grouped in a series of impulses, rush into the neurohypophysis and cause the axon terminals to produce a wave-like release of oxytocin into the bloodstream and, accordingly, a reflex contraction of the alveoli, increasing the pressure and velocity of milk excretion. At the same time, the characteristic feature of the sequence of reflex peaks at all stages of each woman is preserved.

4.2.4 Maternal Reflex of Milk Secretion

Milk removed from the alveolar-ductal system of the breast in a nursing woman with a child or by any other method (milk can be expressed manually or using a breast pump), refills the alveoli due to another neuroendocrine reflex—the maternal reflex of milk secretion. The experimental data accumulated to date indicate that the maternal reflex of milk secretion has similarities and differences in morphofunctional characteristics with the reflex of milk excretion. Let us focus briefly in the introduction to the section on the main links of the milk secretion reflex. For Fig. 4.35 the general scheme of the maternal reflex of milk secretion is presented. The initial afferent link for the secretion reflex and the milk excretion reflex is common-sensory receptors of the areolar-nipple area of the breast. Impulse activity that occurs in response to adequate stimulation in the afferent nerve fibers that form these receptors enters the somatosensory cortex, where a woman has a sense of mechanical action on the areolar-nipple area of the gland. However, the main central link of the maternal reflex of milk secretion as well as the reflex of milk excretion is the hypothalamus. Initially, it was suggested by Jofri Harris (1948) that special cells of the hypothalamus, in response to adequate stimuli, secrete a "special factor" into the bloodstream that stimulates the release of the hormone prolactin and is called prolactin-releasing hormone. But despite the fact that the "search" for prolactin-releasing hormone has been conducted since the time when it was assumed to exist (Harris 1948), there is still no convincing evidence of its existence (Freeman et al. 2000; Ben-Jonathan and Hnasko 2001; Crowley 2015; Grattan 2015).

At the same time, prolactin—an inhibitory factor—dopamine (DA), which is contained in special cells of the hypothalamus, was discovered and studied in detail. These cells are mainly concentrated in the arcuate and periventricular nuclei of the

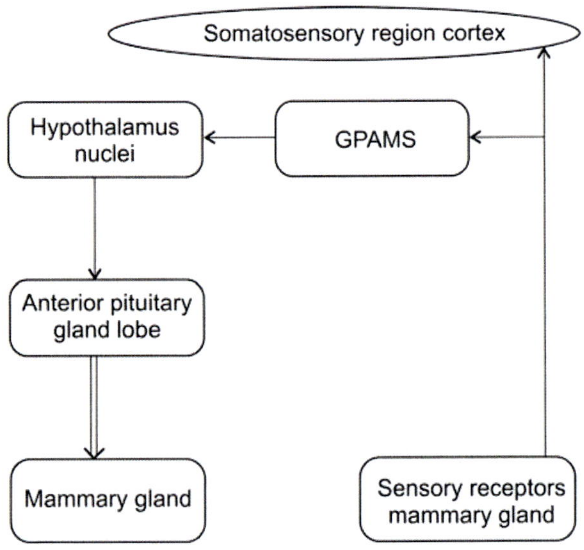

Fig. 4.35 Scheme of maternal reflex of milk secretion. *GISM* generator of pulse activity of milk secretion

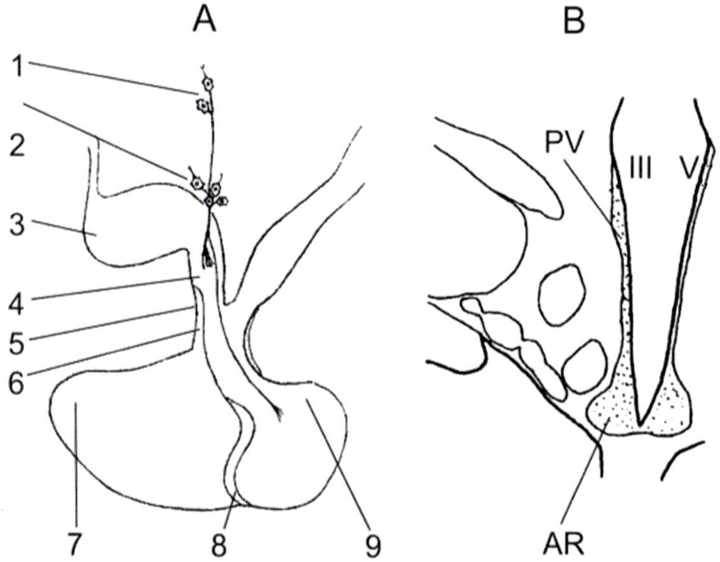

Fig. 4.36 Human pituitary structures involved in the formation of milk secretion reflex. (**a**) Schematic representation of the human pituitary gland showing the course of neuronal axons from arcuate and periventricular nuclei. 1—neurons of the periventricular nucleus, 2—neurons of the arcuate nucleus, 3—optic chiasm, 4—median eminence, 5—pituitary stalk, 6—tuberal part of the pituitary gland, 7—anterior part of the pituitary gland, 8—intermediate part of the pituitary gland, 9—posterior part of the pituitary gland. (**b**) Histological section pattern through human hypothalamus. IIIV—the third ventricle of the brain, PV—the periventricular nucleus, AR—the arcuate nucleus (from Spencer et al. 1985)

hypothalamus (Fig. 4.30b). Dopaminergic neurons similarly to oxytocinergic neurons have spontaneous activity. Efferent nerve fibers of dopaminergic neurons pass into the median eminence of the pituitary gland (Fig. 4.36a) and secrete dopamine from the terminals into the portal bloodstream, which is then transferred through the blood vessels to the cells located in the anterior lobe of the pituitary (adenohypophysis)—lactotrophs. Lactotrophs secrete prolactin continuously into the systemic bloodstream through exocytosis. Dopamine through dopamine receptors (D2) localized in lactotrophs blocks the release of prolactin. Afferent impulse activity from breast receptors in addition to oxytocinergic cells goes to dopaminergic cells, but in contrast to oxytocinergic neurons, it is believed that it does not increase the frequency of dopaminergic neurons, but reduces. In this regard, during feeding, the impulse activity that occurs in the afferent nerve fibers of the areola and nipple receptors in response to SSE, apparently will reduce the secretion of dopamine and, accordingly, increase the secretion of prolactin in lactotrophs. The release of prolactin into the bloodstream in a woman while feeding a child is wavy (McNeilly et al. 1983). Moreover, just as in the case of reflex release of oxytocin into the bloodstream, the time characteristics do not coincide with the time patterns of impulse activity from the areolar-nipple region receptors. Therefore, by analogy with the milk excretion reflex, it can be assumed that the impulse activity from the areolar–nipple complex receptors passes not "directly" to the dopaminergic cells, but through a special neuronal network localized in the hypothalamus and surrounding brain structures. This neuronal formation selects afferent impulse activity coming from breast receptors and forms periodic "bursts" of action potentials that provide, respectively, periodic inhibition of dopamine release into the bloodstream. In the diagram, this structure is designated as a generator of periodic pulse activity of milk secretion (GPAMS) (Fig. 4.35).

With the blood flow from the adenohypophysis, prolactin reaches the alveolar-ductal section of the breast. Effectors of the maternal reflex of milk secretion are secretory cells of the breast alveoli, in which prolactin provides the formation of milk. Large experimental material indicates that prolactin is the main hormone for milk secretion in women, as well as for other mammals (Freeman et al. 2000; Grattan 2015; Lacasse et al. 2016). However, in early studies (Cowie 1974), it was noted that the degree of significance of prolactin for milk secretion in mammalian animals differs. Prolactin is crucial for nonruminant animals during the entire lactation period. In ruminants, prolactin is needed at the beginning of lactation, and in established lactation in ruminants, prolactin has little or no effect on milk secretion (Cowie 1974). However, it was later experimentally proved that prolactin is necessary for effective milk secretion in ruminants during the entire lactation period (Lacasse et al. 2016).

It should be noted that prolactin is also known as oxytocin, a multifunctional hormone (Ben-Jonathan and Hnasko 2001; Freeman et al. 2000; Grattan 2015); however, it received its name in connection with its lactation function. Its name is made up of the initial parts of the English words Promote LACTation, i.e., promotes lactation. The gene encoding prolactin was found in all vertebrates. In humans, it is localized on chromosome 6 (Owerbach et al. 1981). It should be noted that human prolactin (hPRL) circulating in the blood is heterogeneous in nature and exists in several forms (Fig. 4.37B). The main active forms of hPRL contained in the human adenohypophysis are monomeric variants (Fig. 4.37A). Monomeric hPRL consists of a single chain of 199–200 amino acids and has a molecular weight of 23 kD. The chain contains 4 spiral sections collected as a bundle of spirals (Teilum et al. 2005; Brooks 2012) A variant of monomeric hPRL is glycosylated hPRL with a molecular weight of 25 kD. In addition to monomer forms, there is a dimeric (big) form with a mass of 50–60 kD, which is less than 20% of the total amount of hPRL in the blood serum. There is also a polymer (big, big) form of hPRL with a molecular weight of more than 100 kD. Its content is less than 5% in the blood serum. In addition, the so-called fragmented prolactin with a molecular weight of 16 or less kD fragments is observed in the blood serum (Smith and Norman 1990).

In contrast to the reflex of milk excretion, the reflex of milk secretion in a woman is carried out only when the sensory receptors of the nipple and the areola of the breast are stimulated by the child's SSE, manually or by a breast pump, i.e., due to unconditional stimuli. Clinic observations of breastfeeding women found that they did not show a change in the concentration of prolactin in the blood on the appearance of the child or on its cry, as was found for oxytocin (Noel et al. 1974;

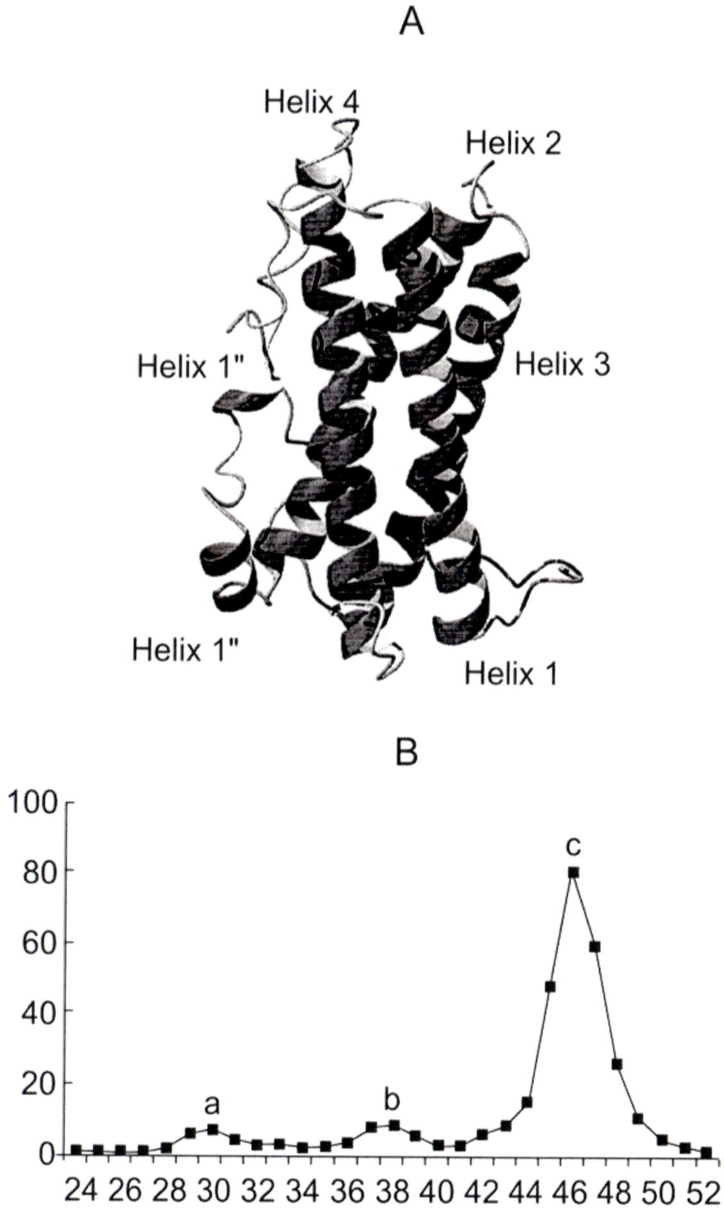

Fig. 4.37 Structure of human monomeric prolactin (**A**) (from Teilum et al. 2005) and distribution of human prolactin fractions in blood serum (from Beltran et al. 2008). A. Human monomeric prolactin consisting of a sequence of 199–200 amino acids. The chain contains 4 helical sections (Helix 1,2,3,4), assembled in the form of a bundle of spirals. (**B**) Chromogram illustrating the distribution of immunoreactive forms of human prolactin in blood serum. a, b, c are respectively big, big-form, big- form, monomer form. Ordinate axis the concentration of prolactin in mMIM/l, abscissa axis the fraction number

McNeilly et al. 1983). Note that the mechanisms of the milk secretion reflex in comparison with the mechanisms of the milk excretion reflex have been studied to a lesser extent, both for women and other mammals.

The formation of the milk secretion reflex begins with adequate stimulation of the sensory receptors of the areolar-nipple area of the breast by the child's oral apparatus. As already noted, the sensory receptors are the same for the reflex of milk excretion and for the reflex of milk secretion in a woman. Therefore, this information will not be provided in this section. Recall that the sensory receptors of the areolar-nipple area of the breast during lactation perform a double function: (1) perform the reception of external stimulus and (2) participate in the formation and maintenance of lactation. In this regard, there are two ways to transmit sensory information. Along the first path, information is quickly transmitted to the brain about the amplitude, gradient, duration, and location of the impact on the mammary gland of extrasensory stimuli. The final link of the first pathway is the somatosensory region of the cortex, where a woman is believed to have a sense of the impact of various extrasensory stimuli, including the child's oral stimulation on the areolar-nipple area of the breast when feeding it. For the second pathway, the final structures are the neuroendocrine cells of the hypothalamus.

Afferent pathways of the milk secretion reflex as well as the woman milk excretion reflex remain unexplored. The question of afferent pathways of milk excretion and secretion reflexes was studied in laboratory animals, and to a lesser extent in farm animals (Grachev and Alekseev 1980; Tsingotjidou and Papadopoulos 2008; Crowley 2015). In morphophysiological studies, it was found that the afferent pathway of milk excretion and secretion reflexes to the cells of the hypothalamus differs from the lemniscal system of conducting pathways to the somatosensory cortex and is carried out along the spinocervical pathway. Nerve fibers pass ipsilateral to the lateral cervical nucleus. The nerve cells of this nucleus are activated when milk is excreted by the pups. Further, part of the nerve fibers makes a cross, passes through the ventrolateral part of the medulla oblongata, and reaches the neuronal network—GPIA and GPAMS, localized in the lateral covering of the midbrain (Crowley 2015). It is believed that at the level of the medulla oblongata, there is a divergence of nerve pathways that control the release of oxytocin and prolactin. However, detailed information about the course of neural pathways to the neural secretory cells of the hypothalamus is currently not available even for laboratory animals (Crowley 2015). Undoubtedly, because of species differences, women's milk secretion reflex pathways have their own characteristics. However, it can be assumed that in general they coincide with the pathways of other mammals, and GPAMS as well as the female GPIA is located in the lateral covering of the midbrain.

According to available experimental data, the initial part of the efferent pathway of the milk secretion reflex in women, as well as in other mammals, is the dopaminergic nerve cells (Spencer et al. 1985). These cells are localized in the arcuate nucleus (ArN) of the hypothalamus, which is often also called the infundibular nucleus, indicating its location in the infundibulum wall (funnel) of the pituitary (Fig. 4.36). The core surrounds the side and back of the funnel in the form of a horseshoe. Above, the core has a continuation in the form of a narrow periventricular (PrN) core. The morphometric characteristics of ArN and PvN, as well as its neuronal composition in women, are less studied than the hypothalamus SON and PVN (Spencer et al. 1985; Braak and Braak 1992; Swaab et al. 1993; Dudas et al. 2010). The size of the soma of dopaminergic neurons ArN and PvN is within 15–20 microns. Uni- or bipolar neurons are closely adjacent to each other, with the long axes of the soma neurons located mainly along the walls of the funnel. Along the course of the nerve fibers that depart from the soma of neurons, extensions are observed (Bergland and Torack 1969a, b; Seyama et al. 1980a; Dudas et al. 2010), which contain secretory granules with a diameter of 50–100 nm and microvesicles with a diameter of 30–50 nm. Part of the nerve fibers of dopaminergic neurons remain in the ArN and PvN, and form thin branches at the end sites. Another part of the nerve fibers passes into the median eminence (Fig. 4.36a) of the pituitary gland and is also divided into thin processes with extensions along the course of the processes (Braak and Braak 1992). In addition, there is evidence that the nerve fibers of the ArN and PvN neurons also pass into the neurohypophysis (Fig. 4.36a) (Seyama et al. 1980a; Braak and Braak 1992). These fibers were identified as dopaminergic B-fibers in contrast to oxytocinergic A-fibers (Seyama et al. 1980a). However, they occur in the neurohypophysis much less often than in the median eminence.

Numerous studies have shown that the release of dopamine (DA) from nerve endings in various structures of the peripheral and central nervous system is associated with the electrical activity of the neurons. Therefore, it is of great interest to register the electrical activity of DA neurons ArN and PvN in women. However, just as in the case of oxytocinergic neurons PVN and SON, for methodological and ethical reasons, these examinations have not been carried out in women to date. Such experiments were carried out on DA neurons of ArN and PvN rats (Lyons et al. 2010, 2012; Lyons and Broberger 2014; Romanò et al. 2013, Le Tissier et al. 2016). However, in all cases, the experiments were performed not on the whole animal, but on sections of the brain of males, virgin and lactating females. Spontaneous activity from DA neurons was registered, which was divided into three types according to the temporary organization (Fig. 4.38A): DA neurons with regular impulse activity (Fig. 4.38Aa), with impulses in the form of individual bursts (Fig. 4.38Ab) and with irregular impulse activity (Fig. 4.38Ac). It was noted that in lactating rats there was a tendency to decrease the percentage of neurons with impulses in the form of individual bursts (Fig. 4.38B) (Romanò et al. 2013). Simultaneously with the pulse activity of DA neurons, using the amperometric method, it was possible to register a change in the concentration of DA near the neurons generating the pulse activity. A clear relationship was found between the change in the frequency of action potentials DA neurons and the concentration of DA (Fig. 4.38B) (Romanò et al. 2013). It is of great interest how the impulse activity of DA neurons changes with adequate stimulation of breast receptors. However, no such experiments have been conducted so far.

Thus, as a result of the generation of pulse activity of DA neurons from the nerve endings of their axons through exocytosis, DA exits into the intercellular space and

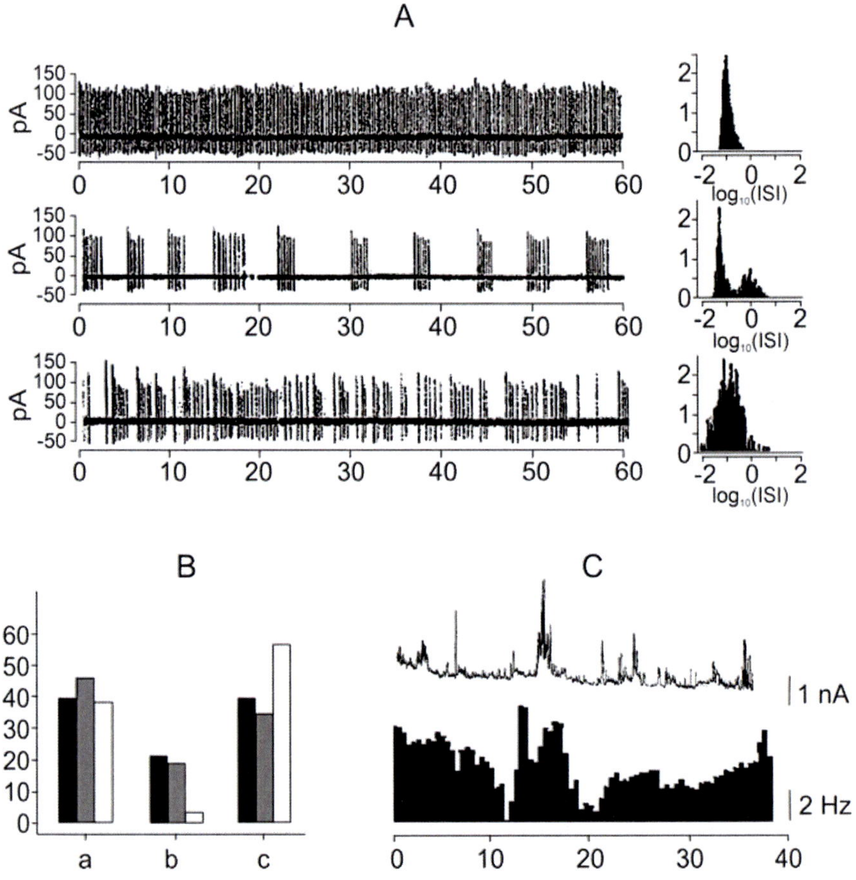

Fig. 4.38 Spontaneous electrical activity of rat hypothalamus dopaminergic neurons and associated change in dopamine concentration in extracellular media (from Romanò et al. 2013). (**A**) Types of spontaneous activity of tuberoinfundibular neurons. The upper oscillogram is regular activity, the middle is explosive activity, the lower is irregular impulse activity. For all oscillograms: abscissa axis—time s, ordinate axis—current pA. On the right is the inter-pulse interval (ISI) distribution in $\log_{10}$ (ISI). (**B**) Percentage of cells showing a different type of impulse activity: a—regular activity, b—explosive, b—irregular. For all histograms of c ordinate, the number of cells in percent. Dark columns are neurons of male rats, gray columns are non-lactating females, light-lactating females. (**C**) Combined recording of change in DA secretion (upper oscillogram) and histogram of electrical activity of tuberoinfundibular neuron (lower oscillogram) at 30 s bin in nonactive rats. Current calibration for upper oscillogram—1 nA. Calibration of pulse activity frequency for lower histogram—2 Hz. For all oscillograms at the bottom time in min

then enters the capillary network of the median eminence pituitary gland, where the nervous efferent link of the milk secretion reflex ends.

The blood supply to the human pituitary gland is fairly well studied at the light-microscopic level. First of all, we should note the work of McConnell (1953) and Xuereb and coworkers (1954a, b). Morphological studies of pituitary vascularization

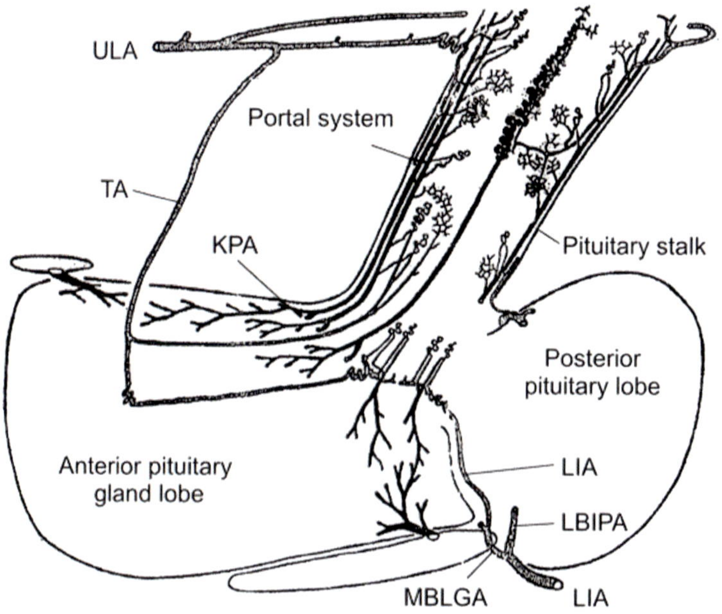

Fig. 4.39 Schematic diagram of the main arterial blood vessels of the human pituitary gland. (from Xuereb et al. 1954a). ULA—upper left pituitary artery, TA—trabecular artery, KPA—capillary plexus of adenohypophysis, LIA—lower infundibular artery, IPA—inferior pituitary artery, LBIPA—lateral branch of inferior pituitary artery, MBLGA—medial branch lower hypophyseal artery

and neighboring areas of the hypothalamus have shown that arterial blood supply to the pituitary is carried out by two groups of vessels: the right and left upper and lower pituitary arteries, which are branches of internal carotid arteries (McConnell 1953; Xuereb et al. 1954a, b) (Fig. 4.39). Each upper left and right pituitary artery is divided into two main branches that run to the front and back sides of the pituitary stalk. In some cases, two upper pituitary arteries immediately branch off from the carotid arteries: anterior and posterior.

Here it should be pointed out that the vessels surrounding the pituitary gland from the outside and passing inside have a very complex branching (Xuereb et al. 1954a, b) and it is not possible to draw it on the same scale. Therefore, the vascular supply of the pituitary gland is represented by a scheme taken from the works of Xuereb G.P. with coworkers (1954a, b). Branches of the pituitary upper arteries form numerous anastomoses in various parts of the upper part of the pituitary stalk, repeatedly branching forming a capillary network in the outer and inner part of the median eminence and the upper part of the pituitary stalk. It is noted that when branching, capillaries form "amazing shapes." In addition to the regular network of capillaries in the walls of the median eminence and the upper part of the pituitary stalk, there are scattered loops of capillaries of various degrees of complexity in different directions, which are twisted into straight branches of various lengths

(Xuereb et al. 1954a, b). This is partially shown in Fig. 4.39. The median eminence capillaries belong to sinusoid capillaries. The peculiarity of this type of capillaries is the presence of intercellular and transcellular holes in their endothelial cells with a diameter of 0.5–3 mcm and fenestra with a diameter of 60–80 nm, which are usually formed in the form of sieve-like plates. The basal membrane of this type of capillaries is almost completely absent. All this facilitates the transfer of various substances, including DA from the intercellular fluid to the blood and vice versa.

From the left and right branches of the anterior upper arteries branch additional branches called the trabecular arteries (Xuereb et al. 1954a). In most cases, they are held down at a noticeable distance from the stalk of the pituitary gland and enter the anterior pituitary. Less often, the trabecular arteries can pass in the immediate vicinity of the pituitary stalk (Fig. 4.39). The trabecular arteries enter the adenohypophysis and then travel to the lower part of the pituitary stalk in the form of parallel branches, while they do not give branches to the epithelial tissue through which they pass. There are numerous anastomoses between the branches of the trabecular arteries and the branches of the upper pituitary arteries passing along the pituitary stalk. In addition, branches of the trabecular arteries give anastomoses with branches of the lower pituitary arteries.

The lower pituitary arteries branch off from the carotid arteries, two on each side. The arteries have a sinuous shape. In the border area between the neurohypophysis and the adenohypophysis, they are divided into two main branches: medial and lateral (Fig. 4.39). The contralateral medial arteries form anastomoses with each other. The lateral branches run along the border between the adeno and the neurohypophysis and then turn in the medial direction at the base of the pituitary stalk and connect with the opposite lateral branches. Thus, the neurohypophysis is surrounded by an arterial ring of branches of the lateral and medial lower pituitary arteries. In the system of the lower pituitary arteries, another main artery was found, which was called the artery of the "lower part of the infundibular process" (Xuereb et al. 1954a) (Fig. 4.39). Usually, this artery originates from the medial branch of the right or left lower pituitary artery, but sometimes it originates from the lateral branches. In most cases, the artery of the lower part of the infundibular process forms anastomoses with the trabecular artery (Fig. 4.39).

As shown by experiments with the filling of pituitary vessels (Xuereb et al. 1954a), the epithelial tissue of the adenohypophysis is not directly supplied with arterial blood. The adenohypophysis has a capillary plexus, which communicates with the capillary plexus of the median eminence through a special portal system of vessels (Fig. 4.39). In the area where the leaf-like part (pars tuberalis) of the pituitary passes into the adenohypophysis, portal vessels branch out and contact the capillary system of the adenohypophysis. Experimental studies have concluded that blood in the portal system can flow in both directions (Xuereb et al. 1954b). Just as in the median eminence, the capillaries of the adenohypophysis are of the sinusoid type. The outflow of blood from the sinusoid capillary plexus of the adenohypophysis is carried out through thin venous vessels of various diameters to the venous sinuses surrounding the pituitary gland, from where blood flows to larger diameter veins (Xuereb et al. 1954a).

Through the portal blood flow to the endocrine cells of the adenohypophysis—lactotrophs that produce prolactin, biologically active substances, including DA, are received, which modulate the release of prolactin into the systemic bloodstream.

Immunohistochemical studies (Lloyd et al. 1988). It was shown that lactotrophs make up approximately 32–55% of the cellular composition of the woman adenohypophysis. The highest density of lactotrophs in the adenohypophysis is observed in its posterolateral sections (Horvath and Kovacs 1988; Elster 1993). In electron microscopic studies, it was found that the type of lactotrophs on sections varies so that their identification is difficult even for a specialist (Horvath and Kovacs 1988). Under normal conditions, most lactotrophs are small elongated cells. Sometimes they have a multisided form. The nuclei of cells often have protrusions. The membranes of the rough endoplasmic reticulum and Golgi apparatus are well developed. Rare spherical secretory granules have a diameter of 150–250 nm. Extrusion of granules with prolactin is observed in various sections of lactotrophs (Horvath and Kovacs 1988).

As has been demonstrated in numerous studies on various cellular structures and in particular the nerve endings of oxytocinergic neurons, the process of hormone exocytosis is associated with an increase in the concentration of calcium ions inside the terminals. In turn, the intake of calcium ions into the cell is due to an increase in the permeability of the cell membrane to calcium ions and changes in its membrane potential. In this regard, it is of great interest to register the electrical activity of lactotrophs during prolactin extrusion. However, just as in the case of oxytocinergic neurons, it has not yet been possible to do this on women's lactotrophs for methodological and ethical reasons. However, such experiments were carried out on lactotrophs of the rat adenohypophysis. Research was started in the 1990s and is currently conducted mainly on lactotrophs isolated from the adenohypophysis in the method of tissue culture (Stojilkovic 2006). Microelectrode studies have shown that the membrane potential of lactotrophs is not stable, but is characterized by oscillations from the level of the membrane potential of about −60 mV (Fig. 4.40a). It is interesting to note that oscillations of the membrane potential are also observed in other endocrine cells of the adenohypophysis (Fig. 4.40b). When depolarization oscillations reach a certain (threshold) value of the membrane potential, action potentials are generated, resembling action potentials that occur in nerve cells. However, the amplitude of the action potentials of lactotrophs does not reach the values of the reversion potentials. In addition, the duration of action potentials of lactotrophs is longer than the duration of action potentials of nerve cells and fibers. Oscillations of the membrane potential of lactotrophs in culture often have the form of a plateau on which action potentials are generated (Fig. 4.40a). These data are consistent with the results of electrophysiological studies conducted on sections of the mouse adenohypophysis, i.e., in situ (Bonnefont and Mollard 2003), and suggest that lactotrophs in tissue culture mainly preserve the properties of lactotrophs located in the intact animal's adenohypophysis. Registration of spontaneous electrical activity of lactotrophs, intracellular Ca^{+2} content, and basal extracellular level of prolactin (Fig. 4.40d) indicate a close relationship between the electrical activity generated in lactotrophs caused by the entry of Ca^{+2} into cells through special channels and the

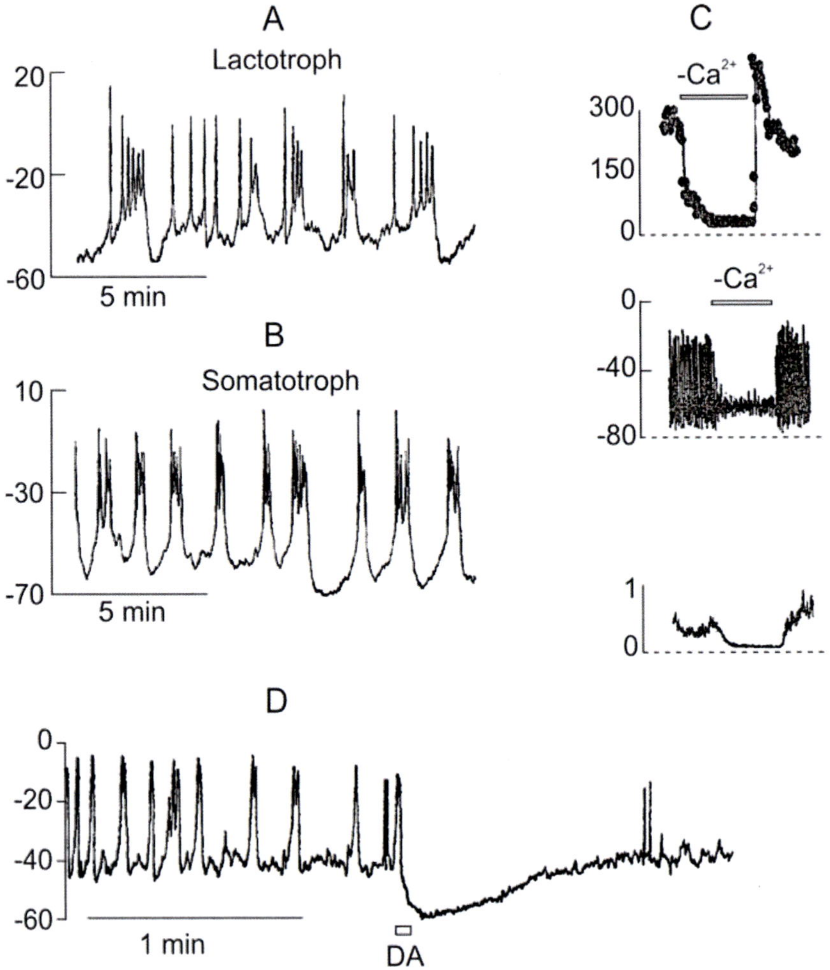

Fig. 4.40 Spontaneous electrical activity of lactotrophs (**A**) and somatotrophs (**B**) (from Stojilkovic 2006), its change when calcium ions are removed from the external environment of the washing lactotroph (**C**) (from Van Goor et al. 2001) and the effect of DA on the electrical activity of lactotrophs (**D**) (from Oxford and Tse 1993). (**A**, **B**) Ordinate: the membrane potential of lactotrophs and somatotrophs in mV, horizontally the time mark −5 min; (**c**) Effect of calcium ion removal (shown by a horizontal line) on the basal concentration of prolactin (on the ordinate—ml) of the upper panel, on the membrane potential (on the ordinate—mV) the middle panel and the intracellular concentration of calcium ions the lower panel (ordinate—μM); (**D**) The effect of dopamine (DA) is marked by a light strip on the electrical activity of lactotroph. Ordinate: the membrane potential in mV, the horizontal line—time 1 min

secretion of prolactin. In addition, the formation of the membrane potential plateau and overlapping action potentials bursts play a crucial role in Ca^{+2}-activated K^{+} channels (Stojilkovic et al. 2005). Here it is important to note that in experiments not only on lactotrophs, but also on other endocrine cells of the adenohypophysis, it was found that with patterns of impulse activity in the form of bursts, exocytosis of hormones significantly increases. Prolactin from lactotrophs enters the capillary plexus of the adenohypophysis, then into the thin venous vessels of various diameters and into the venous sinuses surrounding the pituitary gland, from where the blood flows into the systemic circulation. Thus, the presence of background impulse activity in lactotrophs by analogy with neurosecretory oxytocinergic cells suggests the presence of a background (base) concentration of prolactin in the blood. However, as mentioned above, hypothalamic neurons also have spontaneous electrical activity. As a result, DA neurons secrete DA into the median eminence, which through the portal system enters the capillary plexus of the adenohypophysis and affects the lactotrophs. Experiments on rat adenohypophysis lactotrophs in tissue culture have shown that DA causes hyperpolarization of the membrane and reduces the electrical activity of lactotrophs (Oxford and Tse 1993) (Fig. 4.40c). Therefore, it can be assumed that the spontaneous electrical activity of lactotrophs in the whole body is less than in isolated lactotrophs and, consequently, the output of prolactin in the bloodstream is reduced. The measurement of the base concentration of prolactin in the blood of nonpregnant women could normally range from 10 to 50 ng/ml (Fig. 2.9).

At the beginning of pregnancy, the concentration of prolactin begins to increase (Fig. 2.13) and by the end of pregnancy can reach 200 ng/ml (Tyson et al. 1972; Rigg et al. 1977). During pregnancy, the volume of the adenohypophysis is believed to almost double due to an increase in the number and size of lactotrophs. Therefore, the increase in the base concentration of prolactin can be attributed to a quantitative factor. However, experimental data available to date, although obtained in rats, indicate that the regulation of prolactin content depending on the state of lactation function can be provided by the plasticity of the nervous system (Romanò et al. 2013). It was found that DA neurons of the hypothalamic nuclei contain receptors for prolactin. An increase in the content of prolactin in the blood as a result of action on prolactin receptors causes an increase in the impulse activity of DA neurons (Lyons et al. 2012, 2014) (Fig. 4.41) and, accordingly, an increase in the output of DA to the capillary plexus of the median eminence. Through the portal system, this is transmitted to the capillary plexus of the adenohypophysis, and through the D2 receptors of lactotrophs, it reduces the exocytosis of prolactin. As a result, it is possible for a nonpregnant animal to avoid the undesirable phenomenon of hyperprolactinemia. During the transition to the state of pregnancy and lactation, the positive feedback is violated, and an increase in the concentration of prolactin does not increase the release of DA. Moreover, the electrical properties of DA neurons do not change. The decrease in DA output is due to a decrease in phosphorylation of tyrosine hydroxylase, a key enzyme that provides DA synthesis. It is possible that this mechanism is used to create lactation hyperprolactinemia in women.

Along with the inhibitory system for regulating prolactin secretion using the dopamine system, there is no possibility, as in other endocrine cells of the adenohypophysis, of the existence of mechanisms for regulating prolactin secretion using releasing factors—releasing hormones. The most studied in this regard is thyrotropin-releasing hormone (TRH) (Grachev and Galantsev 1973; Freeman

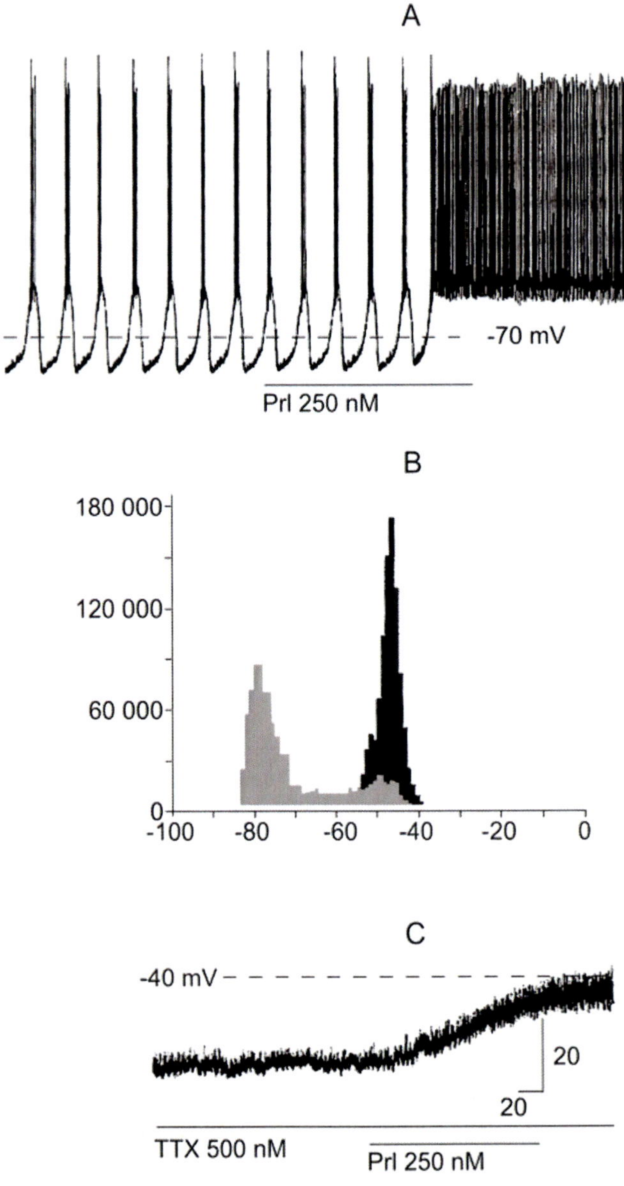

Fig. 4.41 Effect of prolactin on the electrical activity of tuberoinfundibular DA neurons (from: Lyons et al. 2012). (**a**) Change of spontaneous electrical activity of lactotrophs under the action of prolactin. Time of 250 nM prl action is marked by a horizontal line. A dashed line indicates the level of the membrane potential of the lactotroph. Vertical calibration 20 mV, horizontal calibration 20 s. (**b**) Distribution of the values of the membrane potential of lactotrophs in the control (gray color) of the histogram (two-phase) and under the influence of prolactin (black color) (monophase). Ordinate: number of measurements, abscissa membrane potential—mV. (**c**) Alteration of the membrane potential of lactotroph under the influence of prolactin. Action potentials are blocked by 500 nM tetrodotoxin. The effects of prolactin and tetrodotoxin are indicated by horizontal lines. Vertical calibration 20 mV, horizontal calibration 20 s

et al. 2000). Note that TRH was originally isolated as a pituitary factor that stimulates the release of thyroid-stimulating hormone (TSH) from the endocrine cells of the adenohypophysis. The available data on the possible use of TRH as a releasing hormone for prolactin is contradictory. Thus, there is a number of experimental data in favor of the possible participation of TRH in the regulation of prolactin secretion (Freeman et al. 2000). In particular, lactotrophs of the anterior pituitary have receptors for TPH. In addition, intravenous administration of TRN in physiological concentrations increased both thyrotropin and prolactin secretion in lactating women (Jacobs et al. 1971; Gautvik et al. 1973; Tyson et al. 1975; Rossmanith et al. 1995) (Fig. 4.42). Undoubtedly of decisive importance are data about the change in the concentration of TPH while feeding the baby. In studies on women, with the exception of one case (Dawood et al. 1981), during the removal of their milk by the child, it was found that the concentration of prolactin in the blood increased, and the level of thyrotropin remained unchanged. (Gautvik et al. 1973; Tyson et al. 1975; Gehlbach et al. 1989) (Fig. 4.42). These data allow us to conclude that TRH is not directly involved in the implementation of the reflex of milk secretion in women. The high sensitivity of lactotrophs to TRH indicates that TRH in certain cases can have a modulating effect on the secretion of prolactin.

It should be noted here that in addition to TRH, a number of hormones and biologically active substances were detected (Freeman et al. 2000), although only in experiments in mammalian animals, which in physiological concentrations could modulate (increase or decrease) the secretion of prolactin by lactotrophs: oxytocin, vasopressin, vasoactive intestinal polypeptide, somatostatin, calcitonin, holicystokinin, endothelin.

Measuring the level of prolactin in the blood of lactating women in the process of milk excretion by the child showed that, as in the case of oxytocin, the graphical representation of changes in the concentration of prolactin depended on the time intervals at which blood samples were taken. If prolactin was detected in blood samples from nursing women taken after an interval of 10–15 min (Noel et al. 1974; Howie et al. 1980; Uvnäs-Moberg et al. 1990; Tay et al. 1996) (Fig. 4.43a), the prolactin concentration curve gradually increased, reached a maximum, and then slowly decreased. However, if blood samples were taken after 25–30 s (McNeilly et al. 1983), the change in the concentration of prolactin was impulsive (Fig. 4.43b). It is interesting to note that when determining the concentration of prolactin in the blood of rats during the feeding of pups, in the case of blood sampling every 1–2 min (Higuchi et al. 1983), the patterns of dynamics of the prolactin content also had a pulse appearance. Comparing the patterns of peaks of oxytocin and prolactin in the blood of rats during curbs feeding allowed the authors (Higuchi et al. 1983) to make the assumption that the pulse rise in the concentration of oxytocin and prolactin is provided by the operation of a GPIA. Given that when feeding a child, the change in the concentration of oxytocin and prolactin is also impulsive (Fig. 4.43b), we can assume that this assumption is true for a lactating woman.

Unique surveys to determine the effect of infant feeding frequency on basal prolactin levels were performed by Tay and colleges (1996). Blood samples were taken from lactating women through a catheter inserted into a vein every 10 min

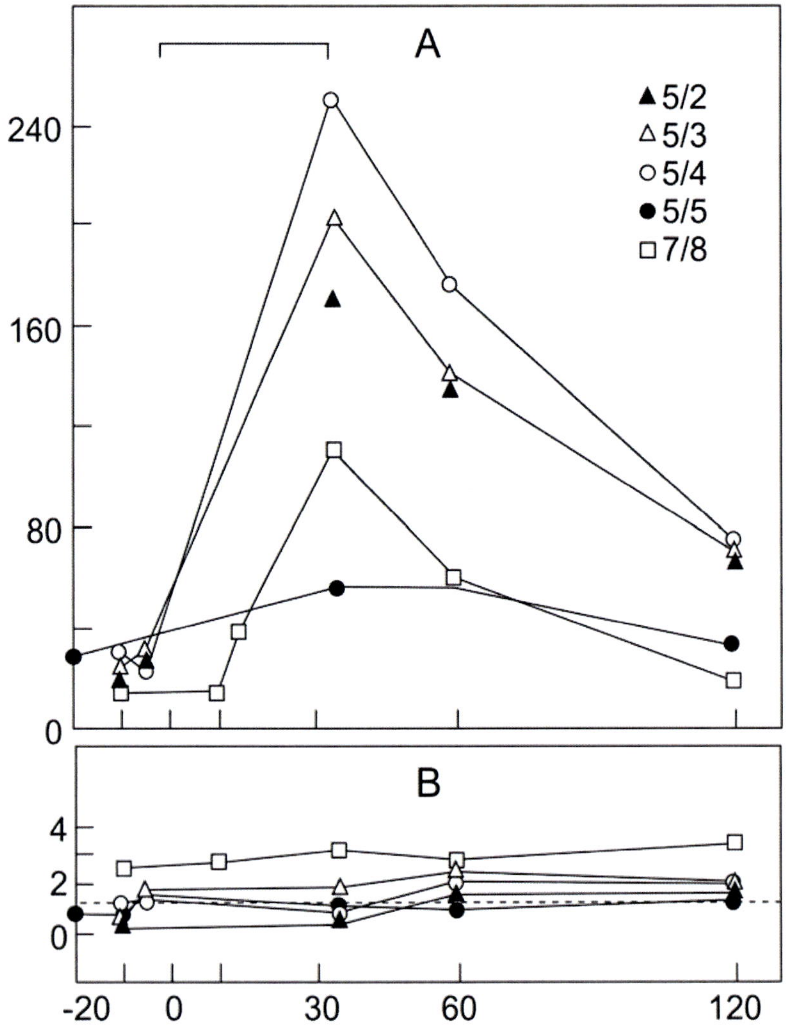

Fig. 4.42 Dynamics of the concentration of prolactin (**a**) and thyroid hormone (**b**) in the blood of a lactating woman for 5 days when milk is excreted by a child (from Gautvik et al. 1973). In (**a**), the horizontal line shows the time of milk withdrawal by the child. Various symbols indicate the days of measuring the concentration of hormones when milk is excreted by a child. Concentration of prolactin on (**a**) ordinate ng/ml, concentration of thyroid hormone on (**b**) ordinate µIU/ml, abscissa: time for all graphs—min. The arrow at the top indicates the first cry of the child. The periods of the child's effects on the areolar-nipple region of the gland are hatched with rectangles. The ordinate: concentration of oxytocin in ng/l (upper graph) and prolactin in mIE/l (lower graph), abscissa: time in min. (**b**) Change in blood prolactin levels in individual lactating women according to infant feeding rate (blood samples were taken after 10 min) (from Tay et al. 1996). The hatched rectangles show the night time. At this time, the light in the room was turned off. Asterisks on top indicate feeding episodes. Ordinate: the concentration of prolactin in mIE/L, abscissa: time of day

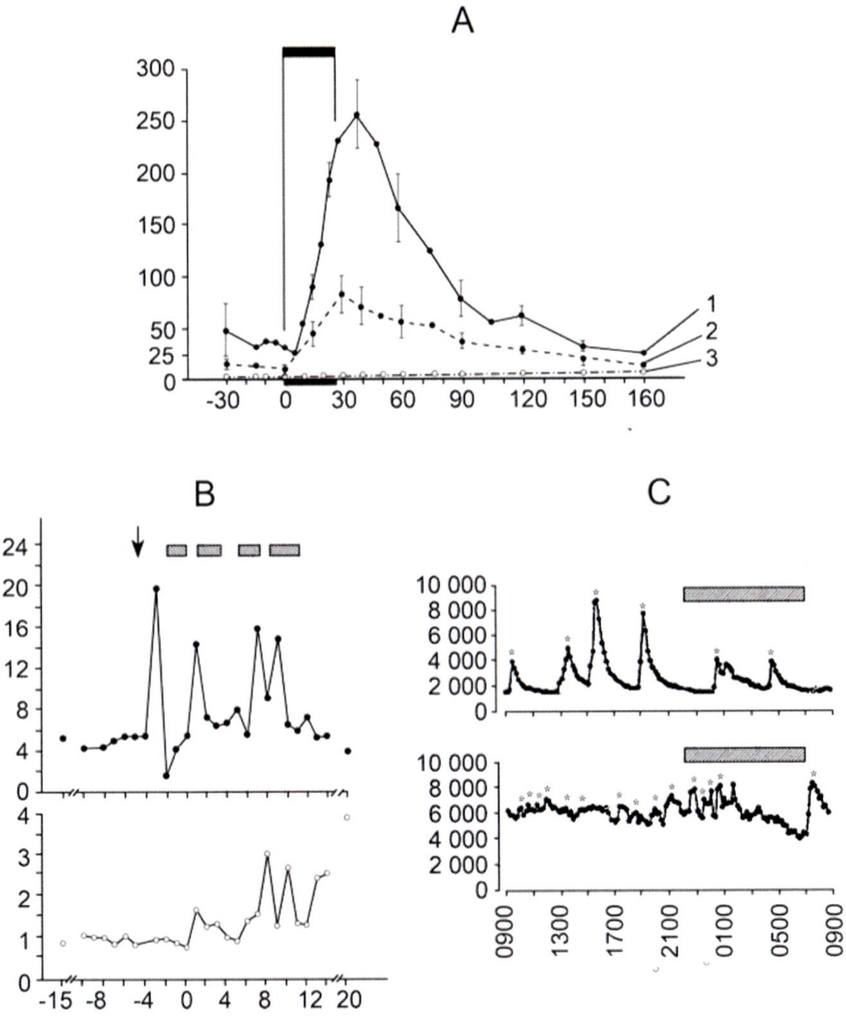

Fig. 4.43 (**a**) Change in blood prolactin concentration in lactating women when milk is ejected by the child. The periods of the child's effects on the areolar-nipple region of the gland are hatched with rectangles. (1) 8–41 days of lactation, (2) 63–104 day of lactation, (3) growth hormone concentration during in this period. The ordinate: concentration of prolactin and growth hormone in ng/l. abscissa: time in min. (**b**) Change in the concentration of oxytocin (upper graph) and prolactin (lower graph) in the blood of a woman when feeding a baby (blood samples were taken at 25–35 s) (by: McNeilly et al. 1983). The arrow at the top indicates the first cry of the child, the periods of the child's effects on the areolar-nipple region of the gland are hatched with rectangles. The ordinate axis is the concentration of oxytocin in ng/l (upper graph) and prolactin in mIU/l (lower graph). On the abscissa axis for graphs time in min. (**c**) Change in blood prolactin levels in individual nursing women according to infant feeding rate (blood samples were taken after 10 min) (Tay et al. 1996). The hatched rectangle shows night time. At this time, the light in the room was turned off. Asterisks from above indicate episodes of feeding. Ordinate: prolactin concentration in mIU/L, abscissa: time of day

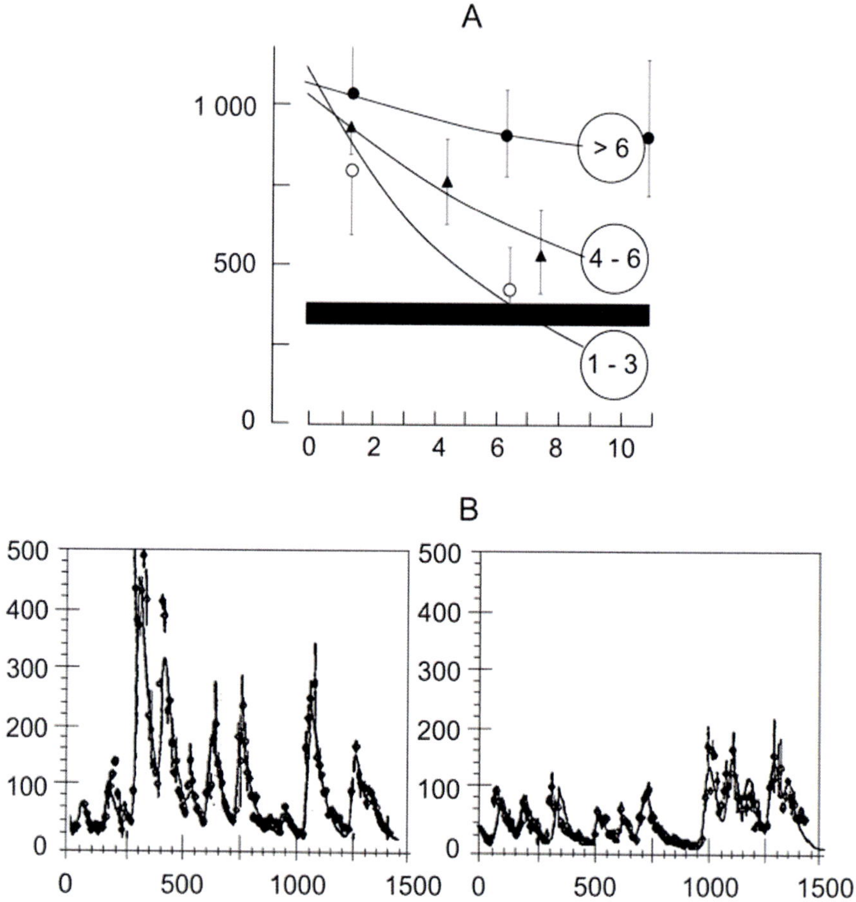

Fig. 4.44 Change in the basal blood concentration of prolactin in a woman (**a**) (from Delvoye et al. 1977b) and during milk excretion by a child (**b**) (from Nunley et al. 1991) during lactation. (**a**) In circles, numbers indicating the frequency of feeding a child per day. A gray horizontal line indicates the basal level in non-acting women. Ordinate: the concentration of prolactin in μIU/ml, abscissa: time in months. (**b**) Change in prolactin concentration during infant feeding. Blood samples were taken 10 min later. Left graph—3 weeks after delivery. Right graph is 3 months after childbirth. Ordinate for all plots: the concentration of prolactin ng/ml, abscissa, min

during the day (Fig. 4.43c). If the intervals between feeding the child were 2–4 h, the maximum value of the concentration of prolactin increased during each feeding decreased to the original initial level (Fig. 4.43c, top graph). However, when the interval between feedings was reduced to 40 min, the basal level significantly increased (Fig. 4.43c, lower graph). At the same time, the maximum amplitude of increasing the concentration of prolactin in response to each feeding significantly decreased.

The basal concentration of prolactin (Figs. 4.1c and 4.44a) in the blood of a lactating woman changes significantly during the entire lactation period. It reaches its maximum at the time of delivery (Fig. 4.1c). During the first 1.5–2 weeks in nursing women, the basal level of prolactin changes slightly. However, then the concentration of prolactin begins to decrease (Delvoye et al. 1977a, b; Martin et al. 1980; Battin et al. 1985). Moreover, the rate of decline depended on the frequency of feeding the child: the more the frequency of feeding, the slower the decrease in basal prolactin levels occurred (Fig. 4.44a) (Delvoye et al. 1977b). In a non-nursing woman, the basal concentration of prolactin was increased during the first 3–4 days, but then within 2 weeks, it decreased to the level of blood content in the woman before pregnancy. In addition to a decrease in basal prolactin levels during lactation, there was also a decrease in the values of prolactin levels during milk excretion by the child (Noel et al. 1974; Johnston and Amico 1986; Nunley et al. 1991). (Fig. 4.44b).

We can assume that this dynamics of prolactin in the postpartum period has a great physiological meaning. In particular, prolonged basal hyperprolactinemia will adversely affect a woman's body. However, under the influence of a high concentration of prolactin by the time of lactopoesis, the number of lobules of type 3–4 (Fig. 2.11), which have the highest secretory activity (Russo and Russo 2004a, b) reaches the maximum value. Therefore, the norm for maintaining secretory activity is apparently not required for the increased basal prolactin levels. The amount of milk ejection by the child is replenished to a sufficient extent by a short-term increase in the level of prolactin during feeding. In this case, the frequency of application of the child becomes a determining factor for the level of prolactin in the blood of a nursing woman.

Prolactin released as a result of extrusion from the lactotrophs of the adenohypophysis through the bloodstream reaches the alveolar-ductal region of the breast. The main effector of the maternal reflex of milk secretion is the secretory (epithelial) cells of the breast alveoli. A necessary condition for the action of prolactin on the structures of the alveolar-ductal department of the breast is the presence of prolactin receptors (PrlR) in them. Unfortunately, data on the localization of PrlR in breast tissue of a lactating woman are not available in the literature. As already mentioned, there are only two immunohistochemical studies on the distribution of PrlR in the normally functioning mammary gland of nonpregnant women (Gill et al. 2001; Ueda et al. 2011). It is interesting to note that the authors (Ueda et al. 2011) expected the appearance of color on PrlR on the basal part of epithelial cell membranes, which are suitable for capillaries carrying prolactin. However, the main immunohistochemical reaction was observed on the apical parts of epithelial cells that exit into the cavity of the alveolar buds and ducts. Similar localization of PrlR was observed in non-lactating mammary glands of dogs (Michel et al. 2012). The distribution of PrlR depending on the functional state of the gland was traced in rats. In non-lactating rats, PrlR receptors were found in the cytosol and nuclei of gland epithelial cells. However, in pregnant rats, PrlR was observed already on the membrane of apical and basal parts of epithelial cells (Camarillo et al. 2001). It can be assumed that in pregnant and lactating women, the number of PrlR receptors in epithelial (secretory) cells for childbirth will also increase, including on the membrane of their basal area.

PrlR exist in several isoforms (Brooks 2012). To date, 7 human PrlR isoforms are known. The effect of prolactin on the cells of the alveolar-duct region of the breast of a lactating woman is mainly carried out through the PrlR isoform with a long chain of amino acid sequences-long PrlR (lPrlR). All isoforms of PrlR consist of three sites-domains: extracellular domain, transmembrane domain, and intracellular domain. lPrlR includes 598 amino acids, of which 210 are located in the extracellular domain, 24 in the transcellular domain, and 364 in the intracellular domain.

It should be noted that the study of the structure and function of prolactin receptors, as well as signaling intracellular pathways that transmit the action of prolactin to the epithelial cell nucleus, where transcription of the corresponding genes is initiated, has been actively conducted in the last decade. However, a number of issues remain to be resolved on this subject. Nevertheless, it can be considered established that the action of prolactin on lPrlR begins with the binding of one prolactin molecule to two monomeric sites of the extracellular PrlR domain, which leads to their dimerization. Prolactin-induced structural changes in the extracellular domain are transmitted via the transmembrane domain to the intracellular domain causing its activation. In turn, when the intracellular domain is activated, it interacts with molecules of signaling systems that modulate cellular processes. In particular, the proximal region of the intracellular domain is associated with a tyrosine kinase called Janus kinase 2 (Jak2). When the intracellular domain is activated, Jak2 phosphorylates, which increases their catalytic activity. Next, the JAk2-lPrlR complex phosphorylates the main target: a protein called signal transmitter and activator of transcription (STAT). The STAT family in vertebrates includes seven proteins. Four of them, STAT1, STAT3, and mainly STAT5a and STAT5b, were identified as transmitters of prolactin receptors (Freeman et al. 2000; Clevenger et al. 2009; Binart et al. 2010; Brooks 2012). Phosphorylated STAT 5a molecules are dimerized and transported to the nucleus, where they bind to the corresponding regulatory sequences of genes and in particular to protein genes, triggering their transcription.

In the development of the breast and the processes of milk secretion in women, in addition to oxytocin and prolactin, a number of reproductive and metabolic hormones are also taken (Neville et al. 2002). Surveys were conducted to study changes in the concentration of estrogen, progesterone, follicle-stimulating and luteinizing hormones, placental lactogen, growth hormone, insulin, gastrin, calcitonin, vasopressin, adrenocorticotropic hormone (ACTH), cortisol, thyroid hormones, and catecholamines in the blood of lactating women during milk excretion (Dawood et al. 1981; Widström et al. 1984; Chiodera et al. 1991; Amico et al. 1994; Nissen et al. 1996). It turned out that from the above-mentioned hormones during the excretion of milk by the child, i.e., adequate stimulation of the areolar-nipple receptors of the mother's breast, the secretion of insulin, adrenocorticotropic hormone, and cortisol changes.

In particular, there was a gradual increase in the concentration of insulin in the blood of the nursing mother during milk withdrawal (Widström et al. 1984). Insulin stimulates the formation of proteins in the secretory cells of a woman's breast (Lawrence and Lawrence 1999). However, an analysis of the data obtained led to the conclusion that an increase in the concentration of insulin may occur indirectly through a reflex increase in the concentration of prolactin (Gustafson et al. 1980; Widström et al. 1984).

The concentration of ACTH and cortisol in the mother's blood during the excretion of milk by the child decreased (Chiodera et al. 1991; Amico et al. 1994; Nissen et al. 1996). ACTH and cortisol are known to be stress hormones. Therefore, a decrease in the concentration of ACTH and cortisol in the blood of the nursing mother will help to establish a calm state in the woman. In other words, breastfeeding is an anti-stress defense for a lactating woman (Nissen et al. 1996). Research on the mechanisms of reducing ACTH and cortisol during infant feeding showed that the decrease in the concentration of these hormones was also mediated, but via oxytocin. The question of the mechanisms of action of oxytocin on the output of ACTH and cortisol is not fully resolved. Available data to date indicate that oxytocin may have an inhibitory effect on the release of ACTH from adenohypois, as well as on the release of cortisol from the adrenal glands (Legros et al. 1988; Chiodera et al. 1991; Legros 2001).

References

Abraira VE, Ginty DD (2013) The sensory neurons of touch. Neuron 79:618–639

Akre J (1989) Infant feeding. The physiological basis. Bull World Health Organ 67(Suppl):109 p

Alberts JR, Pickler RH (2012) Evolution and development of dual ingestion systems in mammals: notes on a new thesis and its clinical implications. Int J Pediatr 2012:1–12

Alekseev NP (2009) Visceral sensory system. In: Altman JA, Kulikov GA (eds) Physiology of sensory systems and higher nervous activity, vol 1. Academy, pp 236–283. (in Russian)

Alekseev NP, Ilyin VI (2016) The mechanics of breast pumping: compression stimuli increased milk ejection. Breastfeed Med 1:370–375

Alekseev NP, Grachev II, Velling VA (1975) Investigation of electrical activity of mechanoreceptor units of the rat mammary gland nipple skin. Physiol Zh SSSR 61:738–743. (in Russian)

Alekseev NP, Grachev II, Velling VA (1976) Functional characteristics of rapidly adapting mechanoreceptor units of the rat mammary gland nipple. Physiol Zh SSSR 62:375–384. (in Russian)

Alekseev NP, Ilyin VI, Yaroslavsky VK, Gaidukov SN, Tikhonova TK, Spesivtsev YA, Omelianyuk EV, Tkachenko NN (1994) Role of vacuum and tactile stimuli in the process of removing milk from a woman's breast. Physiol Zh SSSR 89:67–84. (in Russian)

Alekseev NP, Ilyin VI, Yaroslavski VK, Gaidukov SN, Tikhonova TK, Specivcev YA, Omelyanjuk EV, Tkachenko NN (1998) Compression stimuli increase the efficacy of breast pump function. Eur J Obstet Gynecol Reprod Biol 77(2):131–139

Alekseev NP, Omelyanyuk EV, Talalaeva NE (2000) The dynamics of the reflexes of removing milk during continuous rhythmic stimulation of areola-nipple complex of the breast. Russian Physiol Zh 86:711–719. (in Russian)

Alekseev NP, Omelyanyuk EV, Talalaeva NE (2003) The changing of time parameters of stimulus compression in the breeding of milk during breastfeeding. Russian Physiol Zh 89(10):1403

Alekseev NP, Il'in VI, Schegolkova AV (2010) Prevention and elimination of pathological engorgement in breast-feeding women. J Obstet Women's Dis LIX:95–99. (in Russian)

Alekseev NP, Il'in VI, Talalaeva NE (2014) Contribution of the compression component of the executive mechanism of the milk-producing apparatus to the process of milk excretion in lactating women. J Obstet Women's Dis LXIII:91–97. (in Russian)

Amico JA, Johnston JM, Vagnucci AH (1994) Suckling-induced attenuation of plasma cortisol concentrations in postpartum lactating women. Endocr Res 1:79–87

Ardran GM, Kemp FH (1959) A correlation between suckling pressures and the movements of the tongue. Acta Paediatr 48(3):261–272
Ardran GM, Kemp FH, Lind J (1958a) A cineradiographic study of bottle feeding. Br J Radiol 361:11–22
Ardran GM, Kemp FH, Lind J (1958b) A cineradiographic study of breast feeding. Br J Radiol 363:156–162
Battin DA, Marrs RP, Fleiss PM, Mishell DR Jr (1985) Effect of suckling on serum prolactin, luteinizing hormone, follicle-stimulating hormone, and estradiol during prolonged lactation. Obstet Gynecol 65(6):785–788
Ben-Jonathan N, Hnasko R (2001) Dopamine as a prolactin (PRL) inhibitor. Endocr Rev 22 (6):724–763
Beltran L, Fahie-Wilson MN, McKenna TJ, Kavanagh L, Smith TP (2008) Serum total prolactin and monomeric prolactin reference intervals determined by precipitation with polyethylene glycol: evaluation and validation on common immunoassay platforms. Clin Chem 54 (10):1673–1681
Berga SE (1984) Electrical potentials and cell-to-cell dye movement in mouse mammary gland during lactation. Am J Phys 247:20–25
Bergland RM, Torack RM (1969a) Microtubules and neurofilaments in axons of the human pituitary stalk. Exp Cell Res 54(1):132–134
Bergland RM, Torack RM (1969b) An electron microscopic study of the human infundibulum. Z Zellforsch Mikrosk Anat 99(1):1–12
Binart N, Bachelot A, Bouilly J (2010) Impact of prolactin receptor isoforms on reproduction. Trends Endocrinol Metab 6:362–368
Bonnefont X, Mollard P (2003) Electrical activity in endocrine pituitary cells in situ: a support for a multiple-function coding. FEBS Lett 548(1-3):49–52
Bourque CW (1990) Intraterminal recordings from the rat neurohypophysis in vitro. J Physiol 421:247–262
Braak H, Braak E (1992) Anatomy of the human hypothalamus (chiasmatic and tuberal region). Prog Brain Res 93:3–14
Brake SC, Tavana S, Myers MM (1986) A method for recording and analyzing intra-oral negative pressure in suckling rat pups. Physiol Behav 36:575–578
Brooks CL (2012) Molecular mechanisms of prolactin and its receptor. Endocr Rev 33(4):504–525
Brown CH, Bains JS, Ludwig M, Stern JE (2013) Physiological regulation of magnocellular neurosecretory cell activity: integration of intrinsic, local and afferent mechanisms. J Neuroendocrinol 25:678–710
Bu'Lock F, Woolridge MW, Baum JD (1990) Development of co-ordination of sucking, swallowing and breathing: ultrasound study of term and preterm infants. Dev Med Child Neurol 32:669–678
Bukowska M, Essick GK, Trulsson M (2010) Functional properties of low-threshold mechanoreceptive afferents in the human labial mucosa. Exp Brain Res 201:59–64
Burgess PR, Perl E (1973) Cutaneous mechanoreceptorsand noiciceptors. In: Handbook of sensory physiology, vol 2. Springer, Heidelberg, pp 29–78
Camarillo IG, Thordarson G, Moffat JG, Van Horn KM, Binart N, Kelly PA, Talamantes F (2001) Prolactin receptor expression in the epithelia and stroma of the rat mammary gland. J Endocrinol 171(1):85–95
Chambers MR, Andres KH, von Duering M, Iggo A (1972) The structure and function of the slowly adapting type II mechanoreceptor in hairy skin. Q J Exp Physiol Cogn Med Sci 57(4):417–445
Chiodera P, Salvarani C, Bacchi-Modena A, Spallanzani R, Cigarini C, Alboni A, Gardini E, Coiro V (1991) Relationship between plasma profiles of oxytocin and adrenocorticotropic hormone during suckling or breast stimulation in women. Horm Res 35(3-4):119–123
Clevenger CV, Gadd S, Zheng J (2009) New mechanisms for PRLr action in breast cancer. Trends Endocrinol Metab 20:223–229

Cobo E (1993) Characteristics of the spontaneous milk ejecting activity occurring during human lactation. J Perinat Med 2:77–85
Cobo E, De Bernal MM, Gaitan E, Quintero CA (1967) Neurohypophyseal hormone release in the human. II. Experimental study during lactation. Am J Obstet Gynecol 97:519–529
Cornelissen L, Fabrizi L, Patten D, Worley A, Meek J, Boyd S, Slater R, Fitzgerald M (2013) Postnatal temporal, spatial and modality tuning of nociceptive cutaneous flexion reflexes in human infants. PLoS 8:1–10
Cowie AT (1974) Overview of the mammary gland. J Invest Dermatol 63:2–9
Crowley WR (2015) Neuroendocrine regulation of lactation and milk production. Compr Physiol 5 (1):255–291
Dawood MY, Khan-Dawood FS, Wahi RS, Fuchs F (1981) Oxytocin release and plasma anterior pituitary and gonadal hormones in women during lactation. J Clin Endocrinol Metab 52 (4):678–683
de Vries JIP, Visser GHP, Prechtl HFR (1984) Fetal motility in the first half of pregnancy. Clin Dev Med 94:46–64
Delvoye P, Delogne-Desnoeck J, Nyampeta U, Robyn C (1977a) Time course of physiological hyperprolactinaemia during two years lactation. Clin Endocrinol 7(3):257–259
Delvoye P, Demaegd M, Delogne-Desnoeck J (1977b) The influence of the frequency of nursing and of previous lactation experience on serum prolactin in lactating mothers. J Biosoc Sci 9 (4):447–451
Doucet S, Soussignan R, Sagot P, Schaal B (2009) The secretion of areolar (Montgomery's) glands from lactating women elicits selective, unconditional responses in neonates. PLoS 4:e7579
Doucet S, Soussignan R, Sagot P, Schaal B (2012) An overlooked aspect of the human breast: areolar glands in relation with breastfeeding pattern, neonatal weight gain, and the dynamics of lactation. Early Hum Dev 88:119–128
Dreifuss JJ, Tribollet E, Mühlethaler M (1981) Temporal patterns of neural activity and their relation to the secretion of posterior pituitary hormones. Biol Reprod 24(1):51–72
Drewett RF, Woolridge M (1979) Sucking patterns of human babies on the breast. Early Hum Dev 3:315–321
Du Vigneaud V (1955) Oxytocin, the principal oxytocic hormone of the posterior pituitary gland: its isolation, structure, and synthesis. Experientia (Suppl 2):9–22
Dudas B, Baker M, Rotoli G, Grignol G, Bohn MC, Merchenthaler I (2010) Distribution and morphology of the catecholaminergic neural elements in the human hypothalamus. Neuroscience 171(1):187–195
Elad D, Kozlovsky P, Blum O, Laine AF, Po MJ, Botzer E, Dollberg S, Zelicovich M, Ben Sira L (2014) Biomechanics of milk extraction during breast-feeding. Proc Natl Acad Sci U S A 111 (14):5230–5235
Elster AD (1993) Modern imaging of the pituitary. Radiology 187(1):1–14
Enomoto K, Furuya K, Maeno T, Edwards C, Oka T (1987) Mechanically induced electrical responses in murine mammary epithelial cells in primary culture. FEBS Lett 223(1):82–86
Enomoto K, Furuya K, Yamagishi S, Oka T, Maeno T (1994) The increase in the intracellular Ca2+ concentration induced by mechanical stimulation is propagated via release of pyrophosphorylated nucleotides in mammary epithelial cells. Pflugers Arch 427(5-6):533–542
Essick G, Guest S, Martinez E, Chen C, McGlone F (2004) Site-dependent and subject-related variations in perioral thermal sensitivity. Somatosens Mot Res 21:159–175
Febo M (2011) A bold view of the lactating brain: functional magnetic resonance imaging studies of suckling in awake dams. J Neuroendocrinol 23:1009–1019
Finan DS, Barlow SM (1998) Intrinsic dynamics and mechanosensory modulation of non-nutritive sucking in human infants. Early Hum Dev 52(2):181–197
Fliers E, Swaab DF, Pool CW, Verwer RW (1985) The vasopressin and oxytocin neurons in the human supraoptic and paraventricular nucleus; changes with aging and in senile dementia. Brain Res 342(1):45–53

Freeman ME, Kanyicska B, Lerant A, Nagy G (2000) Prolactin: structure, function, and regulation of secretion. Physiol Rev 80(4):1523–1631

Fuchs A-R, Romero R, Keefe D, Parra M, Oyarzun E, Behnke E (1991) Oxytocin secretion and human parturition: pulse frequency and duration increase during spontaneous labor in women. Am J Obstet Gynecol 165:1515–1523

Furuya K, Enomoto K (1990) Real-time imaging of intracellular calcium change with simultaneous single channel recording in mammary epithelial cells. Brain Res Bull 25(5):779–781

Furuya K, Enomoto K, Furuya S, Yamagishi S, Edwards C, Oka T (1989) Single calcium-activated potassium channel in cultured mammary epithelial cells. Pflugers Arch 414(2):118–124

Gairns FW (1953) Sensory nerve endings other than taste buds in the human tongue. J Physiol 121 (2):65–71

Garbin CP, Sakalidis VS, Chadwick LM, Whan E, Hartmann PE, Geddes DT (2013) Evidence of improved milk intake after frenotomy: a case report. Pediatrics 132(5):413–147

Gardner H, Kent JC, Lai CT, Mitoulas LR, Cregan MD, Hartmann PE, Geddes DT (2015) Milk ejection patterns: an intra- individual comparison of breastfeeding and pumping. BMC Pregnancy Childbirth 15:156

Gardner H, Kent JC, Prime DK, Lai CT, Hartmann PE, Geddes DT (2017) Milk ejection patterns remain consistent during the first and second lactations. Am J Hum Biol 00:22960

Gautvik KM, Weintraub BD, Graeber CT, Maloof F, Zuckerman JE, Tashjian AH Jr (1973) Serum prolactin and TSH: effects of nursing and pyro Glu-His-ProNH2 administration in postpartum women. J Clin Endocrinol Metab 37(1):135–139

Gebhart GF (2000) Visceral pain-peripheral sensitization. Gut 47(Suppl. IV):54–55

Geddes DT (2009a) Ultrasound imaging of the lactating breast: methodology and application. Int Breastfeed J 4:1–17

Geddes DT (2009b) The use of ultrasound to identify milk ejection in women - tips and pitfalls. Int Breastfeed J 4(5):1–7

Geddes DT, Kent JC, Mitoulas LR, Hartmann PE (2008) Tongue movement and intra-oral vacuum in breastfeeding infants. Early Hum Dev 84:471–477

Gehlbach DL, Bayliss P, Rosa C (1989) Prolactin and thyrotropin responses to nursing during the early puerperium. J Reprod Med 34(4):295–298

Gill S, Peston D, Vonderhaar BK, Shousha S (2001) Expression of prolactin receptors in normal, benign, and malignant breast tissue: an immunohistological study. J Clin Pathol 54:956–960

Goldfield EC, Buonomo C, Fletcher K, Perez J, Margetts S, Hansen A, Smith V, Ringer S, Richardson MJ, Wolff PH (2010) Premature infant swallowing: patterns of tongue-soft palate coordination based upon videofluoroscopy. Infant Behav Dev 33(2):209–218

Goldfield EC, Smith V, Buonomo C, Perez J, Larson K (2013) Preterm infant swallowing of thin and nectar-thick liquids: changes in lingual-palatal coordination and relation to bolus transit. Dysphagia 28(2):234–244

Goudsmit E, Hofman MA, Fliers E, Swaab DF (1990) The supraoptic and paraventricular nuclei of the human hypothalamus in relation to sex, age and Alzheimer's disease. Neurobiol Aging 11:529–536

Grachev II, Galantsev VP (1973) Physiology lactation. General and comparative. L., "Science," 591 p. (in Russian)

Grachev II, Alekseev NP, Velling VA (1977) Slow-adapting mechanoreceptor units of the mammary nipple of a guinea pig. Physiol Zh SSSR 63:391–400

Grachev II, Alekseev NP (1980) Role of receptors in lactation regulation. L. Nauka

Grachev II Skopichev VG, Popov SM (1976a) Cytophysiology of milk secretion. L. Nauka (in Russian)

Grachev II, Alekseev NP, Velling VA (1976b) Properties of mechanoreceptors of the guinea pig mammary gland nipple (rapidly adapting mechanoreceptor). Physiol Zh SSSR 62:885–891

Grattan DR (2015) 60 years of neuroendocrinology: the hypothalamo-prolactin axis. J Endocrinol 226(2):101–122

Green JD (1948) The histology of the hypophysial stalk and median eminence in man with special reference to blood vessels, nerve fibers and a peculiar neurovascular zone in this region. Anat Rec 100(3):273–295

Grynkiewicz G, Poenie M, Tsien RY (1985) A new generation of Ca2+ indicators with greatly improved fluorescence properties. J Biol Chem 260(6):3440–3450

Gunter M (1945) Sore nipples; causes and prevention. Lancet 2(6378):590–593

Gustafson AB, Banasiak MF, Kalkhoff RK, Hagen TC, Kim HJ (1980) Correlation of hyperprolactinemia with altered plasma insulin and glucagon: similarity to effects of late human pregnancy. J Clin Endocrinol Metab 51(2):242–246

Hafström M, Lundquist C, Lindecrantz K, Larsson K, Kjellmer I (1997) Recording non-nutritive sucking in the neonate. Description of an automatized system for analysis. Acta Paediatr 86:82–90

Harris GW (1948) Neural control of the pituitary gland. Physiol Rev 28(2):139–179

Hartshorne DJ, Ito M, Ikebe M (1989) Myosin and contractile activity in smooth muscle. Adv Exp Med Biol 255:269–277

Hendrik HD (2013) Sucking-pads and primitive sucking reflex. J Neonatal Perinatal Med 6 (4):281–283

Hewer EE (1935) The development of nerve endings in the human foetus. J Anat 69:369–379

Higuchi T, Honda K, Fukuoka T, Negoro H, Hosono Y, Nishida E (1983) Pulsatile secretion of prolactin and oxytocin during nursing in the lactating rat. Endocrinol Jpn 30(3):353–359

Higuchi T, Tadokoro Y, Honda K, Negoro H (1986) Detailed analysis of blood oxytocin levels during suckling and parturition in the rat. J Endocrinol 110(2):251–256

Hofman MA, Goudsmit E, Purba JS, Swaab DF (1990) Morphometric analysis of the supraoptic nucleus in the human brain. J Anat 172:259–270

Horvath E, Kovacs K (1988) Fine structural cytology of the adenohypophysis in rat and man. J Electron Microsc Tech 8(4):401–432

Howie PW, McNeilly AS, McArdle T, Smart L, Houston M (1980) The relationship between suckling-induced prolactin response and lactogenesis. J Clin Endocrinol Metab 50(4):670–673

Ilyin VI, Alekseev NP (2014) Influence of vacuum stimuli on the length of the areolar-nipple complex of the breast of lactating women. Russian Physiol Zh 100(2):232–237. (in Russian)

Iriki A, Nozaki S, Nakamura Y (1988) Feeding behavior in mammals: corticobulbar projection is reorganized during conversion from sucking to chewing. Dev Brain Res 44:189–196

Jackson KM, Nazar AM (2006) Breastfeeding, the immune response, and long-term health. J Am Osteopath Assoc 106:203–207

Jacobs LS, Snyder PJ, Wilber JF, Utiger RD, Daughaday WH (1971) Increased serum prolactin after administration of synthetic thyrotropin releasing hormone (TRH) in man. J Clin Endocrinol Metab 33(6):996–998

Jacobs LA, Dickinson JE, Hart PD, Doherty DA, Faulkner SJ (2007) Normal nipple position in term infants measured on breastfeeding ultrasound. J Hum Lact 23(1):52–59

Janbu T, Koss KS, Thoresen M, Wesche J (1985) Blood velocities to the female breast during lactation and following oxytocin injections. J Dev Physiol 7(6):373–380

Johansson RS, Vallbo AB (1983) Tactile sensory coding in the glabrous skin of the human hand. Trend Neurosci 6:27–32

Johansson RS, Trulsson M, Olsson KA, Westberg KG (1988) Mechanoreceptor activity from the human face and oral mucosa. Exp Brain Res 72:204–208

Johnson KO, Hsiao SS (1992) Neural mechanisms of tactual form and texture perception. Annu Rev Neurosci 15:227–250

Johnston JM, Amico JA (1986) A prospective longitudinal study of the release of oxytocin and prolactin in response to infant suckling in long term lactation. J Clin Endocrinol Metab 62:653–657

Jonas K, Johansson LM, Nissen E, Ejdebäck M, Ransjö-Arvidson AB, Uvnäs-Moberg K (2009) Effects of intrapartum oxytocin administration and epidural analgesia on the concentration of plasma oxytocin and prolactin, in response to suckling during the second day postpartum. Breastfeed Med 4(2):71–82

Jourdain P, Israel JM, Dupouy B, Oliet SH, Allard M, Vitiello S, Theodosis DT, Poulain DA (1998) Evidence for a hypothalamic oxytocin-sensitive pattern-generating network governing oxytocin neurons in vitro. J Neurosci 18(17):6641–6649

Karyakin MG, Alekseev NP (1992) The influence of tactile stimulation of nipples on the milk ejection reflex in the rat. J Comp Physiol A 170:645–650

Kimura T, Tanizawa O, Mori K, Brownstein MJ, Okayama H (1992) Structure and expression of a human oxytocin receptor. Nature 356(6369):526–529. (Erratum in: Nature 1992, 357 (6374):176, 1992.)

Kimura T, Ito Y, Einspanier A, Tohya K, Nobunaga T, Tokugawa Y, Takemura M, Kubota Y, Ivell R, Matsuura N, Saji F, Murata Y (1998) Expression and immunolocalization of the oxytocin receptor in human lactating and non-lactating mammary glands. Hum Reprod 13:2645–2653

Koutcherov Y, Mai JK, Ashwell KW, Paxinos G (2000) Organization of the human paraventricular hypothalamic nucleus. J Comp Neurol 423(2):299–318

Lacasse P, Ollier S, Lollivier V, Boutinaud M (2016) New insights into the importance of prolactin in dairy ruminants. J Dairy Sci 99(1):864–874

Lagercrantz H, Changeux JP (2009) The emergence of human consciousness: from fetal to neonatal life. Pediatr Res 65:255–260

Lawrence RA, Lawrence RM (1999) Breastfeeding: a guide for medical profession, 5th edn. Mosby

Le Tissier P, Campos P, Lafont C, Romanò N, Hodson DJ, Mollard P (2016) An updated view of hypothalamic-vascular-pituitary unit function and plasticity. Nat Rev Endocrinol 13 (5):257–267

Lederis K (1961) Vasopressin and oxytocin in the mammalian hypothalamus. Gen Comp Endocrinol 1:80–89

Legros JJ (2001) Inhibitory effect of oxytocin on corticotrope function in humans: are vasopressin and oxytocin ying-yang neurohormones? Psychoneuro 26(7):649–655

Legros JJ, Chiodera P, Geenen V (1988) Inhibitory action of exogenous oxytocin on plasma cortisol in normal human subjects: evidence of action at the adrenal level. Neuroendocrinology 48 (2):204–206

Lincoln DW (1984) The postery pituitary. In: Austin CR, Short RV (eds) Hormonal control of reproduction, 2th edn. Cambridge University Press, Cambridge, pp 21–51

Lincoln DW, Paisley AC (1982) Neuroendocrine control of milk ejection. J Reprod Fertil 65:571–586

Lincoln DW, Wakerley JB (1975) Factors governing the periodic activation of supraoptic and paraventricular neurosecretory cells during suckling in the rat. J Physiol 250(2):443–461

Lloyd RV, Anagnostou D, Cano M, Barkan AL, Chandler WF (1988) Analysis of mammosomatotropic cells in normal and neoplastic human pituitary tissues by the reverse hemolytic plaque assay and immunocytochemistry. J Clin Endocrinol Metab 66:1103–1110

Lollivier V, Marnet PG, Delpal S, Rainteau D, Achard C, Rabot A, Ollivier-Bousquet M (2006) Oxytocin stimulates secretory processes in lactating rabbit mammary epithelial cells. J Physiol 570:125–140

Lucas A, Drewett RB, Mitchell MD (1980) Breast-feeding and plasma oxytocin concentrations. Br Med J 281(6244):834–835

Luther EC, Arballo JC, Sala NL, Cordero Funes JC (1974) Suckling pressure in humans: relationship to oxytocin-reproducing reflex milk ejection. J Appl Physiol 36:350–353

Lyons DJ, Broberger C (2014) TIDAL WAVES: network mechanisms in the neuroendocrine control of prolactin release. Neuroendocrinology 35(4):420–438

Lyons DJ, Horjales-Araujo E, Broberger C (2010) Synchronized network oscillations in rat tuberoinfundibular dopamine neurons: switch to tonic discharge by thyrotropin-releasing hormone. Neuron 65(2):217–229

Lyons DJ, Hellysaz A, Broberger C (2012) Prolactin regulates tuberoinfundibular dopamine neuron discharge pattern: novel feedback control mechanisms in the lactotrophic axis. J Neurosci 32 (23):8074–8083

Lyons DJ, Broberger C (2014) TIDAL WAVES: network mechanisms in the neuroendocrine control of prolactin release. Neuroendocrinology 35(4):420–438

Macefield VG (2005) Physiological characteristics of low-threshold mechanoreceptors in joints, muscle and skin in human subjects. Clin Exp Pharmacol Physiol 32:135–144

Manaye KF, Lei D, Tizabi Y, Dávila-García MI, Mouton PR, Kelly PH (2005) Selective neuron loss in the paraventricular nucleus of hypothalamus in patients suffering from major depression and bipolar disorder. J Neuropathol Exp Neurol 64(3):224–229

Martin RH, Glass MR, Chapman C, Wilson GD, Woods KL (1980) Human alpha-lactalbumin and hormonal factors in pregnancy and lactation. Clin Endocrinol 13(3):223–230

McClellan HL, Sakalidis VS, Hepworth AR, Hartmann PE, Geddes DT (2010) Validation of nipple diameter and tongue movement measurements with B-mode ultrasound during breastfeeding. Ultrasound Med Biol 36(11):797–807

McConnell EM (1953) The arterial blood supply of the human hypophysis cerebri. Anat Rec 115 (2):175–203

McGlone F, Reilly D (2010) The cutaneous sensory system. Neurosci Biobehav Rev 34 (2):148–159

McManaman JL, Neville MC (2003) Mammary physiology and milk secretion. Adv Drug Deliv Rev 55(5):629–641

McNeilly AS, Robinson IC, Houston MJ, Howie PW (1983) Release of oxytocin and prolactin in response to suckling. Br Med J 286:257–259

Meyer-Lindenberg A, Domes G, Kirsch P, Heinrichs M (2011) Oxytocin and vasopressin in the human brain: social neuropeptides for translational medicine. Nat Rev Neurosci 12(9):524–538

Michel E, Feldmann SK, Kowalewski MP, Bley CR, Boos A, Guscetti F, Reichler IM (2012) Expression of prolactin receptors in normal canine mammary tissue, canine mammary adenomas and mammary adenocarcinomas. BMC Vet Res 8:1–8

Miller JL, Sonies BC, Macedonia C (2003) Emergence of oropharyngeal, laryngeal and swallowing activity in the developing fetal upper aerodigestive tract: an ultrasound evaluation. Early Hum Dev 71(1):61–87

Minderaa RB, Volkmar FR, Hansen CR, Harcherik DF, Akkerhuis GW, Cohen DJ (1985) Snout and visual rooting reflexes in infantile autism. J Autism Dev Disord 15:409–416

Mizuno K, Ueda A (2001) Development of sucking behavior in infants with Down's syndrome. Acta Paediatr 90:1384–1388

Monaghan P, Moss D (1996) Connexin expression and gap junctions in the mammary gland. Cell Biol Int 20:121–125

Montalbetti N, Dalghi MG, Albrecht C, Hediger MA (2014) Nutrient transport in the mammary gland: calcium, trace minerals and water soluble vitamins. J Mammary Gland Biol Neoplasia 19:73–90

Morris M, Stevens SW, Adams MR (1980) Plasma oxytocin during pregnancy and lactation in the cynomolgus monkey. Biol Reprod 23:782–787

Mortazavi NS, Hassiotou F, Geddes D, Hassanipour F (2015) Mathematical modeling of mammary ducts in lactating human females. J Biomech Eng 137(7):1086–1091

Morton A (1969) A quantitative analysis of the normal neuron population of the hypothalamic magnocellular nuclei in man and of their projections to the neurohypophysis. J Comp Neurol 136:143–157

Naccarato AG, Viacava P, Bocci G, Fanelli G, Aretini P, Lonobile A, Montruccoli G, Bevilacqua G (2003) Definition of the microvascular pattern of the normal human adult mammary gland. J Anat 203(6):599–603

Nakano H, Furuya K, Furuya S, Yamagishi S (1997) Involvement of P 2-purinergic receptors in intracellular Ca2+ responses and the contraction of mammary myoepithelial cells. Pflugers Arch 1:1–8

Nakano H, Furuya K, Yamagishi S (2001) Synergistic effects of ATP on oxytocin-induced intracellular Ca2+ response in mouse mammary myoepithelial cells. Pflugers Arch 442 (1):57–63
Neville MC (1990) The physiological basis of milk secretion. Part I. Basic physiology. Ann N Y Acad Sci 586:1–11
Neville MC, Allen JC, Archer PC, Casey CE, Seacat J, Keller RP, Lutes V, Rasbach J, Neifert M (1991) Studies in human lactation: milk volume and nutrient composition during weaning and lactogenesis. Am J Clin Nutr 54(1):81–92
Neville MC, Morton J, Umemura S (2001) Lactogenesis. The transition from pregnancy to lactation. Pediatr Clin N Am 48:35–52
Neville MC, McFadden TB, Forsyth I (2002) Hormonal regulation of mammary differentiation and milk secretion. J Mammary Gland Biol Neoplasia 7:49–66
Neville MC, Anderson SM, McManaman JL, Badger TM et al (2012) Lactation and neonatal nutrition: defining and refining the critical questions. J Mammary Gland Biol Neoplasia 17:167–188
Newton ER (2004) Breastmilk: the gold standard. Clin Obstet Gynecol 47(3):632–642
Nguyen DA, Parlow AF, Neville M (2001) Hormonal regulation of tight junction closure in the mouse mammary epithelium during the transition from pregnancy to lactation. J Endocrinol 170:347–356
Nissen E, Uvnäs-Moberg K, Svensson K, Stock S, Widström AM, Winberg J (1996) Different patterns of oxytocin, prolactin but not cortisol release during breastfeeding in women delivered by caesarean section or by the vaginal route. Early Hum Dev 45(1-2):103–118
Noel GL, Suh HK, Frantz AG (1974) Prolactin release during nursing and breast stimulation in postpartum and nonpostpartum subjects. J Clin Endocrinol Metab 38(3):413–423
Nowak AJ, Smith WL, Erenberg A (1995) Imaging evaluation of breast-feeding and bottle-feeding systems. J Pediatr 126:130–134
Nunley WC, Urban RJ, Kitchin JD, Bateman BG, Evans WS, Veldhuis JD (1991) Dynamics of pulsatile prolactin release during the postpartum lactational period. J Clin Endocrinol Metab 72 (2):287–293
Ochoa J, Torebjörk E (1983) Sensations evoked by intraneural microstimulation of single mechanoreceptor units innervating the human hand. J Physiol 342:633–654
Oder AL, Stalling DL, Barlow SM (2013) Short-term effects of pacifier texture on NNS in neurotypical infants. Int J Pediatr 2013:1–8
Olins GM, Bremel RD (1982) Phosphorylation of myosin in mammary myoepithelial cells in response to oxytocin. Endocrinology 110(6):1933–1938
Olins GM, Bremel RD (1984) Oxytocin-stimulated myosin phosphorylation in mammary myoepithelial cells: roles of calcium ions and cyclic nucleotides. Endocrinology 114 (5):1617–1626
Owerbach D, Rutter WJ, Cooke NE, Martial JA, Shows TB (1981) The prolactin gene is located on chromosome 6 in humans. Science 212(4496):815–816
Oxford GS, Tse A (1993) Modulation of ion channels underlying excitation-secretion coupling in identified lactotrophs and gonadotrophs. Biol Reprod 48(1):1–7
Pang WW, Hartmann PE (2007) Initiation of human lactation: secretory differentiation and secretory activation. J Mammary Gland Biol Neoplasia 12(4):211–221
Penfield W, Boldrey E (1937) Somatic motor and sensory representation in the cerebral cortex of man as studied by electrical stimulation. Brain 60:389–443
Phillips JR, Johansson RS, Johnson KO (1992) Responses of human mechanoreceptive afferents to embossed dot arrays scanned across fingerpad skin. J Neurosci 12:827–839
Pinelli J, Symington A (2005) Non-nutritive sucking for promoting physiologic stability and nutrition in preterm infants. Cochrane Database Syst Rev 19(4):CD001071
Prieto CR, Cardenas H, Salvatierra AM, Boza C, Montes CG, Croxatto HB (1996) Sucking pressure and its relationship to milk transfer during breastfeeding in humans. J Reprod Fertil 108:69–74

Prieto CR, Cardenas H, Croxatto HB (1999) Variability of breast sucking, associated milk transfer and the duration of lactational amenorrhoea. J Reprod Fertil 115:193–200

Prime DK, Geddes DT, Hepworth AR, Trengove NJ, Hartmann PE (2011) Comparison of the patterns of milk ejection during repeated breast expression sessions in women. Breastfeed Med 6(4):183–190

Ramsay DT, Kent JC, Owens RA, Hartmann PE (2004) Ultrasound imaging of milk ejection in the breast of lactating women. Pediatrics 113(2):361–367

Ramsay DT, Mitoulas LR, Kent JC, Cregan MD, Doherty DA, Larsson M, Hartmann PE (2006) Milk flow rates can be used to identify and investigate milk ejection in women expressing breast milk using an electric breast pump. Breastfeed Med 1(1):14–23

Raymond K, Cagnet S, Kreft M, Janssen H, Sonnenberg A, Glukhova MA (2011) Control of mammary myoepithelial cell contractile function by α3β1 integrin signalling. EMBO J 30 (10):1896–1906

Rigg LA, Lein A, Yen SS (1977) Pattern of increase in circulating prolactin levels during human gestation. Am J Obstet Gynecol 129:454–456

Riordan J (2005) Anatomical and biological imperatives. In: Riordan J (ed) Breastfeeding and human lactation, 3rd edn. Jones and Bartlett Publishers, pp 65–93

Romanò N, Yip SH, Hodson DJ, Guillou A, Parnaudeau S, Kirk S, Tronche F, Bonnefont X, Le Tissier P, Bunn SJ, Grattan DR, Mollard P, Martin AO (2013) Plasticity of hypothalamic dopamine neurons during lactation results in dissociation of electrical activity and release. J Neurosci 33(10):4424–4433

Rossmanith WG, Hohl B, Lüttke B (1995) Stimulated thyrotropin and prolactin secretion in lactating and non-lactating women. Gynecol Endocrinol 9(3):181–188

Rothemund Y, Schaefer M, Grüsser SM, Flor H (2005) Localization of the human female breast in primary somatosensory cortex. Exp Brain Res 164:357–364

Roudaut Y, Lonigro A, Coste B, Hao J, Delmas P, Crest M (2012) Touch sense: functional organization and molecular determinants of mechanosensitive receptors. Channels (Austin) 6:234–245

Russo J, Russo IH (2004a) The breast as a developing organ, Chapter 2. In: Molecular basis of breast cancer. Prevention and treatment. Springer, pp 1–48

Russo J, Russo I (2004b) Development of the human breast. Maturitas 49:2–15

Sakalidis VS, Williams TM, Garbin CP, Hepworth AR, Hartmann PE, Paech MJ, Geddes DT (2013) Ultrasound imaging of infant sucking dynamics during the establishment of lactation. J Hum Lact 29(2):205–213

Sandholm L-E (1968) The effect of intravenous and intranasal oxytocin on intramammary pressure during early lactation. Acta Obst Gynecol Scandinav 47:145–154

Santoro W Jr, Martinez FE, Ricco RG, Jorge SM (2010) Colostrum ingested during the first day of life by exclusively breastfed healthy newborn infants. J Pediatr 156:29–32

Satogami N, Miki Y, Koyama T, Kataoka M, Togashi K (2010) Normal pituitary stalk: high-resolution MR imaging at 3T. Am J Neuroradiol 31(2):355–359

Schaal B (2010) Mammary odor cues and pheromones: mammalian infant-directed communication about maternal state, mammae, and milk. Vitam Horm 83:83–136

Scheithauer BW, Horvath E, Kovacs K (1992) Ultrastructure of the neurohypophysis. Microsc Res Tech 20(2):177–186

Segami Y, Mizuno K, Taki M, Itabashi K (2013) Perioral movements and sucking pattern during bottle feeding with a novel, experimental teat are similar to breastfeeding. J Perinatol 33(4):319–323

Seyama S, Pearl GS, Takei Y (1980a) Ultrastructural study of the human neurohypophysis. I. Neurosecretory axons and their dilatations in the pars nervosa. Cell Tissue Res 205(2):253–271

Seyama S, Pearl GS, Takei Y (1980b) Ultrastructural study of the human neurohypophysis. III. Vascular and perivascular structures. Cell Tissue Res 206(2):291–302

Shennan DB, Peaker M (2000) Transport of milk constituents by the mammary gland. Physiol Rev 80(3):925–951

Shuleikina KV (1967) Food reactions of a newborn in the first days of life. Obstet Gynecol 10:28–34. (in Russian)

Sica-Blanco Y, Mendez-Bauer C, Sala N, Cabot H, Caldeyro-Barcia R (1959) Nuevo metodo para el estudio de la funcionalidad mamaria en el mujer. Arch Ginec Obstet Montevideo 17:63–70

Smith CR, Norman MR (1990) Prolactin and growth hormone: molecular heterogeneity and measurement in serum. Ann Clin Biochem 27:542–550

Smith WL, Erenberg A, Nowak A, Franken EA Jr (1985) Physiology of sucking in the normal term infant using real-time US. Radiology 156:379–381

Soloff MS, Chakraborty J, Sadhukhan P, Senitzer D, Wieder M, Fernstrom MA, Sweet P (1980) Purification and characterization of mammary myoepithelial and secretory cells from the lactating rat. Endocrinology 106(3):887–897

Spencer S, Saper CB, Joh T, Reis DJ, Goldstein M, Raese JD (1985) Distribution of catecholamine-containing neurons in the normal human hypothalamus. Brain Res 328(1):73–80

Stirling JW, Chandler JA (1976a) The fine structure of normal, resting terminal ductal-lobular unit of the female breast. Virchows Arch A Path Anat Histol 372:205–226

Stirling JW, Chandler JA (1976b) Ultrastructural studies of the female breast: I. 9 + 0 cilia in myoepithelial cells. Anat Rec 186(3):413–416

Stirling JW, Chandler JA (1977) The fine structure of ducts and subareolar ducts in the resting gland of the female breast. Virchows Arch A Path Anat Histol 373:119–132

Stojilkovic SS (2006) Pituitary cell type-specific electrical activity, calcium signaling and secretion. Biol Res 39(3):403–423

Stojilkovic SS, Zemkova H, Van Goor F (2005) Biophysical basis of pituitary cell type-specific Ca2 + signaling-secretion coupling. Trends Endocrinol Metab 16(4):152–159

Summerlee AJ, Lincoln DW (1981) Electrophysiological recordings from oxytocinergic neurones during suckling in the unanaesthetized lactating rat. J Endocrinol 90(2):255–265

Swaab DF, Hofman MA, Lucassen PJ, Purba JS, Raadsheer FC, Van de Nes JA (1993) Functional neuroanatomy and neuropathology of the human hypothalamus. Anat Embryol (Berl) 187:317–330

Tagliavini A, Tabak J, Bertram R, Pedersen MG (2016) Is bursting more effective than spiking in evoking pituitary hormone secretion? A spatiotemporal simulation study of calcium and granule dynamics. Am J Physiol Endocrinol Metab 310(7):E515–E525

Tanaka S, Kogo M, Chandler SH, Matsuya T (1999) Localization of oral-motor rhythmogenic circuits in the isolated rat brain stem preparation. Brain Res 821:190–199

Takei Y, Seyama S, Pearl GS, Tindall GT (1980) Ultrastructural study of the human neurohypophysis. II. Cellular elements of neural parenchyma, the pituicytes. Cell Tissue Res 205(2):273–287

Tay CCK, Glasier AF, McNeilly AS (1996) Twenty-four hour patterns of prolactin secretion during lactation and the relationship to suckling and the resumption of fertility in breast-feeding women. Hum Reprod 11(5):950–955

Teilum K, Hoch JC, Goffin V, Kinet S, Martial JA, Kragelund BB (2005) Solution structure of human prolactin. J Mol Biol 351(4):810–823

Theodosis DT, Dreifuss JJ, Orci L (1978) A freeze-fracture study of membrane events during neurohypophysial secretion. J Cell Biol 78(2):542–553

Thibonnier M, Conarty DM, Preston JA, Plesnicher CL, Dweik RA, Erzurum SC (1999) Human vascular endothelial cells express oxytocin receptors. Endocrinology 140(3):1301–1319

Thoresen M, Wesche J (1988) Doppler measurements of changes in human mammary and uterine blood flow during pregnancy and lactation. Acta Obstet Gynecol Scand 67:741–745

Tolkunov Yu A, Markov AG (2005) The physiology of the alveoli of the breast. Nauka, Saint Petersburg, p 219 (in Russian)

Trulsson M, Essick GK (1997) Low-threshold mechanoreceptive afferents in the human lingual nerve. J Neurophysiol 77(2):737–748

Trulsson M, Essick GK (2010) Sensations evoked by microstimulation of single mechanoreceptive afferents innervating the human face and mouth. J Neurophysiol 103:1741–1747
Trulsson M, Johansson RS (2002) Orofacial mechanoreceptors in humans: encoding characteristics and responses during natural orofacial behaviors. Behav Brain Res 135:27–33
Tsingotjidou AS, Papadopoulos GC (2003) Anatomic organization of the ascending branch of the milk-ejection reflex in sheep: primary afferent neurons. J Comp Neurol 460:66–79
Tsingotjidou AS, Papadopoulos GC (2008) The milk-ejection reflex in the sheep: an anatomical study on the afferent pathway. Anat Histol Embryol 37:245–250
Tyson JE, Hwang P, Guyda H, Friesen HG (1972) Studies of prolactin secretion in human pregnancy. Am J Obstet Gynecol 113(1):14–20
Tyson JE, Khojandi M, Huth J, Andreassen B (1975) The influence of prolactin secretion on human lactation. J Clin Endocrinol Metab 40(5):764–773
Ueda EK, Huang K, Nguyen V, Ferreira M, Andre S, Walker AM (2011) Distribution of prolactin receptors suggests an intraductal role for prolactin in the mouse and human mammary gland, a finding supported by analysis of signaling in polarized monolayer cultures. Cell Tissue Res 346:175–189
Uvnäs-Moberg K, Widström AM, Werner S, Matthiesen AS, Winberg J (1990) Oxytocin and prolactin levels in breast-feeding women. Correlation with milk yield and duration of breast-feeding. Acta Obstet Gynecol Scand 69(4):301–306
Van Goor F, Li Y-X, Stanko S, Stojilkovic SS (2001) Paradoxical role of large-conductance calcium–activated K^+ (BR) channels in controlling action potential–driven Ca^{2+} entry in anterior pituitary. J Neurosci 21(16):5902–5915
Vallbo AB, Johansson RS (1984) Properties of cutaneous mechanoreceptors in the human hand related to touch sensation. Hum Neurobiol 3:3–14
Varendi H, Porter RH, Winberg J (1994) Does the newborn baby find the nipple by smell?
Vorontsov IM, Fateeva EM, Hasenson LB (1993) Breastfeeding children. PPMI, SPb., 198 p. (in Russian)
Weber F, Woolridge MW, Baum JD (1986) An ultrasonographic study of the organisation of sucking and swallowing by newborn infants. Dev Med Child Neurol 28:19–24
Weitzman RE, Leake RD, Rubin RT, Fisher DA (1980) The effect of nursing on neurohypophyseal hormone and prolactin secretion in human subjects. J Clin Endocrinol Metab 51:836–839
West CP, McNeilly AS (1979) Hormonal profiles in lactating and non-lactatin women immediately after delivery and their relationship to breast engorgement. Br J Obstet Gynecol 86:501–506
Weston GC, Cann L, Rogers PA (2003) Myometrial microvascular endothelial cells express oxytocin receptor. BJOG 110(2):149–156
Widström AM, Winberg J, Werner S, Hamberger B, Eneroth P, Uvnäs-Moberg K (1984) Suckling in lactating women stimulates the secretion of insulin and prolactin without concomitant effects on gastrin, growth hormone, calcitonin, vasopressin or catecholamines. Early Hum Dev 10 (1-2):115–122
Wolff PH (1968) The serial organization of sucking in the young infant. Pediatrics 42:943–956
Woolridge MW (1986) The 'anatomy' of infant sucking. Midwifery 2(4):164–171
Woolridge MW, How TV, Drewett RF, Rolfe P, Baum JD (1982) The continuous measurement of milk intake at a feed in breast-fed babies. Early Hum Dev 6:365–373
Xuereb GP, Prichard MM, Daniel PM (1954a) The arterial supply and venous drainage of the human hypophysis cerebri. Q J Exp Physiol Cogn Med Sci 39(3):199–217
Xuereb GP, Prichard MM, Daniel PM (1954b) The hypophysial portal system of vessels in man. Q J Exp Physiol Cogn Med Sci 39(3):219–230
Zafeiriou DI (2004) Primitive reflexes and postural reactions in the neurodevelopmental examination. Pediatr Neurol 31:1–8
Zagorodnyuk VP, Brookes SJ, Spencer NJ (2010) Structure-function relationship of sensory endings in the gut and bladder. Auton Neurosci 153(1–2):3–11
Zanardo V, Straface G (2015) The higher temperature in the areola supports the natural progression of the birth to breastfeeding continuum. PLoS One 10(3):1–7

Zanardo V, Volpe F, de Luca F, Straface G (2017) A temperature gradient may support mother-infant thermal identification and communication in the breast crawl from birth to breastfeeding. Acta Paediatr 106(10):1596–1599

Zimmerman E, Barlow SM (2008) Pacifier stiffness alters the dynamics of the suck central pattern generator. J Neonatal Nurs 14(3):79–86

The Period of Established Lactation: Lactogenesis III

5

Abstract

During the period of steady-state lactation (lactogenesis III), the content of milk components, the dynamics of their concentrations, and the rate of secretion in mature milk change with the time of lactation. Each mother produces milk specific to the functional state of the child. The volume of milk depends on the blood content of the monomeric form of prolactin. An analysis of the dependence of increased breast productivity on the use of various galactagogues showed that the most effective pharmacological galactagogues is domperidone, increasing prolactin in the blood. A decrease in the secretory activity of the gland and its volume is observed, soon after the child reduces its milk consumption. This is accompanied by an involution of the breast.

5.1 Composition and Dynamics of Concentrations of Components of Mature Milk

Lactogenesis II in breastfeeding women lasts on average 10–12 days. During this time, the composition and volume of milk change relatively quickly (Figs. 4.3 and 4.4). Transition milk acquires the status of "mature milk." Lactogenesis II is replaced by a period of established lactation, which is referred to as lactogenesis III or lactopoiesis. Mature female milk consists of several phases: the water phase, which contains mineral and organic substances dissolved in water (87%), the colloidal phase with protein dispersion (0.3%), and the emulsion phase with fat globules (4%) and phase including live cells (Picciano 2001). The qualitative and quantitative composition of women's milk has been intensively studied since the 1990s of the last century, since data on the components of breast milk, along with fundamental interest, are of great practical importance, in particular for the development of the composition of milk formula (Allen et al. 1991; Vorontsov et al. 1993; Zanardo et al. 2001; Picciano 2001; Gossage et al. 2002; Newton 2004; Lawrence

N. P. Alekseev, *Physiology of Human Female Lactation*,
https://doi.org/10.1007/978-3-030-66364-3_5

and Lawrence 1999; Mastroeni et al. 2006; Ballard and Morrow 2013; Smilowitz et al. 2013; Bode et al. 2014; Chollet-Hinton et al. 2014; Andreas et al. 2015; Martin et al. 2016; Witkowska-Zimny and Kaminska-El-Hassan 2017; Cacho and Lawrence 2017). However, it should be noted that the qualitative and quantitative composition of mature woman's milk is not fully defined. As a result of better chemical analysis methods, such as nuclear magnetic resonance (NMR), more and more components are found in women's milk (Smilowitz et al. 2013; Andreas et al. 2015).

The components of breast milk can be divided into several categories (Picciano 2001) (Table 5.1).

The presence of cells and microorganisms in milk makes it a special liquid. According to the figurative expression of I. Bode (Bode et al. 2014), milk is a "living liquid." When looking at the categories of women's milk in more detail, it was found that some categories of mature women's milk consist of several hundred components. Thus, more than 200 fatty acids and 400 different proteins were found in milk (Andreas et al. 2015). Thus, the total number of components of mature milk can be very high. Experimental studies have shown that the composition of milk during lactopoiesis is not constant, but changes throughout lactation (Neville et al. 1991; Allen et al. 1991; Mitoulas et al. 2002a, b).

Long-term (12–17 months) surveys were conducted to study the composition of milk in lactating women (Neville et al. 1991; Allen et al. 1991; Mitoulas et al. 2002a; Perrin et al. 2017). Figures 5.1 and 5.2 present data on the dynamics of concentrations and rates of secretion of certain organic and mineral components of milk of lactating women during lactopoiesis (Allen et al. 1991). It turned out that the content of a number of milk components significantly changed over time. So, the concentration of protein, sodium, potassium, and citrate decreased by 25% between 1 and 6 months of lactation, and the concentration of lactose, ionized calcium, and glucose in these terms increases by 10%. It should be noted that there are discrepancies between different authors about the fat content in milk in the period of lactopoiesis. Analysis of available data in the literature indicates that the main reasons for the variation in the values of fat in the milk of a lactating woman are her nutrition and the volume of fat reserves in the body of a nursing woman. Thus, as a result of insufficient nutrition, the amount of fat in milk may decrease during lactation, in particular between 1 and 3 months of lactation from 3.5 to 3.1% (Prentice et al. 1983; Brown et al. 1986). With a sufficiently high-calorie diet, the fat content of milk in women is higher and between 4 and 6 months increased from 4.2 to 5.6% (Fig. 5.1).

Determination of the concentrations of milk components in the second year of lactation (Perrin et al. 2016) showed that the concentration of total protein increases. Particularly noted is the increase in the content of antimicrobial proteins: lysozyme, lactoferrin, and IgA. The level of lactose and fat did not change. The study of the mineral composition of milk during this period found a decrease in the concentration of zinc and calcium. It should be noted that changes in the concentration of milk components during lactopoiesis occurred more slowly than during lactogenesis II.

Intercellular contacts of secretory cells by the time of lactopoiesis are closed. In this regard, it can be assumed that the qualitative and quantitative composition of

Table 5.1 Components of human milk

Protein and nonprotein nitrogen	
Proteins	Nonprotein nitrogen
α-Lactalbumin	α-Amino nitrogen
α-Lactoglobulin	Creatinine
Caseins	Creatine
Enzymes	Glucosamine
Growth factors	Nucleic acids
Hormones	Nucleotides
Lactoferrin	Polyamines
Lysozyme	Urea
Secretory IgA	Uric acid
Other immunoglobulins	
Fats	
Fatty acids	
Phospholipids	
Sterols	
Triglycerides	
Carbohydrates	
Lactose	
Oligosaccharides	
Glycopeptides	
Bifidus factors	
Vitamins	
Fat-soluble vitamins	Water-soluble vitamins
Vitamin A	Biotin
Vitamin E	Choline
Vitamin K	Folate
Vitamin D	Inositol
Carotenoids	Niacin
	Pantothenic acid
	Riboflavin
	Thiamine
	Vitamin B12
	Vitamin B6
	Vitamin C
Mineral and ionic constituents	
Mineral macronutrients	Mineral micronutrients
Bicarbonate of	Chromium
Calcium	Cobalt
Chlorine	Copper
Citrate	Fluoride
Magnesium	Iodine
Phosphorus	Iron
Potassium	Manganese

(continued)

Table 5.1 (continued)

Mineral and ionic constituents	
Mineral macronutrients	Mineral micronutrients
Sodium	Molybdenum
Sulfur	Nickel
	Selenium
	Zinc
	Aluminum
	Barium
	Bohr
	Cadmium
	Lead
	Strontium
	Titan
Cells	Bacteria
Myoepithelial cells	*Staphylococcus*
Lactocytes	*Acinetobacter*
Stem cells	*Streptococcus*
Leukocytes	*Pseudomonas*
Lymphocytes	*Lactococcus*
Macrophages	*Enterococcus*
Neutrophils	*Lactobacillus*

milk in the alveolar-ductal system during lactopoiesis is mainly regulated by intracellular processes occurring in secretory cells. Through the capillary blood flow to the secretory cells are delivered milk precursors, some of which through the basal and apical parts of the membrane of secretory cells (III, IV pathways, Fig. 4.2) during intracellular control directly enter the alveolar cavity. Other precursors passing through the basement membrane inside the secretory cell are included in synthetic intracellular processes, and then the newly synthesized products are transported through the apical membrane to the alveolar cavity (I, II, Fig. 4.2), thereby establishing the necessary homeostasis of the milk composition of the nursing mother for a certain period of child development. Large experimental material indicates that the change in the composition of milk in a nursing mother occurs according to the age of the child, providing him with optimal physical and mental development, and is genetically determined. Recent research has found that the composition of a mother's milk is affected even by the sex of the child. The energy value of milk from mothers who fed boys was 25% higher than that of mothers who fed girls (Powe et al. 2010a, b).

The composition of the mother's milk depends on the infective status of the child. Thus, in the case of a child's disease, an increase in the number of white blood cells, macrophages, and TNF-α was detected in the mother's milk (Riskin et al. 2012).

Thus, each mother produces milk specific to her child, including its functional state, which is probably the main reason for differences in the composition of milk in

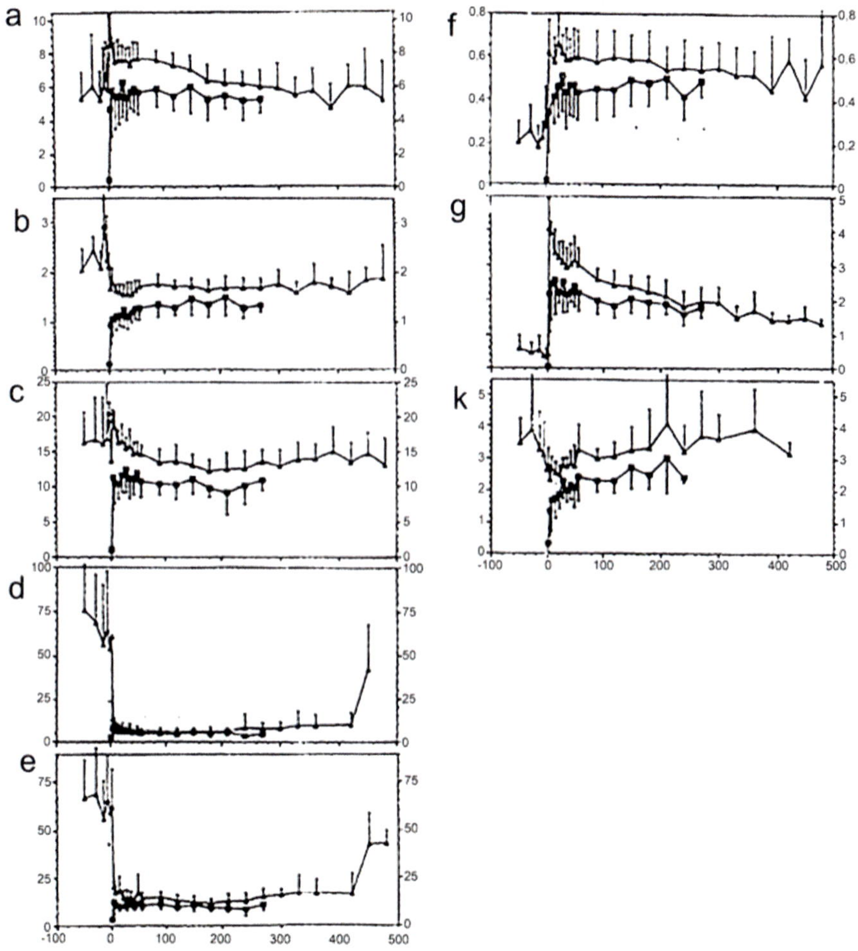

Fig. 5.1 Concentration (triangles) and secretion rate (squares). Some mineral components of milk, during lactopoiesis (from Allen et al. 1991). (**a**) Ca, (**b**) Mg, (**c**) K, (**d**) Na, (**e**) Cl, (**f**) free PO4-, (**g**) citrate, (**k**) Ca^{+2} ionized. Ordinate for all plots on the right is mM/L concentration, on the left for all plots is secretion rate. Abscissa: time—days after childbirth. Mean values for milk composition were determined by pooling across individuals. The secretion rate of each component for these days was derived for each individual by multiplying her time-corrected volume by time-correcting concentration of that component. Mean secretion rate was then derived by pooling across individuals

nursing women (Bravi et al. 2016). Additionally, the cause of variations in the composition of milk is the mother's nutrition. However, as already noted in the case of fat content of milk, the influence of the mother's nutrition on the composition of her milk is apparent with significant and long-term changes in the mother's diet.

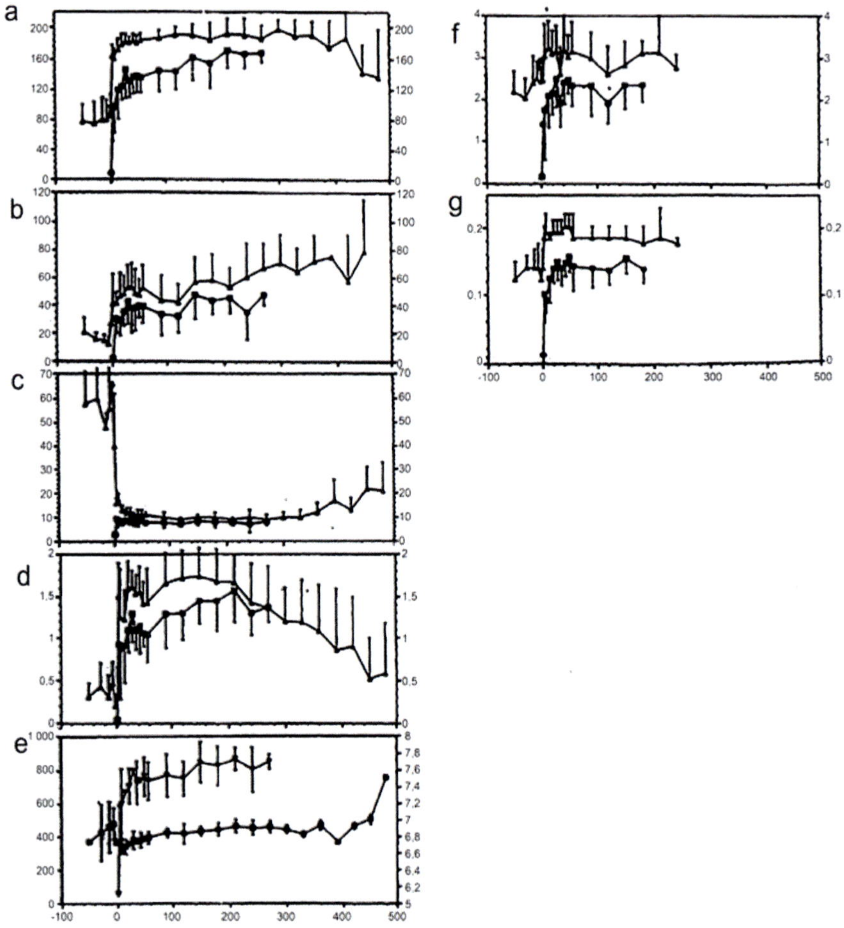

Fig. 5.2 Concentration (triangles) and rate of secretion (squares) of certain organic milk components, milk volume (triangles), and pH (squares) during lactopoiesis (from Allen et al. 1991). (**a**) Lactose, ordinate axis: mM/l concentration on the left, secretion rate on the right, (**b**) fat, ordinate axis: on the left concentration g/l, secretion rate on the right, (**c**) protein, ordinate axis: on the left is a concentration of g/l, secretion rate on the right, (**d**) glucose, on the ordinate axis: on the left is the concentration of mM/L secretion rate on the right, (**e**) the volume of milk and pH, ordinate axis: on the left is milk volume mL/day, on the right is pH, (**f**) urea, ordinate: mM/L concentration on the left, secretion rate on the right, (**g**) creatinine, ordinate: mM/L concentration on the left, secretion rate on the right. Abscissa: time—days after childbirth. Mean values for milk composition were determined by pooling across individuals The secretion rate of each component for these days was derived for each individual by multiplying her time-corrected volume by time-correcting concentration of that component. Mean secretion rate was then derived by pooling across individuals

5.2 Productivity of a Lactating Woman During Lactation

Along with the change in the qualitative and quantitative composition of milk components during lactopoiesis with normal growth of the child in the mother, there is an increase to a certain level of total milk volume (Fig. 5.2). A large experimental physiological material shows that the volume of milk during lactopoiesis is influenced by several factors, the main factor of which is the concentration of the hormone prolactin in the blood of lactating women.

5.2.1 Effect of Prolactin on the Volume of Secreted Milk

In the previous chapter, data were presented that the release of prolactin from lactotrophs of the adenohypophysis is regulated by dopamine through the D2 receptors of lactotrophs. A number of pharmacological compounds that block D2 receptors are known (Baron 1999; Reddymasu et al. 2007). In lactation practice, domperidone and metoclopramide are currently used. Oral administration of domperidone causes an increase in plasma prolactin in women (Fig. 5.3A) and, accordingly, an increase in milk production (Fig. 5.3B) (Knoppert et al. 2013), with a dose-dependent effect. At the same time, the use of dopamine agonists 2-bromo-alpha-ergocryptine or cabergoline (Dostinex) (Rolland et al. 1991) caused a decrease in the concentration of prolactin in the blood and a decrease in milk secretion. An increase in women's productivity could be caused by an increase in the concentration of prolactin in the blood using recombinant human prolactin (r-hPRL) (Page-Wilson et al. 2007; Powe et al. 2010a, b, 2011). Subcutaneous administration of r-hPRL was accompanied by an increase in the concentration of prolactin in the blood and an increase in milk during pumping in women with Sheehan syndrome, in which as a result of pituitary necrosis, lactotrophs are damaged and lactation stops (Powe et al. 2010a, b, 2011). It is important to note that in all cases, the increase or decrease in prolactin was accompanied by a corresponding increase or decrease in productivity during various periods of lactation.

However, there are experimental data in the literature (Howie et al. 1980; Hennart et al. 1981; Cox et al. 1996), which allowed us to suggest that prolactin does not have a determining value for the volume of milk, but only has a permissive, initiating action at the beginning of lactation for subsequent lactation (Hennart et al. 1981). As an illustration, we present data from surveys in two groups of lactating women who have the same basal, as well as caused by feeding the child, prolactin levels, in which the amount of milk excreted was significantly different (Howie et al. 1980). There are several reasons for this situation. The first of them may be the difference in the volume of glandular tissue in women with sufficient and insufficient lactation. Here it should be noted that numerous data from the clinic indicate that all lactating women have a difference in the size of the mammary glands and the amount of milk between the mammary glands. At the same time, it is difficult to assume that the concentration of prolactin in the blood of a gland that secretes a larger volume of milk is higher than in the blood of a gland with a smaller volume of milk.

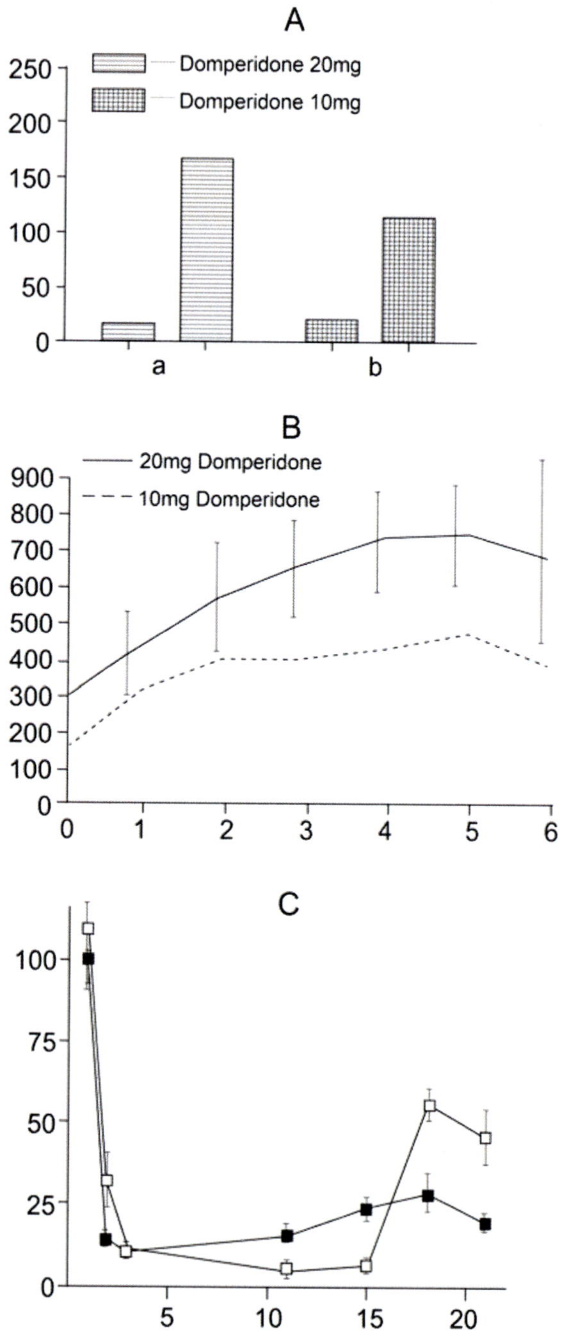

Fig. 5.3 Effects of antagonists (**A**, **B**) (from Knoppert et al. 2013) and agonists (**C**) (from Rolland et al. 1991) Д2 lactotrophic receptors on prolactin (**A**, **C**) content and productivity in lactating women. (**A**) Effect of domperidone on prolactin concentration. The ordinate for all histograms-concentration prolactin in blood plasma μg/L. (**a**) Left basal concentration, right 10 days after taking 20 mg of domperidone three times daily. (**b**) Left basal concentration, right 10 days after taking 10 mg of domperidone three times daily

The difference in the amount of milk at the same concentration of prolactin may be due to different secretory ability of glandular tissue. For example, the mammary glands of women with increased milk volume contain a larger number of type 3 lobes (Fig. 2.11), whose secretory cells produce a larger volume of milk (Russo and Russo 2004a, b).

Milk secretion may be affected by the fractional composition of prolactin. As already noted, the main active forms of prolactin contained in the human adenohypophysis are monomeric variants (Fig. 4.37A). In surveys to determine the relationship between the volume of milk and the concentration of prolactin, total prolactin was determined. Therefore, it is possible that in lactating women with reduced productivity, with an equal total concentration of prolactin with productive women, less active polymer forms predominate in the blood. It can also be noted that milk secretion may be different due to the number of prolactin receptors in secretory cells at the same concentration of prolactin in the blood of lactating women (Marasco 2014).

5.2.2 Effect of Oxytocin on the Volume of Secreted Milk

The other main lactogenic hormone oxytocin also affects the volume of secreted milk. However, the effect of oxytocin is mediated through emptying the alveolar-ductal system of the breast of a lactating woman. Here it should be noted that in contrast to other exocrine glands, the mammary gland has two features (Peaker and Wilde 1996): (1) the secretion continues throughout lactation, and not only during the action of adequate stimuli, and (2) the secret—milk—is not immediately released to the outside, but is initially collected in the alveolar-ductal capacity of the gland. Since the middle and thick milk ducts of the breast of a woman have sensory innervation-mechanosensory and pain, filling the milk ducts will cause stretching of their walls and stimulation of the nerve endings located on the walls of the ducts. Initially, the woman feels "heaviness" in the gland, and with further filling (overflow) of the ductal system, pain begins to appear. Breastfeeding or pumping milk (manually or using a breast pump) eliminates these unpleasant sensations. Undoubtedly, in lactating animals, filling and overflowing of the mammary gland's capacitance system are also accompanied by corresponding "feelings." For example, cows manifest an alarm bellowing if their glands are full of milk because they have not been milked for a long time. Since lactating animals themselves cannot remove excess milk, overfilling the milk storage system can cause damage to the duct walls. In this regard, it has been suggested that the filling of the mammary gland's capacitive system is regulated by a special chemical factor or factors contained in milk (Peaker and Wilde 1996). In initial animal experiments (Wilde et al. 1987), it was found that a fraction of goat milk with a molecular weight of 10–30 KD contains a protein factor(s) that inhibits milk secretion, called Feedback Inhibition Lactation (FIL). The introduction of a solution with this factor(s) into the ducts of the mammary gland of a lactating rabbit inhibited the accumulation of milk in its capacitive system. A similar phenomenon was observed when a solution with this factor (factors) was introduced into the cisternal part of the udder of a goat (Wilde et al. 1988). Subsequently, FIL was found in other mammals (cows, marsupials), including lactating women. Prepared by a method similar to that used in animal experiments (Prentice et al. 1989), the fraction of woman's milk containing protein components with a molecular weight of more than 10 KD when biotesting cultured glandular tissue explants of lactating rabbits inhibited the synthesis of lactose and casein. Unfortunately, a detailed study of the FIL of breast milk was not conducted. The cited work (Prentice et al. 1989) is currently the only experimental confirmation of the presence of FIL in women's milk. However, the results of this survey suggested that women's milk contains a factor(s) that inhibits the secretion of milk volume. It should be noted here that data on animal FIL is also scarce. Thus, the molecular structure of FIL, the dynamics of its concentration during the accumulation of milk in the capacitive system of the gland, and the place of its action on secretory cells are unknown. However, taking into account the available data obtained on animals (Peaker and Wilde 1996), it can be assumed that FIL is formed in the secretory cells of the gland and continuously enters the capacitive system of the gland. An increase in the FIL concentration is possible through the corresponding FIL receptors located on the apical surface of secretory cells which will inhibit the formation of milk. Therefore, in order to maintain the necessary volume of milk secretion, the breast's capacitive system must be sufficiently periodically emptied. This can only be achieved with the help of the hormone oxytocin, which is released from the neurohypophysis during the implementation of the milk withdrawal reflex. As already noted, in the absence of a reflex of milk excretion, and the absence of oxytocin release, the child can withdraw several ml of milk, which is located in the thick ducts in the areolar region. It should be noted here that since the filling of the capacitive system is accompanied by a corresponding feeling, it is possible that in response to mechanical stimulation of the mechanoreceptors and pain receptors of the capacitive system of the breast, biologically active substances may be released from the central nervous system—factors that in addition to FIL will reduce the formation of milk.

The volume of milk sufficient to feed a child is an indicator of successful lactation of nursing women. In this regard, lactation of women of all times and peoples was aimed at achieving the secretion of milk in the quantities necessary for the growth and development of the child. However, as has already been noted, at all times and in all peoples, there have been nursing women whose amount of milk produced by them was not sufficient to feed the child (Wickes 1953a, b, c, d, e). Therefore, for a long time, there have been found lactic remedies that increase the volume of milk in women with insufficient lactation—hypogalactia. Given the importance of this issue for women's lactation, the next paragraph of this chapter is devoted to lactation agents—*galactagogues*—and their mechanisms of action on women's lactation.

Table 5.2 Plants that help increase the volume of milk in hypogalactia

Household name	Latin name
Fenugreek	*Trigonella foenum-graecum*
Goat's-rue	*Galega officinalis*
Common fennel (dill)	*Foeniculum vulgare*
Monk's pepper	*Vitex agnus-castus*
Kiprey	*Epilobium* spp.
Anise	*Pimpinella anisum*
Blessed thistle	*Cnicus benedictus*
Milk thistle (thistle)	*Silybum marianum*
Common nettle	*Urtica dioica*
Cotton seed	*Gossypium* spp.
Common cumin	*Carum carvi*
Common oregano	*Origanum vulgare*
Asparagus	*Asparagus racemosus*
Spur-flower fragrant	*Coleus amboinicus*
Moringa	*Moringa oleifera*
Garlic	*Allium sativum*
Ben oil tree	*Moringa oleifera*

5.2.3 Galactagogues

The first galactagogues that were used since ancient times to increase the milk productivity of lactating women were of plant origin. The so-called herbalists-reference books of medicinal plants of various peoples, which described the healing and medicinal properties of various herbs, and their use in the form of decoctions, infusions, ointments, and powders for the treatment of various diseases, including hypogalactia, have reached our days. Currently, in Russia and other countries of the world, a number of plants that are considered to help a nursing woman in the case of hypogalactia to increase the volume of milk are most often used as lactation agents (Table 5.2):

It should be noted that in some cases, more than one plant is used as a lactation agent, but in combination with other lactation herbs (Mortel and Mehta 2013; Bazzano et al. 2016; Bumrungpert et al. 2018), or extracts from lactation herbs are added to specialized products for nursing women (Gmoshinskaya et al. 2004a, b). It is of interest to what extent each herbal remedy affects the increase in the volume of secreted milk, as well as the mechanisms by which lactation is enhanced. It should be noted that in the scientific literature, information about the milk-bearing properties of plants was presented relatively long ago. However, these data were mostly qualitative "anecdotal" in nature. For example, one study (Jensen, 1945) reported that the use of fenugreek (*Trigonella foenum-graecum*) by Egyptian women during lactation increased their productivity by 900%. Quantitative research on the impact of plant products on the milk productivity of lactating women began to appear since the end of the last century. However, analysis of these studies shows that among them, single surveys represent randomized controlled trials (Mortel and Mehta 2013; Bazzano et al. 2016). Moreover, only three of the most popular plants in medical practice (Bazzano et al. 2016): asparagus (*Asparagus racemosus*), fenugreek (*Trigonella foenum-graecum*), and ben oil tree (*Moringa oleifera*) were examined more than once. The results were statistically reliable (positive effect) only in one case for each plant. In addition, review authors (Mortel and Mehta 2013; Bazzano et al. 2016) additionally note that differences in methods, dosages of plants used, and effects estimates make it difficult to interpret the results.

Difficulties also arise regarding the explanation of the mechanisms of the lactogenic effect of plants. For the plants listed above that have been randomized, it is assumed that the increase in productivity occurs due to an increase in the number of secretory cells of glandular tissue as a result of the action of phytoestrogens contained in these plants (Mortel and Mehta 2013; Bazzano et al. 2016). It should be noted here that difficulties in interpreting the data may be due to the wide range of effects of milk-bearing plants on the human body. Therefore, the lactogenic property of the plants listed above can be mediated through the influence on the functioning of various body systems of lactating women. Thus, the available data do not allow us to consider with sufficient certainty the presence of a lactogenic effect in known plants. To find out the lactogenic properties of plants, it is necessary to increase the number of randomized studies with standardization of survey conditions.

In this section, you should focus on such an important issue as side effects when using herbal milk products. Clinicians believe that teas and infusions from natural milk herbs are harmless to nursing women and their children. However, all herbal preparations contain biologically active substances that, when often consumed by herbal teas, can adversely affect a woman or penetrate milk acting on the child. For example, the milky grass "number 1"—fenugreek (*Trigonella foenum-graecum*)—contains coumarin and nicotinic acid, which can have a strong effect on heart rate, blood pressure, blood sugar, and other body functions (Shawahna et al. 2018). The use of lactic herbs in high concentrations can cause stomach disorders and nausea. Therefore, along with determining the galactogenic effect of plants, it is necessary to consider the side effects that the studied herb may have on the body.

In addition to herbal remedies, pharmacological lactic substances are actively used in lactation practice. There are a number of organic compounds that have a positive effect on increasing the milk productivity of lactating women: human growth hormone (hGH), thyrotropin-releasing hormone (TRH), metoclopramide (Cerucal), and domperidone (Motilium).

hGH Numerous studies of lactation of farm animals have shown that bovine growth hormone (bGH) can increase the milk productivity of animals by more than 30%. Comprehensive research on the use of bGH has allowed it to be used as a lactogenic agent in the industrial production of milk. Studies on lactating women have shown that the introduction of percutaneous hGH was also accompanied by an increase in milk productivity in lactating women (Milsom et al. 1992; Gunn et al. 1996). At the same time, as in animals, the concentration of IGF-1—an insulin-like growth factor—increased in the blood of women. This hormone regulates the processes of cell differentiation and their development and growth and participates in glucose metabolism. It is believed that milk formation is stimulated indirectly through IGF-1. However, for a number of reasons, namely, the need to inject the hormone subcutaneously with a syringe several times a day and the high cost, hGH has not become widespread in clinical practice (Forinash et al. 2012).

TRH The effect of this hormone on the level of prolactin in the blood of lactating women was discussed in detail in Sect. 4.2.4, Chap. 4. A review (Forinash et al. 2012) of studies on the effect of TRH on the milk productivity of lactating women showed that the administration of TRH intranasally, orally, or intravenously in various concentrations increased the blood content of prolactin, triiodothyronine, and thyroxine in women. However, only one out of four studies showed a noticeable increase in milk productivity. Since the duration of examinations was not sufficient to be sure that mothers and children would not develop hyperthyroidism, as well as a slight increase in milk secretion, TRH was not recommended as a galactagogue.

Domperidone and metoclopramide (Cerucal) are organic substances (Baron 1999) that are not hormones. They are widely used in clinical practice for the treatment of dyspepsia in adults and children. In particular, as a result of blocking D2 receptors in the structures of the gastrointestinal tract, as well as in various structures of the central and peripheral nervous system, domperidone (Motilium) and metoclopramide facilitate the formation of peristalsis and emptying of the gastrointestinal tract. In addition, it turned out that these substances as a side effect (Hofmeyr and van Iddekinge 1983; Cann et al. 1983; Maddern 1983) increase the content of prolactin in the blood of lactating women and, accordingly, increase the milk productivity through prolactin. As already noted (Sect. 4.2.1), the increase in prolactin concentration occurs through blocking of D2 receptors of adenohypophysis lactotrophs. Domperidone is most often used because, unlike metoclopramide, it does not penetrate the blood-brain barrier and has a peripheral effect on D2 lactotrophs of the adenohypophysis, located before the blood-brain barrier. Systematic research on the effect of domperidone on the milk productivity of lactating women with insufficient lactation began in 1985. Among them, we can note a number of studies, the results of which indicate a high efficiency of using Motilium to increase the milk productivity of women, especially in preterm birth (Petraglia et al. 1985; Da Silva et al. 2001; Wan et al. 2008; Campbell-Yeo et al. 2010; Knoppert et al. 2013). Domperidone dose dependently increases the content of prolactin in the blood of lactating women and the amount of secreted milk (Fig. 5.3). However, in some cases, there was little or no increase in milk production in nursing women (Grzeskowiak et al. 2013). Perhaps the reason is due to factors that were noted in Sect. 4.2.1, for example, when Motilium is used, there is an increase in the concentration of prolactin, but with the predominance of less active polymer forms.

In clinical practice, domperidone as a lactogenic agent is most often used in a concentration of 10–20 mg orally, three times a day. Calculations show that a child at such doses will receive a small dose of domperidone through the mother's milk

(Anderson 2017; Hondeghem and Logghe 2017), and the risk of toxicity can be ignored. However, for the mother, the question of domperidone safety, according to some authors, remains open. The fact is that domperidone and metoclopramide, in addition to increasing the concentration of prolactin, have another side effect. They cause blocking of hERG (human ether-a-go-go-related gen) of potassium channels involved in repolarization of cardiopotentials (Claassen and Zünkler 2005; Stork et al. 2007). Accordingly, there is an increase in the duration of the QT interval, which can cause cardiac arrhythmia. It should be noted that the first side effect of domperidone in the form of cardiac arrhythmia was reported in 1980 when it was administered intravenously. At the same time, calculations showed that the concentration in blood plasma of domperidone at 10 mg intravenous administration was approximately 80–150 times greater than at oral administration of 20 mg (Betzold 2004). However, further analysis of a large number of patients in the postpartum period showed that oral domperidone increases the percentage of mothers with arrhythmia (Smolina et al. 2016). However, the absolute value of the occurrence of cardiac arrhythmia was very small: 15 patients out of 45,518. Contributing factors to cardiac arrhythmia in patients were the presence of previous cases of arrhythmia and being overweight (Smolina et al. 2016). It is interesting to note the results of surveys on the effect of domperidone on the QT interval in male and female patients aged 19–39 years. It turned out that domperidone at a concentration of 10 mg orally four times a day in women did not significantly change the duration of the QT interval compared to placebo. At the same time, in men, QT increased by 4.2–9.11 ms. It should be added that in clinical practice, doctors who prescribed domperidone to nursing women did not observe cases of cardiac arrhythmia (Grzeskowiak and Amir 2015; Anderson 2017). Thus, the available data indicate that domperidone is the most effective pharmacological galactagogue. However, because of its possible cardiac arrhythmic effect, Motilium should not be used in nursing women with cardiovascular diseases (Da Silva and Knoppert 2004).

In addition to medications, physiotherapy methods are used to increase and preserve the milk productivity of lactating women during the entire lactation period, and the main one is the use of breast pump.

5.2.4 Use of the Breast Pumps to Increase and Preserve the Milk Productivity of Lactating Women

The use of breast pump to establish and maintain lactation in nursing women, as well as to increase their milk productivity, has increased significantly worldwide in recent years (Meier et al. 2016). In some cases, due to a number of circumstances (premature baby, child's illness, mother's illness), a woman is forced to express milk using breast pump for weeks or months. In the literature, the term pump-dependent women has even appeared (Meier et al. 2008). In this regard, it is advisable to focus in more detail in this section on various types of breast pump and their function and contribution to improve the milk productivity of lactating women. As already noted in the Introduction, the problem of milk productivity in nursing women has existed since ancient times. If there was no systematic emptying of the mammary glands in nursing women for various reasons and mainly due to the inability of the child to effectively remove milk, milk productivity decreased. According to the data presented in the previous sections (p. 4.2.1, p. 4.2.2), this occurs for two main reasons: a decrease in the concentration of prolactin in the blood and accumulation within the alveoli of FIL. It should be noted that the algorithm for removing milk from a child in ancient times did not seem to differ from the algorithm for removing milk from a modern child. The child uses two components for milk withdrawal: vacuum and compression (Fig. 4.13). Thus, in order to empty the gland and supply the baby with milk, the woman had to somehow create a vacuum or squeeze—withdraw the milk manually. Apparently, the original method of removing milk from a woman's gland was the method of squeezing, which women use to this day. At the same time, it can be assumed that doctors (healers) of ancient times with inventive abilities tried to simulate the vacuum component of removing the baby's milk. They managed to do this because during archaeological excavations, for example, in the Mediterranean countries, simple devices were found dated to the sixth to fifth century BC.

The first breast pumps found in archaeological excavations in the Mediterranean area (Fig. 5.4a) (Obladen 2012) were ceramic products shaped like teapots. The bottom of the "kettle" was missing, and vice versa, the upper part, where the usual kettle was filled with water, was closed. The device was applied with an open surface to the mammary gland, and through a kind of spout "kettle," someone created a vacuum and the milk from the gland was sucked into the inner volume of the vessel. The short "nose" was inconvenient for sucking milk. Therefore, in other versions, the "nose" was elongated so that the woman herself could suck milk from her breast (Fig. 4.4b). It is interesting to note that this type of device has been used for a long time in lactation practice. For example, in the nineteenth century in the United States (Walker 2005), products made of glass in the form of a cup with a long tube were used to remove milk and correct flat nipples. Extracting milk using a vacuum created in the oral cavity by the woman herself (Fig. 5.4c) or her assistant was a labor-intensive task, so as technology developed, the possibility of creating a vacuum using mechanical devices was sought. In particular, manual piston devices were used for sucking milk, which were connected to a cup placed on the breast (Fig. 5.4d). With the advent of rubber, the vacuum in the cup was created using a rubber balloon (Fig. 5.4e). Here it should be noted that the piston principle and the principle of the rubber balloon proved to be very successful for pumping milk from the mammary glands of lactating women. A detailed description of options for this type of breast pump used at various times in practice can be found in a number of reviews (Lawrence and Lawrence 1999; Walker 2005; Obladen 2012). However, in order to effectively remove milk, it was necessary to set the optimal vacuum amplitude, the frequency of movement of the piston, or compression of the rubber cylinder, i.e., the frequency and duration of vacuum stimuli. This was especially important for women who were going to withdraw milk for a long time. When manually creating a vacuum, it was almost impossible to identify the optimal parameters of vacuum stimuli for the best milk excretion, as well as to compare the effectiveness of

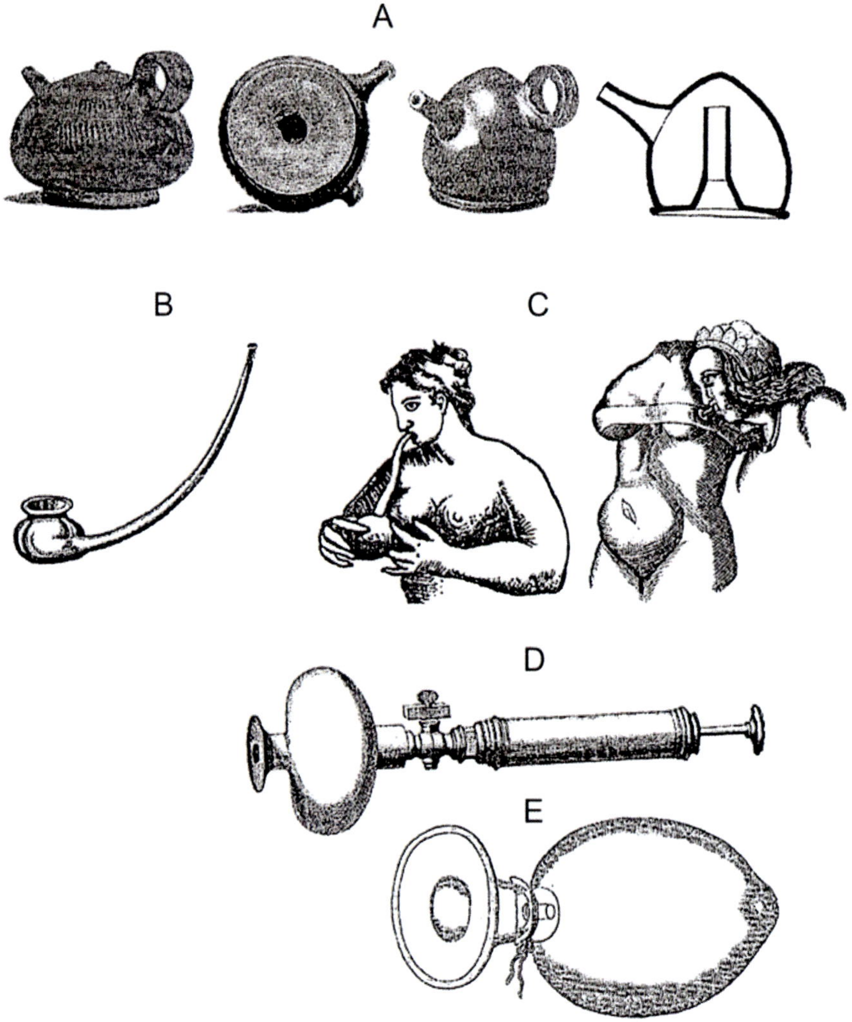

Fig. 5.4 Milk breast pumps of various times and peoples (from Obladen 2012). (**a**) Antique milk breast pumps. (**b**, **c**) Milk breast pumps of the sixteenth to seventeenth centuries. (**d**, **e**) Milk breast pumps of the eighteenth century

different types of breast pumps. In addition, since in the middle of the nineteenth century, mankind learned that the cause of various diseases, including diseases of the mammary glands, is pathogenic microorganisms, and there was a problem of sterilization of breast pumps. An important step in improving breast pumps was the use of electric motors to create a vacuum. Here, first of all, we should note the classic development of Einar Egnell (1956) (Fig. 5.5a). His breast pump consisted of an electric motor, whose rotation through a crank mechanism was converted into reciprocating motion of the piston, creating a pulsating vacuum and pressure. With

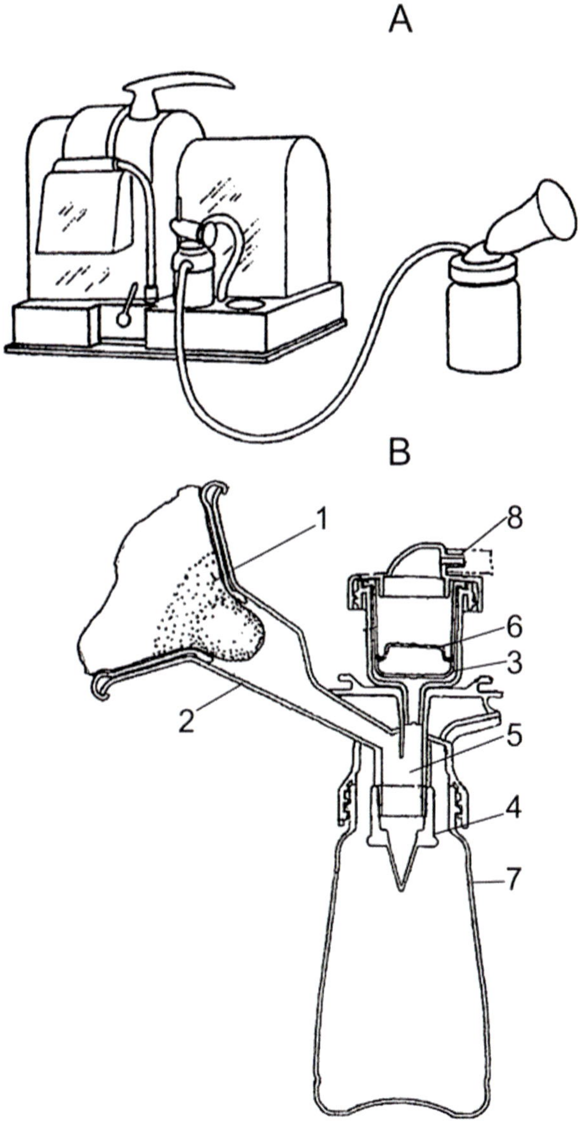

Fig. 5.5 E. Egnell's milk breast pump (**a**) (from Lawrence and Lawrence 1999) and a lining of modern milk breast pumps (**b**). (**b**) (1) Breast nipple and areola, (2) plastic cone, (3) elastic cylinder, (4) check valve, (5) volume inside the plastic cone, (6) compressed elastic cylinder, (7) milk collector, (8) inlet to the elastic cylinder of vacuum stimulus from the compressor

the help of a special valve, the main part of the excess pressure was removed to the atmosphere, and the vacuum was fed through a hose to the milk collector (bottle), connected to a funnel-shaped cup that was placed on the areola with the breast nipple (Fig. 5.5). The motor had a fixed speed of rotation, which in contrast to the manual movement of the piston provided a stable frequency and duration of vacuum stimuli. The amplitude of the vacuum pulses could also be adjusted fairly accurately using a simple device. In this way, the nipple and areola were rhythmically affected by vacuum stimuli representing half of the sinusoid in shape. During the second half of the sinusoid, when the piston created a pressure higher than atmospheric pressure, the pressure inside a funnel-shaped cup rose for a short time by a small amount above atmospheric pressure, and then decreased to atmospheric pressure and followed by a rest phase, during which the pressure inside the lining was equal to atmospheric pressure. A small short increase in pressure was necessary for more efficient operation of the valve, which provides the formation of vacuum pulses. E. Egnell found that the vacuum of 200 mm Hg is the upper limit for the removal of milk by the breast pump. Exceeding this level was often accompanied by pain and in some cases caused damage to the nipple (Biancuzzo 1999). The most effective method for removing milk was the impulse effect of vacuum. At the same time, when the duration of vacuum stimuli is more than 2 s (frequency of 30 imps/min), some women also experienced pain when removing milk. The breast pump created by E. Egnell was the basis for the development of modern breast pumps. As a result of technical progress, it was possible to reduce the size of electric compressors, which led to the appearance of small-sized electric breast pumps for individual use. In addition, the sanitary and hygiene capabilities of breast pumps have been significantly improved. In particular, the funnel-shaped cup was now made of plastic, which allows for sterilization at a temperature of 100 °C, and it was also possible to avoid getting milk into the compressor. Currently, several types of manual and electric breast pumps are used in the practice of breastfeeding, which differ in design and size. However, there are no fundamental differences in the device and functioning. Practice shows that the most effective for removing milk are electric breast pumps. All modern electric breast pumps consist of two main units: a control unit with a compressor and a funnel-shaped cup with a milk collector, which is placed on the breast. In breast pumps used in the clinic, the compressor with the control unit is located separately from the funnel-shaped cup. In individual electric breast pumps, a miniature compressor with control is combined with an overlay. Currently, in both types of electric breast pumps used are of the diaphragm-type compressors. Figure 5.5b shows as an example the scheme of the device, which is used in various modifications in modern clinical or individual electric breast pumps. Breast nipple with part of the areola (1) is located inside plastic cone (2). The result is a closed volume bounded by the nipple with the areola, (1) elastic cylinder, (3) and check valve (4). The vacuum in the volume (5) is where the nipple on the areola is created using the elastic cylinder (3). The inside of the cylinder from the compressor is a pulse of vacuum (8). Since the initial pressure in a limited volume is equal to the atmospheric pressure, when the vacuum increases, the elastic cylinder, depending on the vacuum amplitude, begins to be compressed by atmospheric pressure by a certain amount (6), and its movement resembles that of a piston. There is an increase in the volume inside the cone, as a result of which the pressure in the volume decreases below atmospheric pressure, i.e., a vacuum is formed. The check valve closes and prevents the intake of atmospheric air from the milk collector (7). Due to the resulting pressure difference, the milk begins to flow out of the gland. In addition, under the influence of the vacuum, the nipple and part of the areola are drawn inside the cup. In this case, the skin surface of the nipple and part of the areola of the breast are stretched, and the mechanoreceptors located in it are stimulated. Then the

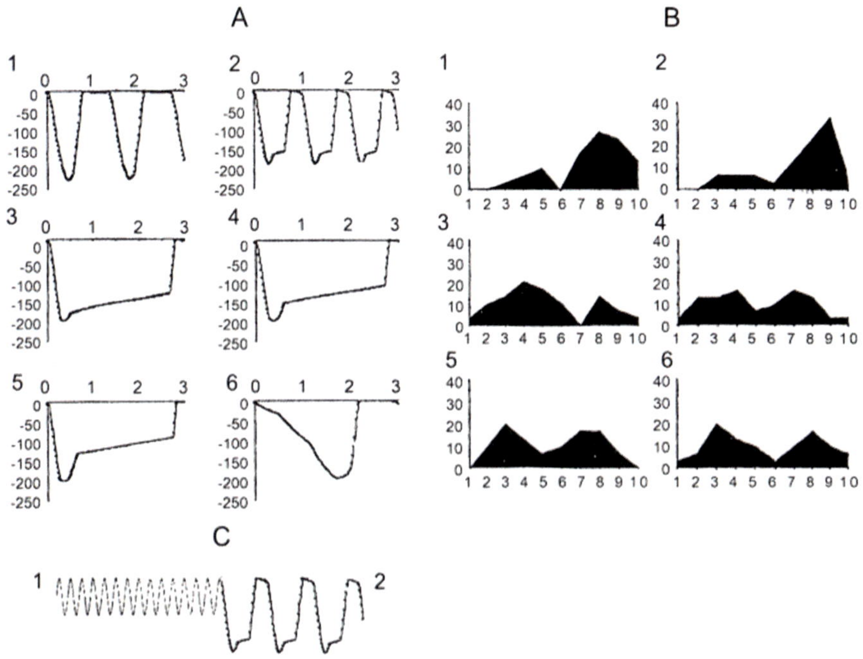

Fig. 5.6 Profiles (**a**) of vacuum stimuli, their "comfort" (**b**) in milk excretion by experimental vacuum pump (from Mitoulas et al. 2002b), and the sequence of vacuum stimuli in the Medela Symphony milk pump. (**a**) Profile of vacuum stimuli. On all plots ordinate: vacuum in mm Hg, abscissa time in s. (**b**) "Comfort" vacuum stimuli. In all ordinate plots, the number of women as a percentage of the total number, abscissa, the comfort coefficient. (**c**) Sequence of stimulating (1) and milk sucking vacuum stimuli (2)

vacuum begins to decrease inside the elastic cylinder and the cylinder spreads out. In order to prevent residual vacuum from accumulating in the volume, the elastic cylinder must take its original shape. Therefore, a small amplitude and short duration of positive pressure were applied inside the cylinder after the vacuum part of the stimulus ended. During the rest phase, the milk is drained through a check valve into the milk collector (5), and then the cycle is repeated. Thus, in this design of the breast pump, milk was excluded from entering the compressor and the control unit. In addition, all parts of the funnel-shaped cup in all types of breast pumps are removable and easy to clean and sterilize.

On electric clinical breast pumps from Medela AG, tests were conducted to optimize the amplitude and shape of vacuum stimuli on 23 lactating women. In amplitude optimization studies (Mitoulas et al. 2002a, b), vacuum stimuli were half-sinusoid in shape and were transmitted into the plastic funnel-shaped cup at a frequency of 47 imps/min (Fig. 5.6). The main indicator of the effectiveness of vacuum stimuli was the amount of milk withdrawn over a certain time. It should be noted that according to the laws of hydrodynamics, the velocity of the liquid in the tube is proportional to the pressure acting on the liquid. Surveys showed that the amount of milk excreted was also proportional to the amount of vacuum stimuli. Since women could experience pain at high vacuum values, a maximum of comfortable vacuum was initially set. Its value fluctuated significantly in various women from −98 to −270 mm Hg and averaged to −190.7 ± mm Hg (Kent et al. 2008), which is consistent with the upper limit of the vacuum set by E. Egnell. For small values of vacuum, −190.7, −143, −125, and −75 mm Hg, the amount of milk withdrawn in 15 min was 118.5 ± 11.4 ml, 90.7 ± 9.4 ml, 81.2 ± 11.2 ml, and 73.2 ± 11.0 ml, respectively. Here it is interesting to note that the number of reflex peaks of milk excretion for 15 min in different women varied significantly (1–12). On average, it was 4.3 and did not depend on the amplitude of the vacuum stimuli used (Kent et al. 2008). Perhaps within the limits of the used values of vacuum stimuli, the stretching of the skin surface of the nipple and areola was sufficient to stimulate the mechanoreceptors and form reflex peaks of milk excretion. Increasing the vacuum in each case additionally increased the rate of milk excretion depending on the vacuum amplitude.

The vacuum stimuli used by the child on the nipple and areola of the breast (Fig. 4.10) or on the nipple (Fig. 4.13) when removing milk are similar to the half of the sinusoid or trapezoid. However, it is possible to choose the form and duration of vacuum stimuli in the breast pump, which could be more effective for removing milk than the vacuum stimuli of the child.

Figure 5.6, a shows profiles of vacuum stimuli in an experimental breast pump from Medela, which were used to remove milk from 30 lactating women (Mitoulas et al. 2002b). It turned out that for each profile, there is a maximum comfortable vacuum. It had the highest value for 1 and 5 profiles, respectively, −191.3 ± 6.5 mm Hg and −182.2 ± 7.4 mm Hg. For profiles 2, 4, 6, and 3, it was less and was respectively −175.9 ± 8.9 mm Hg, −174.4 ± 6.2 mm Hg, −174.1 ± 7.7 mm Hg, and −167.2 ± 6.0 mm Hg. These data allowed the authors to conclude that the profile of vacuum stimuli is important for the effectiveness of breast pumps (Mitoulas et al. 2002b). It is interesting to note that the most effective in removing milk and comfort were vacuum profiles 1 and 2 (Fig. 5.6b), which have the greatest similarity to the steplike vacuum stimuli that a child uses when removing milk from the mother's gland or from the nipple (Figs. 4.10B and 4.13b).

We (Alekseev and Ilyin 2015) conducted detailed studies to optimize the duration of step vacuum stimuli used in breast pumps when expressing milk from the breast of a nursing woman. It is important to point out that, according to the available experimental data, the effectiveness of pumping milk in a woman during the day largely depends on her physiological and psychological state (Lawrence and Lawrence 1999). Therefore, it is possible to estimate with sufficient accuracy the difference in the amount of milk withdrawn using vacuum stimuli of different durations if the milk is expressed in the same session in turn in different containers using a series of vacuum stimuli of different durations. However, there are difficulties associated with the impulsive nature of milk ejection reflexes. We have already mentioned (p. 4.2.3) that under the influence of stimulation of mechanoreceptors of the areola of the breast, the hormone oxytocin is periodically released from the woman's neurohypophysis into the bloodstream, which,

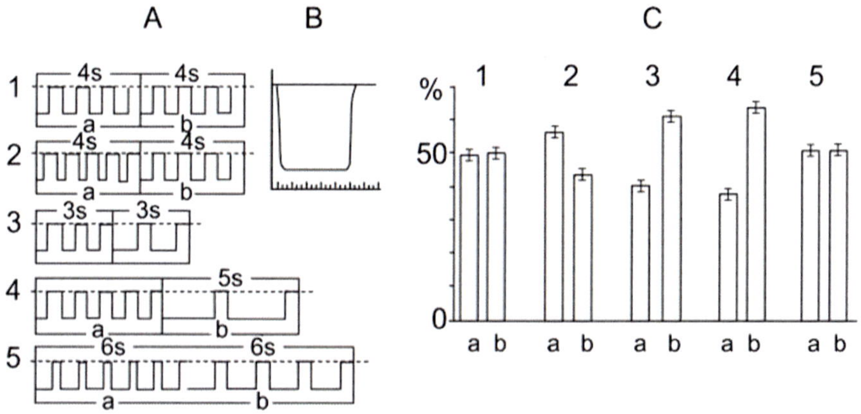

Fig. 5.7 The sequence of vacuum stimuli (**A**), the form of vacuum stimuli (**B**), and the volume of milk in milk collectors a and b excreted by vacuum stimuli of different duration (**B**) (from Alekseev and Ilyin 2015). (**A**) 1–5, horizontal lines on the top indicate the time of the pulse series in s, at the bottom indicate the series of pulses that were supplied to milk collectors a and b. The dotted line indicates the atmospheric pressure level. (**B**) Shape of vacuum pulses. Abscissa: time in ms, the entire scale is 700 ms. Pulse amplitude, −150 mmHg. (**C**) The volume of milk in milk collectors a and b, excreted by vacuum stimuli of various durations in percent of the total amount of milk. Amount of milk in milk collectors excreted by vacuum stimuli with duration of 1–0.5 s (a) and 0.5 s (b), 2–0.5 s (a) and 0.3 s (b), 3–0.5 s (a) and 1 s (b), 4–0.5 s (a) and 2 s (b), 5–0.7 s (a) and 1.05 s (b) pulses. Ordinate: the volume of milk in milk collectors as a percentage of the total amount of excreted milk

accordingly, periodically increases the pressure in the ductal system of the breast (Fig. 4.31), and the peak of pressure increase lasts for 1.5–2 min. The amplitude of peaks and their sequence also change over time. Moreover, these parameters differ markedly for different women (Figs. 4.33 and 4.34). In this regard, alternate pumping of vacuum stimuli with different duration of stimuli in a series into different containers with duration of a series, for example, 2 min, will introduce significant uncertainty in the results. Thus, when pumping milk with vacuum stimuli of one duration and profile during the peak of pressure increase in the alveolar-ductal system of the gland, the amount of milk withdrawn will be greater than when pumping with vacuum stimuli of another duration and profile during the period of lowest pressure in the alveolar-ductal system. Therefore, in order for the effect of the difference in pressure in the ductal system of the gland on the excretion of milk to be minimal, periodically successive series of vacuum pulses with different duration of stimuli in the series should have the shortest exposure time. A technique was developed (Alekseev and Ilyin 2015), which allowed for the removal of milk to act alternately on the nipple and areola with short (3–6 s) series of vacuum stimuli. The effect of steplike vacuum stimuli with a duration of 0.3 s, 0.5 s, 0.7 s, 1 s, and 2 s on milk excretion was studied (Fig. 5.7A). The total duration of the vacuum pulse series on the nipple and gland areola was the same. At the same time, an integer number of periods of both types of vacuum pulses were placed in the series

(Fig. 5.7A). The duration of the milk withdrawal session was 20 min. Thirty-nine lactating women were examined using an experimental device.

In the first part of the work, vacuum stimuli with a duration of 0.5 s, a frequency of 1 imp/s, and an amplitude of 120 mm Hg were selected as the basic ones. Vacuum with an amplitude of 120 mm Hg was comfortable for all women. The amount of milk extracted using 0.5 s vacuum stimuli was compared with the volume of milk extracted using 0.3 s, 1 s, and 2 s vacuum stimuli (Fig. 5.7A). Here it should be pointed out that vacuum stimuli of various durations in our experiments had a rise time of the ascending and fall time of the descending phase of about 0.05 s (Fig. 5.7B). The milk was ejected to two different milk collectors, (a) and (b).

In the first survey (control) (Fig. 5.7A1), the milk collector (a) received milk extracted using 0.5 s series of vacuum pulses of four pulses per series. The tube leading from the funnel-shaped cup where the nipple with the areola was located to the milk collector (b) was blocked at this time. The duration of the series was 4 s, and the frequency of vacuum pulses was 1 imp/s. After the tube connected to the milk collector (a) was closed, the tube leading to the milk collector (b) was opened and the milk was ejected to it, which was also the output using 0.5 s vacuum stimuli of four pulses in a series. The duration of the series was also equal to 4 s (Fig. 5.7A1). Figure 5.7B1 shows histograms of the amount of expressed milk in milk collectors (a) and (b). The volume of milk output in milk collector (a) was 49.5 $\pm$ 2% and in milk collector (b) 50.5 $\pm$ 1.5% ($p \leq 0.05$), i.e., it was almost the same. In the next survey, milk collector (a) received milk using 0.3 s series with vacuum pulses of five per series. The tube leading from the funnel-shaped cup where the nipple with the areola was located to milk collector (b) was blocked at this time. The duration of the series was 4 s, and the frequency of vacuum pulses was 1.25 imp/s. After the tube connected to milk collector (a) was closed, the tube leading to milk collector (b) was opened, and the milk was delivered to it, which was the output using 0.5 s vacuum stimuli of four pulses in a series. The duration of the series was also equal to 4 s, and the frequency of vacuum pulses was 1 imp/s (Fig. 5.7A2). After the end of the 0.5 s series of vacuum stimuli, the tube leading to milk collector (b) was closed, and milk collector (a) was opened and the cycle was repeated. The total duration of the vacuum in the series with 0.5 s pulses exceeded by 0.5 s the total duration of the vacuum in the series 0.3 s (Fig. 5.7A2). Since the milk was removed from the gland only under the action of vacuum stimuli, the ratio between the duration of the vacuum (tv) to the total duration of the series (ts) could be an indicator of the effectiveness of the selected mode of operation of the device: $E = tv/ts$. In the case of 0.3 s of vacuum pulses, the ratio of the sum of the action of 0.3 s of vacuum stimuli, 1.5 s, to the total duration of the series, 4 s, was 0.37 ($E = 0.37$). The ratio of the amount of action 0.5 s of the vacuum, 2 s, stimuli to the total duration of the series, 4 s, was 0.5 ($E = 0.5$). The amount of milk withdrawn during the entire pumping session was 13% higher for 0.5 s of vacuum stimuli than for 0.3 s of vacuum stimuli ($p \geq 0.05$) (Fig. 5.7B2). Since the duration of the series in both cases was the same, the rate of milk withdrawal in the case of 0.5 s of pulses was greater. With a similar method of comparing the action of 3 s series 0.5 s of vacuum stimuli with 3 s series 1 s of vacuum pulses (Fig. 5.7A3), the amount of milk derived using

1 s of vacuum stimuli was about 20% greater than with 0.5 s vacuum pulses ($p \geq 0.05$) (Fig. 5.7B3). The difference of the total time steps of 1 s pulses of vacuum also exceeded 0.5 with a total duration of 0.5 s of stimuli of the vacuum (Fig. 5.7A3). For 0.5 s of stimuli, $E = 0.5$, and for 1 s of stimuli, $E = 0.67$. The frequency of 0.5 s and 1 s of vacuum pulses during the session was equal to 1 imp/s and 0.67 imp/s, respectively. A comparison of the action of 5 s series 0.5 s of vacuum stimuli with 5 s series 2 s of vacuum pulses (Fig. 5.7A4) on the amount of withdrawn milk found that the volume of milk withdrawn using 2 s of vacuum stimuli was 25.8% greater than with 0.5 with vacuum pulses ($p \geq 0.05$) (Fig. 5.7B4). The difference in the total time of action of 2 s pulses exceeded by 1.5 s the total time of action of 0.5 s of the vacuum stimuli (Fig. 5.7A4). For 0.5 s of vacuum stimuli, $E = 0.5$, and for 2 s of stimuli, $E = 0.8$. The frequency of 0.5 s and 2 s of vacuum pulses during the session was equal to 1 imp/s and 0.4 imp/s, respectively. Thus, the conducted surveys have shown that when the values of parameter E increase, the amount and velocity of milk excretion increase. This is due to an increase in the ratio between the duration of the action of the vacuum pulses and the interval (rest time) between the vacuum pulses. The highest value of E can be 1. That is, when the rest time becomes 0 and the nipple and areola are affected by a continuous vacuum. Here it is necessary to indicate that the first models of breast pumps used a permanent vacuum. However, this caused pain in women and severe swelling of the nipple and areola (Lawrence and Lawrence 1999). It should be noted that in our work, when pumping milk in a session with 2 s vacuum pulses in some women, pain was also noted, and swelling of the areolar-nipple area of the breast increased. Despite the absence of a difference in the time of action of 0.3 s–0.5 s and 0.5–1 s of vacuum pulses in the series, there was a large difference in the volume and velocity of the milk output for the second case (Fig. 5.7A2,3). While the difference in the time of action of 1.5 s between the 0.5 s and 2 s series of vacuum stimuli (Fig. 5.7A4) did not show a proportional increase in the amount of milk and velocity for the 2 s series of vacuum stimuli (Fig. 5.7B4), these data suggest that the ductal system of the breast by its biomechanical properties is "tuned" to the excretion of milk by vacuum stimuli of a certain duration. In addition, it can be assumed that a decrease in the efficiency of milk excretion with an increase in the duration of vacuum stimuli to 2 s is associated with the appearance of unpleasant sensations and pain in the nipple-areolar region in women. This may negatively affect the central mechanisms of formation of the milk excretion reflex and consequently reduce the efficiency of pumping (Ueda et al. 1994). Here, our data are consistent with the results of studies on the influence of the form of vacuum stimuli of breast pumps on the process of milk withdrawal (Mitoulas et al. 2002b). It turned out that when evaluating the comfort of the action of vacuum stimuli (Fig. 5.6b) in the case of pumping milk with vacuum pulses of different shapes and duration, the number of women feeling less comfort was at 2–3 s of the duration of the vacuum pulses used in the experimental breast pump (Fig. 5.6b). It can be assumed that increasing the inter-pulse interval (rest time), for example, to 1 s, will reduce the degree of discomfort of long-term vacuum pulses. However, in our case, for 2 s pulses, E will decrease from 0.8 to 0.5 with an interval of 0.67, i.e., it will be the same as for 1 s vacuum pulses. Reducing

the duration of vacuum pulses with a corresponding increase in their frequency in order to increase the total vacuum time is also ineffective. So, for example, in the case of reducing the pulse interval in a series of 0.3 s pulses to 0.3 s, E will increase to 0.5. With a further decrease in the inter-pulse interval, the vacuum pulses will almost merge into a continuous vacuum, which will undoubtedly cause women a sense of discomfort. The above information allows us to assume that the optimal vacuum duration for milk excretion is in the region of 1 s. For testing, 0.7 s vacuum stimuli were selected with an interval between 0.3 s pulses so that the frequency in the series was 1 imp/s. The effectiveness of 0.7 s stimuli was compared with 1.05 s vacuum stimuli with a gap between stimuli 0.45 s and a frequency of 0.67 imp/s (Fig. 5.7A5). The number of 0.7 s vacuum stimuli in the series was 6, and the number of 1.05 s vacuum pulses in the series was 4. The total duration of 0.7 s pulses of vacuum and the total duration of 1.05 from the incentive of the vacuum were the same for 4.2 s. The total duration of series of pulses in both cases was 6 s. For 0.7 s of vacuum stimuli, $E = 0.7$, and for 1.05 s of stimuli, $E = 0.66$. The amount of milk withdrawn during the entire pumping session was for 0.7 s of vacuum stimuli $50.1 \pm 1.8\%$ of the total amount, and for 1.05 s was $49.9 \pm 1.8\%$ (Fig. 5.7B5). There was no statistically significant difference ($p \leq 0.05$). However, women felt less comfortable expressing to 1.05 with the pulses of vacuum. In addition, the puffiness of the areola and nipple was also more pronounced when milk was excreted 1.05 with vacuum stimuli.

Thus, the optimal profile of vacuum stimuli that can be installed in breast pumps, meaning of course electric devices, is trapezoidal (Mitoulas et al. 2002b) or step-shaped (Alekseev et al. 2014) stimuli with a duration of about 0.7 s. However, it was difficult to determine the optimal vacuum amplitude, i.e., the maximum comfortable vacuum amplitude, for all women. Therefore, breast pumps should have a vacuum amplitude regulator so that in each case, depending on the functional state, the nursing woman can set the maximum comfortable vacuum value. Under the action of vacuum stimuli, progressive return movements of the areolar-nipple area of the breast occur. This is clearly visible through the transparent part of the funnel-shaped cup, which is placed on the breast. Stretching the skin of the areola and nipple stimulates mechanoreceptors: Ruffini corpuscles and free nerve endings (Fig. 3.12). As a result, a series of nerve impulses are generated in the nerve fibers forming these receptors (Fig. 4.19b), arriving at neuronal formations, which are supposed to be located in the structures of the midbrain: generators of the impulse activity of the milk ejection reflex, GPIA (Fig. 4.18), and generator of periodic impulse activity of milk secretion, GPAMS (Fig. 4.35). The pulse activity at the output of these generators ensures the release of oxytocin into the bloodstream for the removal of milk and prolactin for secretion of milk, respectively. At the same time, over time, in the area of transition of the funnel-shaped cup to the tubular one (Fig. 5.5b2), there is a sealing of the areola tissues. The movement of the areolar-nipple area is significantly slowed down. Edema may form. It has been suggested that the cause is probably the special viscoelastic properties of the nipple and areola. In connection with this assumption, special studies were conducted on the effect of vacuum stimuli on the areolar-nipple area of the breast in 12 lactating women (Ilyin and Alekseev

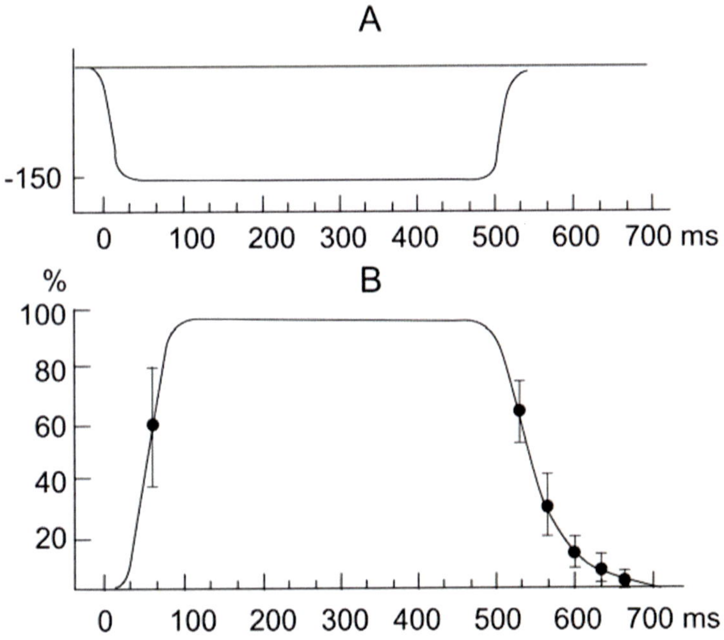

Fig. 5.8 Change of the increase and decline front of the length of the breast areola (from Ilyin and Alekseev 2014). (**a**) A vacuum stimulus. Ordinate axis, amplitude in mm Hg; abscissa, time in ms. (**b**) Change of areola length. Ordinate, the length of the areola as a percentage of the maximum value, which is achieved when the vacuum pulse enters the plateau; abscissa, time in ms

2014). We used vacuum stimuli with a duration of 0.5 s with a rise and fall front of 30 ms, with an amplitude of 140–160 mm Hg, and with intervals between stimuli of 0.5 s. With these parameters of vacuum stimuli, women did not experience pain. Observation of the movements of the nipple and part of the areola in a transparent elastic cup found that there is a stretch mainly of the areolar part. The nipple with the used amplitude of vacuum stimuli practically did not change its size. The increase in the length of the areola varied in different patients from 2 to 6 mm. Figure 5.8 shows the data of elongation and shortening of the areola relative to the maximum value, which did not change when a constant vacuum inside the cup was reached. The lengthening time of the areola exceeded the time of the vacuum stimulus increase (30 ms), but it was still relatively short and was 85–100 ms. It is interesting to note that against the background of noticeable differences in the absolute lengthening of the areola, the time during which the length of the areola reached the maximum value did not differ much in different patients.

Shortening of the areola after switching off the vacuum was less rapid than the lengthening of the areola after switching on the vacuum stimulus (Fig. 5.8b). There are two phases to this process. During the first phase, the length of the areola decreased relatively quickly within 120 ms to a value of about 10% of the maximum value achieved during stretching. In the second phase, there was a slower shortening of the areola during the remaining time until the beginning of the next cycle. The length of the areola at these time parameters of the vacuum stimulus did not return completely to the original level. This appears to be due to the viscous-elastic properties of the areola, which can be modeled as sequentially incorporated viscous and elastic elements. Accordingly, when the tensile force of a certain value was reached, the elastic element reacted first, and then, with some delay, the inertial viscous element and the areola rapidly increased its length. When the vacuum stimulus reached the plateau, due to inertia, the stretching of the viscous element continued. The length of the areola increased slightly during this period. After switching off the vacuum, the elastic element relatively quickly began to return the length of the areola to its original level. At the same time, due to the inertia of the viscous element, shortening of the areola was slowed down, and at this frequency and duration of the vacuum pulses, it was not possible to achieve full achievement of the original length. For each cycle, probably, the residual deformation accumulated and the areolar part in the area of the transition to the nipple gradually thickened, and its stretching decreased (Ilyin and Alekseev 2014). Accordingly, the stimulation of mechanoreceptors decreased, which affected the formation of milk excretion and secretion reflexes. In addition, an increase in tissue density in the internal area could compress the milk ducts, making it difficult to remove milk. Therefore, in some cases, when using breast pumps, you had to remove the cup and massage the areola to eliminate puffiness.

As already noted, after the end of milk excretion by the child, there was also an elongation of the areolar area of the gland, which was located in the child's oral cavity. However, the density of the areolar area adjacent to the nipple did not increase. This could be happening for at least two reasons. Firstly, as shown by the registration of vacuum stimuli created by the child, a series of high- and low-amplitude vacuum stimuli affect the nipple during the process of milk withdrawal. It is assumed that during the intake of milk into the oral cavity of the child, its taste receptors are stimulated, while the frequency of vacuum stimuli in periods decreases and becomes less than 1 imp/s, and their amplitude increases. This effect on the nipple and areola of the woman's breast was designated as "nutritive sucking." Since the output of milk is directly associated with an increase in pressure in the capacitance system of the gland, a decrease in intramammary pressure after the end of the reflex peak reduces the flow of milk to a complete stop and makes unnecessary long-term exposure to high-amplitude vacuum stimuli. In this regard, the patterns of efferent activity of the GPIA coming to the muscles of the orofacial region become shorter. The duration and amplitude of the patterns of vacuum stimuli that are generated in the orofacial region also become smaller, and they begin to perform mainly a stimulating function. This type of exposure to vacuum stimuli was designated as "nonnutritive sucking" (Fig. 4.16). Thus, reducing the time of exposure to high-amplitude vacuum stimuli will reduce the development of puffiness of the areolar area.

Secondly, along with vacuum stimuli for both types of sucking, compression stimuli always act on the areolar area, in which, in addition to speeding up the removal of milk, massaging this area of the gland reduces its swelling, thereby making it soft and more stretchable.

Medela AG on the basis of "nutritive" and "nonnutritive" sucking of a child developed a special sequence of vacuum stimuli for its new vacuum breast pump (Medela Symphony), which allowed to reduce the time of action of high-amplitude vacuum stimuli on the areolar-nipple part of the breast during milk excretion (Fig. 5.6c). The breast pump works as follows. At the beginning of pumping, vacuum stimuli with a frequency of 120 imp/min and an amplitude of 50–150 mm Hg are applied to the nipple and areola. It is believed that at these parameters, the vacuum pulses perform mainly a stimulating effect on the mechanoreceptors of the areola and nipple. Indeed, an increase in the frequency of vacuum stimuli, in particular, the acceleration of the front of rise and fall of stimuli, can further increase the response of rapidly adapting mechanoreceptors of the skin of the areola and nipple (Fig. 4.19). After 2 min, high-frequency stimuli are automatically switched to a suction vacuum with an amplitude of 100–250 mm Hg and a frequency of 40–75 imp/min. At the same time, each woman with the help of a vacuum regulator sets a comfortable vacuum. A series of stimulating and sucking stimuli alternate throughout the entire pumping session. The breast pump allows to manually switch stimulating impulses to suction impulses using the button before the reflex peak of the increase in intramammary pressure begins. Some women may feel the approach of a reflex increase in pressure by tingling in the breast (vascular reaction to an increase in the concentration of oxytocin in the blood). The appearance of a reflex peak of milk excretion can be detected visually at the beginning of milk intake in the milk collector. This algorithm of the breast pump operation complicates its operation in comparison with the previously developed and widely used breast pump in medical practice of the company Medela AG. However, detailed studies of the operation of these two breast pumps (Meier et al. 2008) showed that they do not differ in the efficiency of milk excretion.

In addition to vacuum stimuli, the child during milk excretion affects the areola of the breast simultaneously with compressing stimuli (Figs. 4.11 and 4.13). In this case, compression stimuli additionally effectively stimulate the mechanoreceptors of the areola of the breast (Fig. 4.19b, c) and squeeze milk from the milk ducts localized in the areola area. In contrast to vacuum stimuli, the use of compression stimuli in the breast pump began in the second half of the last century. An analysis of the patent literature has shown that compression stimuli in breast pump are used for two purposes: in some cases, only for tactile stimulation of mechanoreceptors of the skin of the areola of the breast and in others simultaneously for stimulation of mechanoreceptors and excretion of milk. In some breast pumps used in medical practice (e.g., firms: Philips Avent, Medela, Chicco, Embrace Playtex, Canpol Babies), there are devices for tactile mechanical stimulation-massage inserts. Massage inserts are made of elastic material, in most cases, silicone rubber. Their shape follows the shape of a rigid plastic cone. On the external side surfaces of the elastic cone, there are grooves or cavities of various shapes. An example of this most popular massage insert is shown in Fig. 5.9. The elastic insert on the outer surface has grooves in the form of petals (Fig. 5.9a). The insert is placed inside a rigid cone

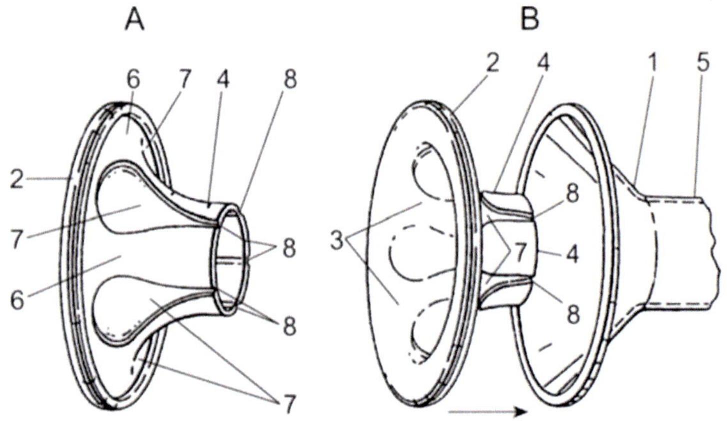

Fig. 5.9 Elastic insert (**a**) and its placement in a rigid plastic cup of a breast pump (**b**) (from Ford 1999). (1) Conical part of rigid plastic cup, (2) elastic insert, (3) conical part of elastic insert, (4) cylindrical part of elastic insert, (5) cylindrical part of rigid plastic cup, (6) internal surface of elastic insert, (7) petal recesses in elastic insert, (8) grooves coming from petal recesses in elastic insert

(Fig. 5.9b) so that cavities are formed limited on one side by the walls of the elastic insert and on the other by the wall of the cone. The cavities are communicated by means of grooves with the volume of a hard cone. The cone with an elastic insert is placed on the breast. Simultaneously with the arrival of vacuum in the cone, the vacuum is also established in the cavities. The walls of the petals are pressed against the walls of a hard cone. After the vacuum is released and air enters the cone, the petal walls are straightened and take their original shape, while they have a tactile effect on the skin of the breast, stimulating the mechanoreceptors. Comparative tests were performed on Medela breast pumps with a rigid plastic cone and Embrace Playtex with an elastic massage insert (Hopkinson and Heird 2009). The elastic insert has wavy protrusions on the conical section that touches the surface of the breast. When vacuum stimuli are applied inside the cone, the walls of the elastic insert move, while the protrusions of the insert perform a periodic tactile effect on the skin surface of the gland and stimulate the skin mechanoreceptors. Additional tactile stimulation during milk withdrawal significantly increased the output of the hormone prolactin in the bloodstream of a woman compared to the cup without a massage insert (Fig. 5.10a).

Previously, similar data were obtained when using breast pumps from White River Concepts, which had an elastic cone. When applying a vacuum stimulus inside the elastic cone, atmospheric pressure compressed the elastic walls of the cone, respectively squeezing the part of the breast located in the cone. At the same time, tactile stimulation of the skin surface mechanoreceptors took place. Comparative tests of the effect of mechanical breast stimulation on prolactin output using breast pump of White River Concepts, Medela, manual pumping and baby sucking (Zinaman et al. 1992) showed that the level of prolactin during milk ejection by

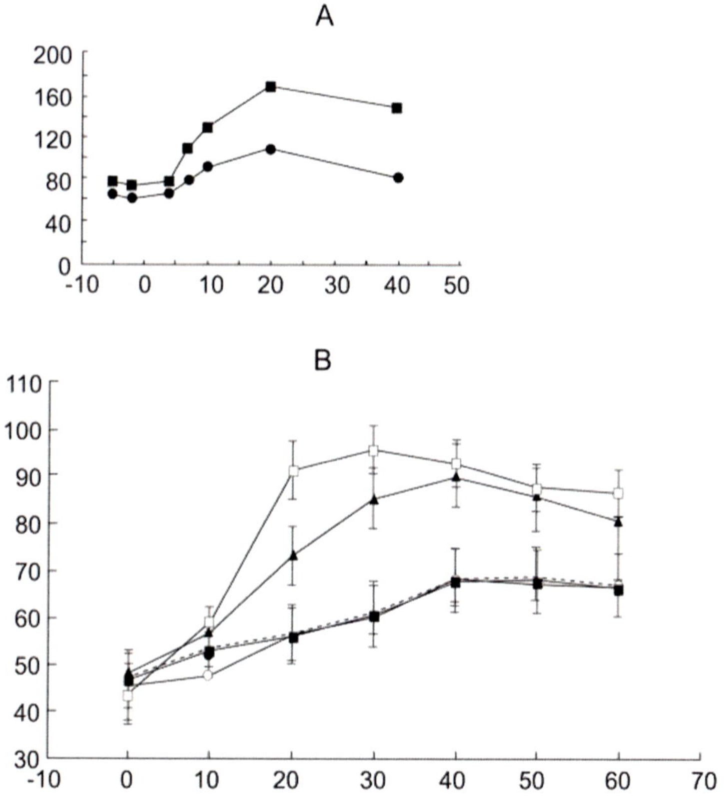

Fig. 5.10 Change in blood prolactin concentration in lactating women with different milk excretion methods. (**a**) Change in blood concentration in lactating women when milk is excreted by Embrace Playtex breast pump with elastic insert (dark squares) and Medela breast pump (dark circles) (from Hopkinson and Heird 2009). Ordinate, prolactin concentration in ng/ml; abscissa, time in min. Prolactin was determined in blood samples taken 5 and 2 min before setting and 4, 7, 10, 20, and 40 min after starting milk excretion. (**b**) Change in the blood concentration of prolactin in lactating women during milk excretion by White River Concepts breast pump (light squares), Medela breast pump (light circles), hand milk excretion (dark squares), and milk excretion by child (dark triangles) (from Zinaman et al. 1992). Ordinate, prolactin concentration in ng/ml; abscissa, time in min. Prolactin was determined in blood samples taken at 10-min interval

the child and the White River Concepts milk pump was approximately the same and significantly exceeded the level of prolactin during manual pumping and pumping by the Medela milk pump (Fig. 5.10b). At the same time, in both studies, there was no significant difference in the concentration of oxytocin in milk excretion by different methods (Zinaman et al. 1992; Hopkinson and Heird 2009).

The most interesting are breast pumps, in which the compression component is used not only to stimulate the mechanoreceptors of the areola of the breast but also to remove milk. It should be noted that in medical practice, more than a dozen different versions of vacuum breast pumps are used. At the same time, there is only one model

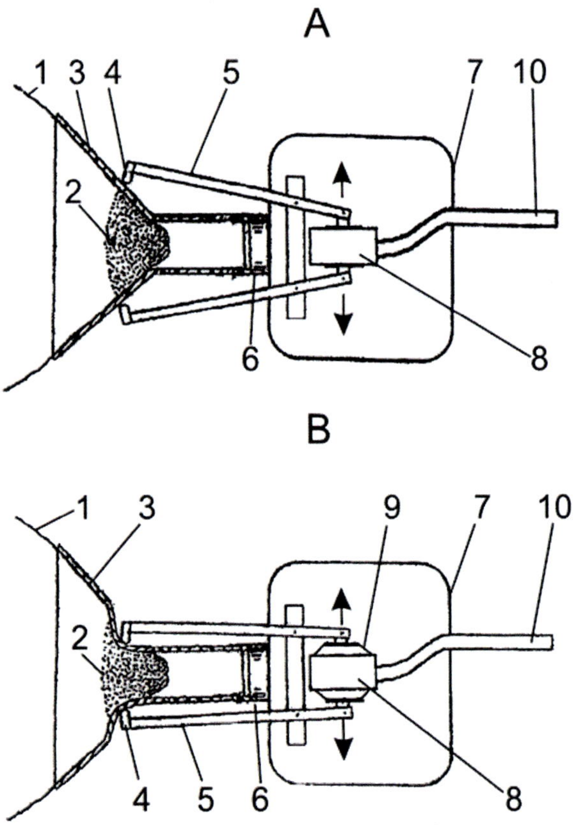

Fig. 5.11 Action of the compression component in the milk breast pump with vacuum and compression components. (**a**) Positive pressure pulse is not turned on. (**b**) Positive pulse is on. (1) Mammary gland, (2) breast areola, (3) elastic cone, (4) rigid plates, (5) levers, (6) movable ledge, (7) body, (8) pneumatic piston, (9) pneumatic pistons membranes, (10) pneumatic hoses

of a breast pump with a compression component used in practice. The breast pump was developed at the Saint Petersburg (Leningrad) State University at the end of the last century by V.I. Ilyin and N.P. Alekseev and was named "Lactopuls." Just like vacuum electric breast pumps, the breast pump "Lactopuls" consists of a control unit with a compressor and a removable funnel-shaped cup that is placed on the breast of a woman. Schematically, the removable funnel-shaped cup is shown in Fig. 5.11a, b. It consists of a conical and cylindrical part (3). However, unlike vacuum breast pumps, the cup is made of an elastic material, such as silicone rubber. The cone part is placed on the breast (1). On two opposite external sides, rigid plates (4) are in contact with the elastic cone, which are fixed at the front ends of the levers (5). The rear parts of the levers are connected to the movable membranes (9) of the pneumatic piston (8), to which excessive pressure pulses are supplied from the control unit through the pneumatic hoses (10). The pneumatic piston is placed in the rigid body (7), to which a cylindrical part of the cup is attached on the outside on a special movable ledge (6). The pump works as follows. Vacuum and compression (overpressure) stimuli are applied in a certain sequence to the removable funnel-shaped cup from the control unit. Just as when milk is excreted by a child (Fig. 4.13) at the beginning, a vacuum acts on the mammary gland, which enters the cup via a pneumatic hose (on Figs. 5.11, 5.12 and 5.13 are shown), and the areola of the breast (2) begins to stretch, stimulating the skin stretching receptors. At the same time, under the influence of vacuum, milk begins to flow out of the ducts. After the amount of vacuum inside the elastic cone tube reaches the maximum value, positive pressure stimuli are applied to the pneumatic piston. The movement of the pneumatic piston membranes is transmitted to the levers (Fig. 5.11b), and the rigid plates compress the conical part cup and the areolar area with milk ducts located in it. It is important to note here that, just as in a child, the amplitude of compression and vacuum stimuli can be independently regulated. Compression stimuli additionally perform tactile stimulation of mechanoreceptors more effectively than the grooves or protrusions of the elastic insert of vacuum breast pumps and consequently increase the output of lactogenic hormones (Alekseev et al. 1994, 1998). Figure 5.12a shows as an example two lactograms of the same lactating woman when expressed by a breast pump with a compression component. On the left graph, the dynamics of changes in the rate of milk excretion by a breast pump with two components is presented. The first peak of the milk excretion reflex due to the release of oxytocin into the blood from the neurohypophysis began about 1.5 min later. On the right graph, the milk was first expressed only by vacuum (marked on the horizontal line at the top of V). During the action of vacuum stimuli for about 4 min, there was no reflex increase in the rate of milk excretion. However, switching on of compression stimuli (marked on the horizontal line at the top of the PV) led to a reflex increase in the rate of milk excretion after 0.5 min. A study of the time dependence of the first reflex peak of increasing the rate of milk excretion in 11 lactating women showed that in the presence of compression stimuli, the first peaks appeared 1–1.5 min after the beginning of pumping (Fig. 5.12b, upper histogram). Turning off the compression stimuli was accompanied by an increase in the latent periods of the appearance of reflex peaks of increasing the rate of milk excretion (Fig. 5.12b, lower histogram). The presented data indicate that additional tactile stimulation of the breast areola accelerates the formation of milk excretion reflexes or, in other words, the output of oxytocin from the neurohypophysis. In addition to the milk excretion reflex, compression stimuli effectively stimulate the release of prolactin from the adenohypophysis. This is well illustrated by the histograms in Fig. 5.12c. Examinations were conducted on 28 lactating women. In each of the three histogram groups, the first histogram represents basal prolactin levels in women who breastfed a child six times a day and accordingly, the second and third histograms in each group—the basal concentration of prolactin when removing milk three times a day using a breast pump with a compression component and three times pumping a day manually. The first group of histograms was based on data obtained on the first day after birth, the second group on the third day after birth, and the third group on the sixth day after birth. The figure clearly shows that when a baby is breastfed, as well as by pumping the breast pump with compression pulses, by day 6 after delivery, the basal level of prolactin significantly increased compared to the basal concentration of prolactin during manual pumping. Moreover, there were no significant differences between

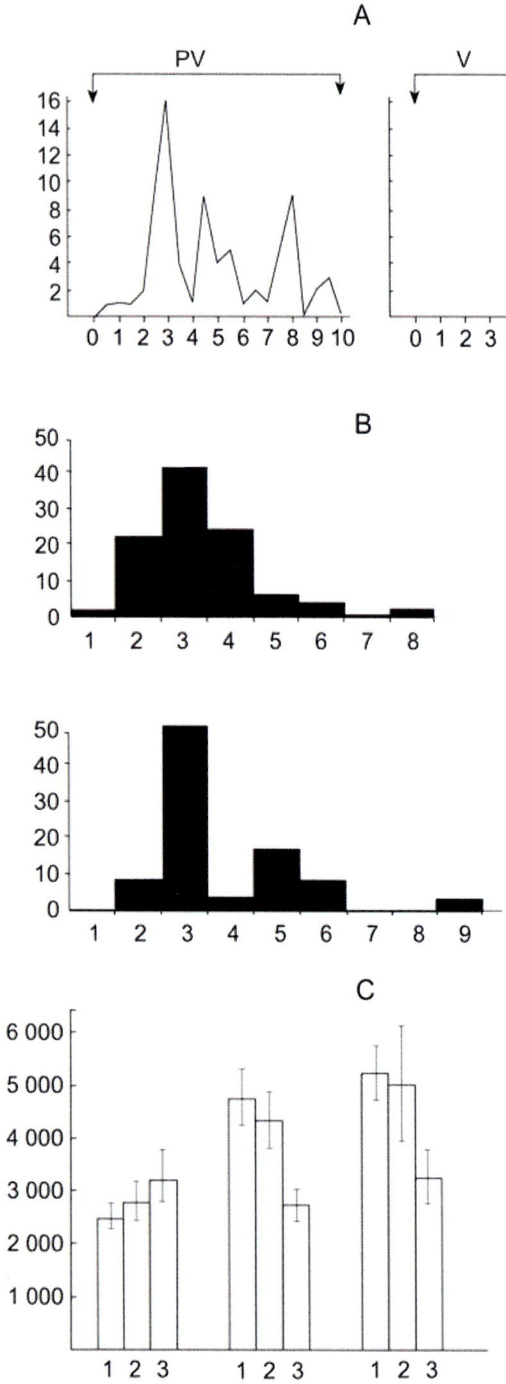

Fig. 5.12 Influence of compression components on formation of milk ejection reflexes (**a**, **b**) and basal level of prolactin (**C**) in lactating women (from Alekseev et al. 1998). (**a**) The rate of milk removal, ml/0.5 min. In each graph, the horizontal line shows the period of milk removal by means

the basal level of prolactin during breastfeeding and milk excretion using a breast pump with compression pulses.

In addition, special examinations have shown that the breast pump with a compression component not only has a high stimulating ability compared to a vacuum breast pump but also more effectively removes milk from the breast of a lactating woman (Alekseev et al. 2014; Alekseev and Ilyin 2016). The survey was conducted on 24 women 5–6 days of lactation using a technique that was used in the work on optimizing the duration of steplike vacuum stimuli used in breast pumps (Fig. 5.7).

According to the method, the nipple and areola were alternately affected by short (4 s) series of 0.5 s vacuum stimuli and 0.5 s vacuum stimuli with compression pulses of 0.27 s. The amplitude of the vacuum pulses was within -120 and -140 mm Hg, i.e., less than the maximum comfortable vacuum amplitude of -191.3 ± 6.5 mm Hg (Mitoulas et al. 2002b). Accordingly, the milk was ejected to various milk collectors: (a) when pumping together with vacuum and compression stimuli and (b) when pumping only with vacuum stimuli (Fig. 5.13A).

When pumping milk alternately with vacuum pulses and vacuum with compression pulses (18 patients), except for one case when the amount of milk was the same, the volume of milk expressed together with the help of vacuum and compression exceeded the amount of milk expressed by a single vacuum. However, the difference in volume varied in different patients in the range of 10%–46%. On average, as follows from the histograms (Fig. 5.13B), the amount of milk expressed by one vacuum was $40.5 \pm 5\%$, and the vacuum with compression stimuli was $59.5 \pm 5\%$. Observations during pumping indicate that the difference in the volume of expressed milk depends on the anatomical characteristics of the breast. According to ultrasound studies, 6–14 milk ducts are suitable for the nipple, which at a distance of 8–9 mm from the base of the nipple have a maximum diameter of 1–5 mm (Prime et al. 2009). However, when the milk ducts pass into the milk ducts of the nipple, they narrow by about five to ten times (Ramsay et al. 2005). Moreover, the diameter of the milk ducts along the length of the nipple varies. So, when approaching the tip of the nipple (1 mm from the tip), the diameter of the ducts is the smallest and is about 0.1 mm. To the outlet, the ducts in most cases expand in the form of a funnel with an increase in

Fig. 5.12 (continued) of the breast pump. PV, breast pump action with both compression and vacuum stimuli; V, breast pump action with compression stimuli switched off (vacuum only). (**b**) Frequency of distribution of the time of occurrence of the first reflex peak of the change in the rate of milk removal (latent period) during milk removal simultaneously with the vacuum and compression stimuli (upper histograms) and only with the vacuum stimuli (lower histograms). Ordinate, number of time intervals (latent periods) as a percentage of the total number of intervals; abscissa, duration of latent periods in min. (**c**) The change in the basal serum prolactin level of lactating women on the first (**a**), third (**b**), and fifth (**c**) days after delivery. (1) women who breastfed their babies six times a day; (2) women whose milk was removed by means of the breast pump with a compression component three times a day from both breasts; (3) women who expressed their milk manually three times a day from both mammary glands. Prolactin concentration, mIU/l

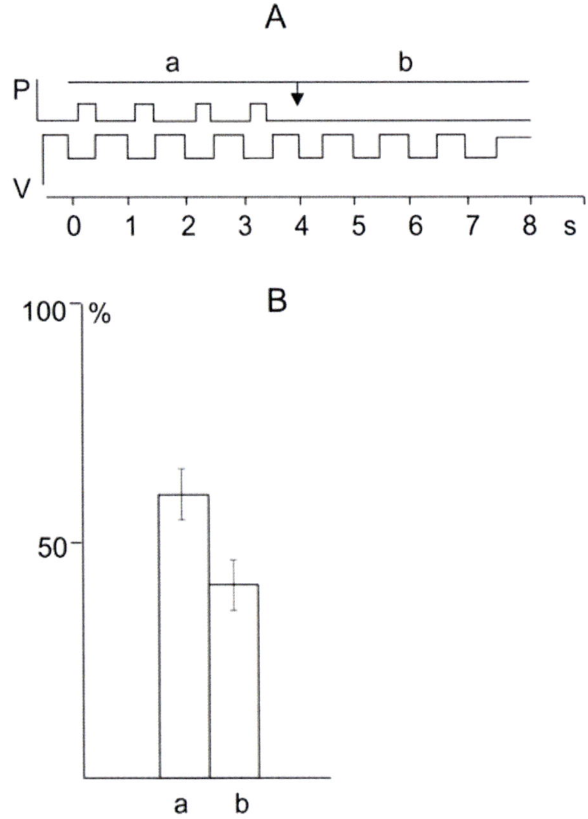

Fig. 5.13 Influence of compression components on the amount of milk removal (from Alekseev et al. 2014). (**A**) Scheme of effects of vacuum and compressive stimuli on the mammary gland. (a) Simultaneous action of vacuum and compression stimuli, milk enters the milk collector (a); (b) action of vacuum stimuli, milk enters the milk collector (b). Ordinate, P (positive pressure), V (vacuum); abscissa, time (s). (**B**) The amount of milk as a percentage of the total volume of milk removal together by means of vacuum and compression pulses and by means of vacuum only (**B**). The ordinate: the amount of milk as a percentage

diameter by 1.5–3 times. When moving deep into the nipple (3.5–4 mm from the tip) in different women, the diameter of the ducts increases and reaches 0.4–0.8 mm. Then the diameter decreases again to an average of 0.4 mm (Going and Moffat 2004; Rusby et al. 2007). Since the resistance to the movement of liquid in the tube is inversely proportional to the area of its cross section, nipples with thin milk ducts will inhibit the output of milk to a greater extent, and in this case, the mammary glands are classified as "tight." In addition, in the first 4 days after birth, when colostrum is present in the ductal system, the viscosity of which is higher than that of transitional and mature milk, difficulties with milk excretion are aggravated. To overcome the "tightness" of the gland, it is necessary to increase the pressure difference between the environment and the milk inside the milk ducts, for example, to increase the vacuum. However, as clinical studies have shown, high vacuum causes pain in the nipple, which inhibits the formation of the milk excretion reflex and consequently slows down the milk output. The child gets out of position by adding compression impulses to the vacuum stimuli so that the overall pressure difference between the environment and the milk inside the milk ducts increases markedly. In particular, when removing milk, the child can create a maximum vacuum of 197 $\pm$ 10 mm Hg. The maximum amplitude of the compression pulses could reach 70 mm Hg. Thus, when added together, the total pressure difference will be about 270 mm Hg, which will significantly increase the efficiency of removing milk from the "tight" gland. At the same time, the woman will not experience discomfort, since when the child creates vacuum and compression stimuli (Fig. 4.13), the values of vacuum and compression for the woman will be comfortable (Kent et al. 2008). The combination of vacuum and compression stimuli in the executive mechanism of the pump was also effective (Alekseev et al. 2014; Alekseev and Ilyin 2016). Surveys have shown that compression stimuli against the background of vacuum stimuli −120 and −140 mmHg can increase the output of milk from the gland by 46%. Compression stimuli made a particularly noticeable contribution when pumping "tight" mammary glands and mammary glands in the first days of lactation filled with colostrum. Here it is interesting to note the results of surveys of A.A. Morton (Morton et al. 2009). In this work, against the background of sucking milk using a Medela Symphony breast pump, the fingers were additionally compressed in the area of the breast in front of the edge of the hard cup. As a result of the combination of vacuum and compression, the amount of milk produced increased significantly (up to 48%). This method was especially effective when removing colostrum from the gland.

In the case of equal volume of milk expressed by vacuum and vacuum with compression pulses, examinations showed that the patient's breast was very "light." Through the transparent cover, it was clearly visible that the milk began to be released from the breast in trickles already under the influence of a vacuum of 60–70 mm Hg, i.e., half the established amount. Moreover, the milk as a result of the reflex of milk excretion began to drip quite intensively from the neighboring breast. In this case, the milk ducts in the nipple probably had a maximum diameter (0.6–0.8 mm), and a vacuum stimulus, before the compressing stimulus took effect, removed most of the milk. In this regard, it is interesting to note the results of a study of the process of removing milk from bottles, the nipples of which have holes for the exit of milk of different diameters (Eishima 1991). It was discovered that the child had the greatest compressive effect tongue and gums on the artificial nipples without holes or artificial nipples with a very small diameter exit hole, which when you turn the bottle in vertical position water is dripping at a speed of 0.04 ml/h. If the hole diameter was increased so that water dripped at a rate of 0.1 ml/sec, the artificial nipple compression was significantly weakened.

Here it should also be noted that when the milk was removed by a breast pump with a compression component, there was no compaction and puffiness of the areolar area of the breast. Massage of the areola with compressive stimuli as the milk is withdrawn, as well as in the case of milk withdrawal by a child, did not allow for compaction and puffiness of the areola. Therefore, the use of the breast pump in clinical practice with a compression component was effective in eliminating postpartum breast engorgement, as well as in the case of elimination of edema in lactostasis (Alekseev et al. 2010, 2015).

Thus, the use of milk-removing devices and especially devices with a compression component is an effective nondrug means of increasing the productivity of lactating women and increasing their lactation period.

5.3 Postlactational Involution of the Mammary Gland

The World Health Organization recommends that breastfeeding women introduce complementary foods from 5–6 months, with a total duration of breastfeeding of 1.5–2 years. The volume of complementary foods increases over time so that the need for milk for the child decreases. In this regard, as shown by numerous surveys, the amount of milk secreted by the mammary glands also decreases. As an example, Fig. 5.14 shows a daily milk production and breast volume in women who started introducing baby complementary food 6 months after birth (Kent et al. 1999). During the first 6 months after delivery, productivity remained virtually unchanged, and then there was a decrease in the amount of secreted milk and the volume of the mammary glands. The decrease in breast productivity probably occurred for several reasons. Common cause of loss of productivity for all of the lobes of the mammary gland was apparently reducing the number of feedings and the time of feedings, resulting in decreased release from the anterior pituitary hormone prolactin, the main hormone in women necessary to produce milk

In addition, due to the decrease in milk consumption by the child, FIL accumulates in the alveoli of the lobules which will inhibit the formation of milk. This occurs primarily, apparently, in the alveoli of the lobes, from which the child less effectively removes milk. For example, the milk ducts in the nipple through which the milk exits the lobe have the smallest diameter and, accordingly, a high resistance to the movement of milk. Therefore, the child will first empty the lobes whose milk ducts in the nipple have a large diameter. The alveoli of the lobes that do not produce milk will overflow with milk, resulting in the process of postlactation involution. The term "involution" is taken from the Latin language (involutio-"clotting") to refer to the process that results in a decrease in the volume of glandular tissue and a decrease in the formation of milk in the secretory cells of the breast. Unfortunately, the processes of postlactation involution of a woman's breast at the cellular and molecular level remain unexplored. The most studied to date among various mammals are the cellular and molecular mechanisms of postlactation involution in the mammary glands of mice due to the use of genetic engineering methods (Watson 2006a, b; Watson and Kreuzaler 2011). In mice, in the nipple of each breast, and they have ten, there is a single duct, which repeatedly divides to the thinnest branches forming alveoli. Multiple mammary glands in mice function independently of each other. In women, each milk lobe has its own duct, which opens at the tip of the nipple. The number of lobes in one breast of women varies from 5 to 12 (King and Love 2006), and they also operate independently from each other. In the study of postlactation involution in the mammary glands of lactating mice, blocking the exit of milk from the gland is used by sealing the exit of the nipple duct of a mammary gland (Li et al. 1997). As a result, milk is not removed from it by the pups, while from other glands, the pups can extract milk without hindrance. This situation is somewhat similar to the above-described inhibition of the removal of milk from the breast lobe of a woman with a thin duct in the nipple. Here it should be noted that the mouse feeds the pups quite often, and the average interval between feedings is 35 ± 2 min (Tolkunov Yu and Markov 2005). Increasing the interval between

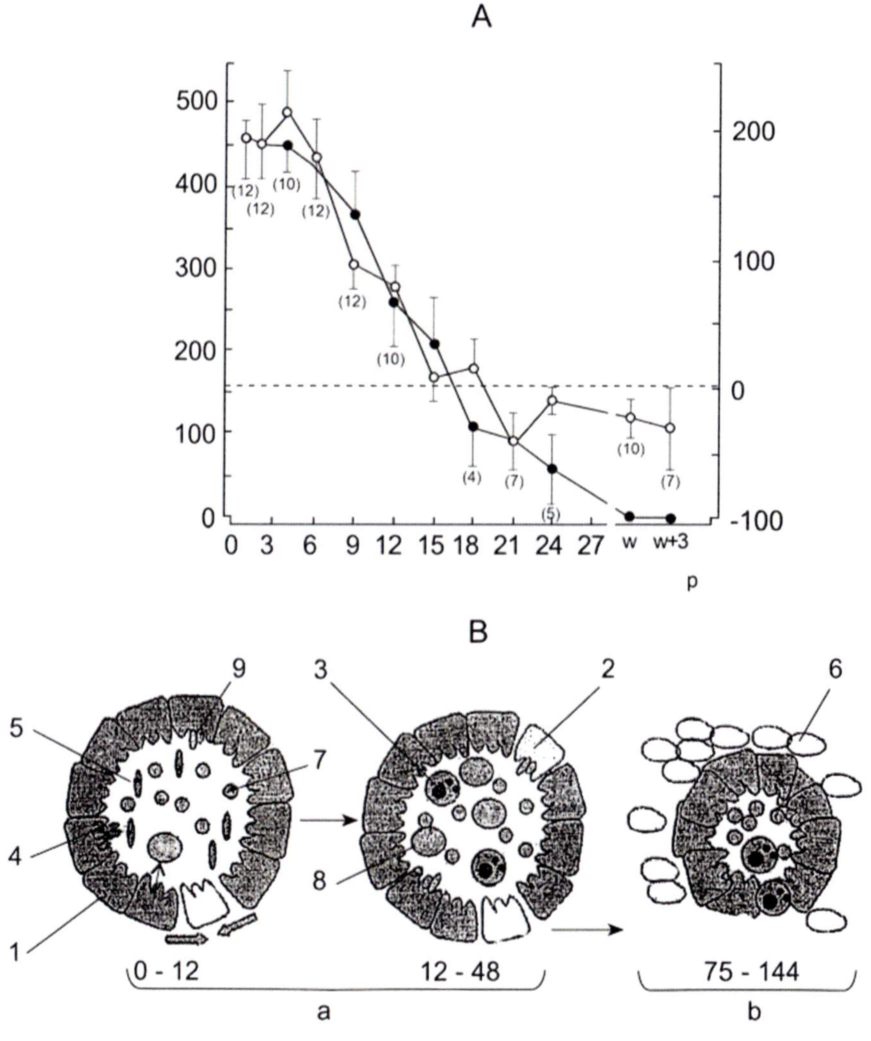

Fig. 5.14 Involution of the breast. (**A**) Daily milk volume (dark circles) and breast volume (light circles) (from Kent et al. 1999). In the graphs under each point, the number of observations. Ordinate: on the left, the daily amount of milk in g; on the right, the volume of mammary glands in ml. The right on the ordinate zero value and the dotted line—the volume of the breast before the onset of pregnancy. Abscissa: time after birth in months, w-time of the end of feeding, w + 3, 3 months after the end of feeding. (**B**) Schematic of mouse alveola involution (from Watson 2006a). (1) Normally functioning secretory cell, (2) dead secretory cell, (3) cell altered by apoptosis, (4) cell death receptor, (5) cell death receptor ligands, (6) adipocyte, (7) milk globules, (8, 9) leukemia inhibitory factor (LIF). Numbers indicate the time of involution in hours. Broad arrows show the direction of closure of normally functioning secretory cells when pushing a dead cell into the alveola cavity

feedings causes overflow of the alveoli of the mammary glands. It is believed that the stasis of the ducts and the overflow of milk from the alveoli and their outgoing ducts is the primary signal for the beginning of involution (Li et al. 1997), during which secretory tissue apoptosis occurs. Since data on postlactation involution of the woman breast at the cellular and molecular level are practically absent in Fig. 5.14B, a brief diagram of mouse alveolus involution is presented as an example (Watson 2006b). In the process of involution of the mammary glands of mice, two phases are distinguished (Fig. 5.14B). In the first 6–12 h of the initial phase, as a result of blocking milk output in the intraalveolar volume, the concentration of specific extracellular ligands and cytokines such as transforming growth factor-β3 (TGF-β3), tumor necrosis factor-α (TNF-α), and leukemia-inhibiting factor (LIF) increases. These agents induce the process of apoptosis through activation of cell death receptors (first pathway) and activation of the proapoptosis transcription factor STAT3 (pathway 2). The dead cells are pushed into the alveolar cavity. In this case, normally functioning neighboring cells move to each other and close the lateral surfaces, thereby preserving the integrity of the alveolus for some time. The first phase of involution is reversible. If the nipple is released after 48 h and the baby starts to withdraw milk, the function of this breast is resumed (Watson 2006b). After the reversible first phase of involution after 48 h of no milk excretion, the second irreversible phase of involution begins (Fig. 5.14B).

As a result of increasing the concentration and activity of such enzymes as matrix metalloproteinases (MMR) in the extracellular medium, the extracellular matrix around the alveoli is destroyed, which increases apoptosis and finally collapses the alveoli. At the same time, the adipocytes surrounding the alveolus are differentiated so that they begin to fill up with fat again. In this phase, the processes of autophagy and heterophagy, that is, self-digestion of breast tissue, are observed. The number of lysosome enzymes—active organelles that digest cellular remains—is growing. The second phase of involution is irreversible, so the alveoli in which it started can no longer lactate again. It is necessary that the processes of differentiation and proliferation of glandular tissue begin again, as it happens during pregnancy. It should be noted that the degree of difference or similarity in molecular and cellular mechanisms between mouse and woman breast involution is unknown. However, in some popular science reviews, the involution of mammary glands in mice and women appears to be identical. However, the available experimental data indicate that, for example, the time parameters of mammary gland involution in mice and women differ significantly. Thus, when a child's feeding stops abruptly, the process of involution in a woman takes up to 42 days (Hartmann and Kulski 1978), while in mice it takes 6–7 days (Watson 2006a), i.e., six to seven times longer than in mice. Perhaps one of the reasons is a slower rate of involution processes, which is due to the peculiarities of woman lactation. In particular, the interval between feedings during the established lactation of a woman is 2.5–3 h, and in mice 0.5 h. So the time of accumulation of milk for the next feeding of a child is five to six times longer than in a mouse for feeding cubs. Accordingly, it seems that under unforeseen circumstances, with a long delay in feeding the child so that the involution of glandular tissue does not begin, the time of the beginning of involution in a woman is shifted to a later date. The process of postlactation involution of the breast in a woman under normal conditions, that is, when feeding to a child and simultaneously reduced consumption of milk by the child, develops smoothly and takes a relatively long period of time (Fig. 5.14A). Moreover, individual lobes of the mammary glands are subjected to involution asynchronously. This is easily determined by the number of streams of milk when squeezing the areola with your fingers. So, at the beginning of involution, compression of the areola of the woman's breast is accompanied by an average of five to ten trickles of milk. In the future, the number of streams begins to decrease, which indicates the cessation of milk formation in some lobes. However, as mentioned above, under special conditions (illness of the child or mother), a woman is forced to abruptly stop feeding. In this case, the process of involution develops much faster and may end within 1–1.5 months.

In addition to reducing the total amount of secreted milk with "slow" and "fast" breast involution, there is a change in the concentration of various components of milk. The introduction of complementary foods and the beginning of involution additionally causes accelerated changes in the composition of milk. Unfortunately, about the impact of breast involution on the content of various components of milk in the literature, there are single works concerning changes in the concentrations of only some components of milk. Thus, when studying the composition of milk in women when introducing complementary foods to children aged 5–7 months (Garza et al. 1983), it was found that there is a gradual change in the concentration of certain components over the next 3 months: an increase in protein concentration from 135 ± 6 to 193 ± 21 mg/100 ml; an increase in fat concentration from 3.6 ± 08 to 7.5 g/100 ml for the first 10 weeks, but then decreased to 4.07 ± 0.7; an increase in iron concentration from 29 ± 6 to 50 ± 5 (μg/1000 ml); an increase in sodium concentration from 136 ± 16 to 297 ± 57 mg/1000 ml; calcium concentration did not change; and a decrease in zinc concentration from 1.2 ± 0.2 to 0.7 ± 0.1 mg/1000 ml.

In subsequent studies (Perrin et al. 2017), the dynamics of concentrations of a number of components of the milk of women who fed children during 11–17 months of lactation were studied. Children received along with milk complementary foods, i.e., the mammary glands of these women were subjected to slow involution. It turned out that during this period, there was an increase in the concentration of total protein from 1.6 ± 0.2 to 1.8 ± 0.3 g/100 ml, proteins that have bactericidal properties and in particular lactoferrin from 180 ± 66 to 280 ± 180 mg/100 ml, lysozyme from 58,000 ± 31,000 μ/ml, and IgA from 21 ± 7.7 to 29 ± 10 mg/100 ml. There was also an increase in the total sodium content from 70 ± 19 to 86 ± 35 μg/ml and simultaneously a decrease in the total concentration of zinc from 560 ± 339 to 420 ± 310 ng/ml and calcium from 200 ± 29 to 180 ± 30 μg/ml. There were no changes in the concentration of lactose, fat, iron, and potassium during this period.

In the literature, there is one work (Hartmann and Kulski 1978) which investigated the dynamics of a number of mineral and organic substances in milk during the "rapid" evolution of the mammary glands of women. Figure 5.15 shows changes in the concentration of various components of milk in women 251–443 days

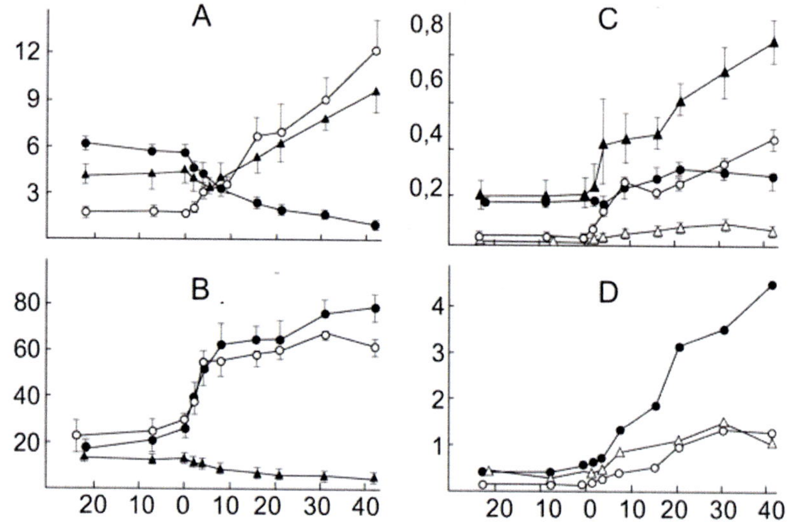

Fig. 5.15 Change in concentration of certain milk components during abrupt stopping of infant feeding (from Hartmann and Kulski 1978). (**a**) Change in concentration: dark circles, lactose; light circles, total protein; dark triangles, fat. Ordinate—concentration in g/100 ml. (**b**) Change in concentration: dark circles, sodium; light circles, chlorine; dark triangles, potassium. Ordinate—concentration in mM. (**c**) Change in concentration: dark circles, α-lactalbumin; light circles, IgM; dark triangles, IgA; light triangles, IgG. Ordinate—concentration in g/100 ml. Abscissa—time in days. (**d**) Change in concentration: dark circles, lactoferrin; light circles, albumin; light triangles, casein. Ordinate—concentration in g/100 ml. For all graphs, abscissa—the time in days

of lactation. In women, breastfeeding was stopped at a certain point so that the milk was not removed from the glands except for small samples (0.5–5.0 ml), which were necessary for analysis. For 45 days, the amount of milk quickly decreased. Comparison of data on changes in the concentration of milk components during "slow" and "fast" involution shows in most cases the unidirectional nature of these changes. Here it should be noted that a number of milk components that were analyzed for "fast" involution were not determined in milk for "slow" involution and vice versa. With "fast" involution, the concentration of total protein, fat, sodium, and chlorine increased, and the concentration of potassium and lactose decreased (Fig. 5.15a, b). The presented data suggest that during involution, there is an increase in the permeability of dense intercellular contacts to sodium, chlorine, potassium, and lactose. As a result, the concentration of these milk components begins to equalize between the extracellular medium and the intraalveolar volume. That is, the situation is the reverse of the transition of the mammary glands from pregnancy to lactation, in which it is believed that the permeability of dense contacts is blocked for 2–3 days after birth. Characteristic of the composition of milk with "slow" and "fast" involution is a noticeable increase in proteins with antimicrobial properties such as alpha-lactalbumin, lysozyme, lactoferrin, and immunoglobulins A, M, and G (Hartmann and Kulski 1978; Perrin et al. 2017) (Fig. 5.15b, d). These proteins are important primarily for women because milk stasis and overflow of the alveolar-ductal system with milk during involution contributes to the growth of bacterial concentrations in milk and the development of inflammatory processes. As clinical practice currently shows, to eliminate possible painful engorgement of some lobes, when a woman after a relatively long period of feeding a child with complementary foods (12–14 months) abruptly stops feeding the child, an additional dopamine agonist—cabergoline (Dostinex)—is used. After removing milk from the swollen lobe against the background of cabergoline action, milk secretion in this lobe is blocked. In addition, in recent studies, however, on the mammary glands of cows (Boutinaud et al. 2017), it was found that during the period of involution of the mammary glands, cabergoline, along with blocking the release of prolactin, increases the exfoliation of epithelial cells of the alveoli and activates matrix metalloproteinases MMR2 and MMR9, i.e., accelerates the second phase of involution (Fig. 5.14B). It is possible that cabergoline accelerates apoptosis in the alveoli of mammary gland of the women. Thus, as a result of postlactation involution, the structure and function of the mammary glands return to the state that is typical for women with a normal menstrual cycle. At the same time, the mammary glands are ready to start lactating again in the event of another pregnancy and childbirth.

References

Alekseev NP, Ilyin VI, Yaroslavski VK et al (1994) Role of vacuum and compression stimuli in the process of milk removal from a woman's breast (1994) Fiziol.zh. USSR 89:67–84. (in Russian)

Alekseev NP, Ilyin VI, Yaroslavski VK et al (1998) Compression stimuli increase the efficacy of breast pump function. Eur J Obstet Gynecol Reprod Biol 77(2):131–139

Alekseev NP, Ilyin VI (2015) Optimization of biomechanical stimuli of vacuum in a device with a compression component for removing milk from the breast of lactating women. Rus J Biomech 19(4):430–438. (in Russian)

Alekseev NP, Ilyin VI, Talalaeva NE (2015) Pathological postpartum breast engorgement: prediction, prevention, and resolution. Breastfeed Med 10(4):203–208

Alekseev NP, Ilyin VI (2016) The mechanics of breast pumping: compression stimuli increased milk ejection. Breastfeed Med 11:370–375

Alekseev NP, Il'in VI, Schegolkova AV (2010) Prevention and elimination of pathological engorgement in breast-feeding women. J obstet Women's Diseases LIX:95–99. (in Russian)

Alekseev NP, Il'in VI, Talalaeva NE (2014) Contribution of the compression component of the executive mechanism of the milk-producing apparatus to the process of milk excretion in lactating women. J Obstet Women's Diseases LXIII:91–97. (in Russian)

Allen JC, Keller RP, Archer P, Neville MC (1991) Studies in human lactation: milk composition and daily secretion rates of macronutrients in the first year of lactation. Am J Clin Nutr 54 (1):69–80

Anderson PO (2017) Domperidone: the forbidden fruit. Breastfeed Med 12:258–260

Andreas NJ, Hyde MJ, Gomez-Romero M, Lopez-Gonzalvez MA, Villaseñor A, Wijeyesekera A, Barbas C, Modi N, Holmes E, Garcia-Perez I (2015) Multiplatform characterization of dynamic changes in breast milk during lactation. Electrophoresis 36:2269–2285

Ballard O, Morrow AL (2013) Human milk composition: nutrients and bioactive factors. Pediatr Clin N Am 60(1):49–74

Baron JA (1999) Domperidone: a peripherally acting dopamine$_2$ – receptor antagonist. Ann Pharmacol 33:429–440

Bazzano AN, Hofer R, Thibeau S, Gillispie V, Jacobs M, Theall KP (2016) A review of herbal and pharmaceutical galactagogues for breast-feeding. Ochsner J 16(4):511–524
Betzold CM. Author replay. Editor (2004) J Midwifery Women's Health 49(2):461
Biancuzzo M (1999) Selecting pumps for breastfeeding mothers. J Obstet Gynecol Neonatal Nurs 28(4):417–426
Bode L, McGuire M, Rodriguez JM, Geddes DT, Hassiotou F, Hartmann PE, McGuire MK (2014) It's alive: microbes and cells in human milk and their potential benefits to mother and infant. Adv Nutr 5(5):571–573
Boutinaud M, Isaka N, Gandemer E, Lamberton P, Wiart S, Taranilla AIP, Sordillo LM, Lollivier V (2017) Inhibiting prolactin by cabergoline accelerates mammary gland remodeling during the early dry period in dairy cows. J Dairy Sci 100(12):9787–9798
Bravi F, Wiens F, Decarli A, Dal Pont A, Agostoni C, Ferraroni M (2016) Impact of maternal nutrition on breast-milk composition: a systematic review. Am J Clin Nutr 104(3):646–662
Brown KH, Robertson AD, Akhtar NA (1986) Lactational capacity of marginally nourished mothers: infants' milk nutrient consumption and patterns of growth. Pediatrics 78(5):920–927
Bumrungpert A, Somboonpanyakul P, Pavadhgul P, Thaninthranon S (2018) Effects of fenugreek, ginger, and turmeric supplementation on human milk volume and nutrient content in breastfeeding mothers: a randomized double-blind controlled trial. Breastfeed Med 13:645–650
Cacho NT, Lawrence RM (2017) Innate immunity and breast milk. Front Immunol 8:1–9
Campbell-Yeo M, Allen AC, Joseph KS, Ledwidge JM, Caddell K, Allen VM, Dooley KC (2010) Effect of domperidone on the composition of preterm human breast milk. Pediatrics 125(1): e107–e114
Cann PA, Read NW, Holdsworth CD (1983) Galactorrhoea as side effect of domperidone. Br Med J 286(6375):1395–1396
Chollet-Hinton LS, Stuebe AM, Casbas-Hernandez P, Chetwynd E, Troester MA (2014) Temporal trends in the inflammatory cytokine profile of human breastmilk. Breastfeed Med 9 (10):530–537
Claassen S, Zünkler BJ (2005) Comparison of the effects of metoclopramide and domperidone on HERG channels. Pharmacology 74(1):31–36
Cox DB, Owens RA, Hartmann PE (1996) Blood and milk prolactin and the rate of milk synthesis in women. Exp Physiol 81:1007–1020
Da Silva OP, Knoppert DC (2004) Domperidone for lactation women. Can Med AssJ 171 (7):725–726
Da Silva OP, Knoppert DC, Angelini MM, Forret PA (2001) Effect of domperidone on milk production in mothers of premature newborns: a randomized, double-blind, placebo-controlled trial. Can Med Ass J 164(1):17–21
Egnell E (1956) The mechanics of different methods of employing the female breast. J Swed Med Assoc 40:1–8
Eishima K (1991) The analysis of sucking behaviour in newborn infants. Early Hum Dev 27 (30):163–173
Ford DR (1999) Breast pump insert. Patent USA №5885246
Forinash AB, Yancey AM, Barnes KN, Myles TD (2012) The use of galactogogues in the breastfeeding mother. Ann Pharmacother 46(10):1392–1404
Garza C, Johnson CA, Smith EO, Nichols BL (1983) Changes in the nutrient composition of human milk during gradual weaning. Am J Clin Nutr 37(1):61–65
Gmoshinskaya MV, Fateeva EM, Shilina NM, Aleshina IV, Kon IY (2004a) Application of specialized products in the nutrition of pregnant and nursing women. Quest Child Nutr 2:7–10. (in Russian)
Gmoshinskaya MV, Aleshina IV, Shilina NM, Gmoshinsky IV, Fandeeva TA et al (2004b) A new special product for pregnant and lactating women. Quest Child Nutr 2(4):46–48. (in Russian)
Going JJ, Moffat DF (2004) Escaping from flatland: clinical and biological aspects of human mammary duct anatomy in three dimensions. J Pathol 203:538–544

Gossage CP, Deyhim M, Yamini S, Douglass LW, Moser-Veillon PB (2002) Carotenoid composition of human milk during the first month postpartum and the response to beta-carotene supplementation. Am J Clin Nutr 76(1):193–197
Grzeskowiak LE, Amir LH (2015) Use of domperidone to increase breast milk supply: further consideration of the benefit-risk ratio is required. J Hum Lact 31(2):315–316
Grzeskowiak LE, Lim SW, Thomas AE, Ritchie U, Gordon AL (2013) Audit of domperidone use as a galactogogue at an Australian tertiary teaching hospital. J Hum Lact 29(1):32–37
Gunn AJ, Gunn TR, Rabone D, Breier BH, Blum WF, Gluckman PD (1996) Growth hormone increases breast milk volumes in mothers of preterm infants. Pediatrics 98(2):279–282
Hartmann PE, Kulski JK (1978) Changes in the composition of the mammary secretion of women after abrupt termination of breast feeding. J Physiol 275:1–11
Hennart P, Delogne-Desnoeck J, Vis H, Robyn C (1981) Serum levels of prolactin and milk production in women during a lactation period of thirty months. Clin Endocrinol 14(4):349–353
Hofmeyr GJ, van Iddekinge B (1983) Domperidone and lactation. Lancet 1(8325):647–651
Hondeghem LM, Logghe NH (2017) Should domperidone be used as a galactagogue? Possible safety implications for mother and child. Drug Saf 40(2):109–111
Hopkinson J, Heird W (2009) Maternal response to two electric breast pumps. Breastfeed Med 4 (1):17–23
Howie PW, McNeilly AS, McArdle T, Smart L, Houston M (1980) The relationship between suckling-induced prolactin response and lactogenesis. J Clin Endocrinol Metab 50(4):670–673
Ilyin VI, Alekseev NP (2014) Influence of vacuum stimuli on the length of the areolar-nipple complex of the breast of lactating women. Russ Physiol Zh 100(2):232–237. (in Russian)
Jensen R (1945) quoted by: Abascal K, Yarnell E (2008) Botanical galactagogues 14(6):288–294
Kent JC, Mitoulas L, Cox DB, Owens RA, Hartmann PE (1999) Breast volume and milk production during extended lactation in women. Exp Physiol 84(2):435–447
Kent JC, Mitoulas LR, Cregan MD, Geddes DT, Larsson M, Doherty DA, Hartmann PE (2008) Importance of vacuum for breastmilk expression. Breastfeed Med 3(1):11–19
King BL, Love SM (2006) The intraductal approach to the breast: raison d'être. Breast Cancer Res 8:206–221
Knoppert DC, Page A, Warren J, Seabrook JA, Carr M, Angelini M, Killick D, Dasilva OP (2013) The effect of two different domperidone doses on maternal milk production. J Hum Lact 29 (1):38–44
Lawrence RA, Lawrence RM (1999) Breastfeeding: a gide for medical profession, 5th edn. Mosby, 966 p
Li M, Liu X, Robinson G, Bar-Peled U, Wagner KU, Young WS, Hennighausen L, Furth PA (1997) Mammary-derived signals activate programmed cell death during the first stage of mammary gland involution. Proc Natl Acad Sci U S A 94(7):3425–3430
Maddern GJ (1983) Galactorrhoea due to domperidone. Med J Aust 2(11):539–540
Marasco LA (2014) Unsolved mysteries of the human mammary gland: defining and redefining the critical questions from the lactation consultant's perspective. J Mammary Gland Biol Neoplasia 19(3–4):271–288
Martin CR, Ling PR, Blackburn GL (2016) Review of infant feeding: key features of breast milk and infant formula. Nutrients 8(5):279–290
Mastroeni SS, Okada IA, Rondó PH, Duran MC, Paiva AA, Neto JM (2006) Concentrations of Fe, K, Na, Ca, P, Zn and Mg in maternal colostrum and mature milk. J Trop Pediatr 52 (4):272–275
Meier PP, Engstrom JL, Hurst NM, Ackerman B, Allen M, Motykowski JE, Zuleger JL, Jegier BJ (2008) A comparison of the efficiency, efficacy, comfort, and convenience of two hospital-grade electric breast pumps for mothers of very low birthweight infants. Breastfeed Med 3(3):141–150
Meier PP, Patel AL, Hoban R, Engstrom JL (2016) Which breast pump for which mother: an evidence-based approach to individualizing breast pump technology. J Perinatol 36(7):493–499
Milsom SR, Breier BH, Gallaher BW, Cox VA, Gunn AJ, Gluckman PD (1992) Growth hormone stimulates galactopoiesis in healthy lactating women. Acta Endocrinol 127(4):337–343

Mitoulas LR, Kent JC, Cox DB, Owens RA, Sherriff JL, Hartmann PE (2002a) Variation in fat, lactose and protein in human milk over 24 h and throughout the first year of lactation. Br J Nutr 88(1):29–37

Mitoulas LR, Lai CT, Gurrin LC, Larsson M, Hartmann PE (2002b) Effect of vacuum profile on breast milk expression using an electric breast pump. J Hum Lact 18(4):353–360

Mortel M, Mehta SD (2013) Systematic review of the efficacy of herbal galactogogues. J Hum Lact 29(2):154–162

Morton J, Hall JY, Wong RJ et al (2009) Combining hand techniques with electric pumping increases milk production in mothers of preterm infants. J Perinatol 29(11):757–764

Neville MC, Allen JC, Archer PC, Casey CE, Seacat J, Keller RP, Lutes V, Rasbach J, Neifert M (1991) Studies in human lactation: milk volume and nutrient composition during weaning and lactogenesis. Am J Clin Nutr 54(1):81–92

Newton ER (2004) Breastmilk: the gold standard. Clin Obstet Gynecol 47(3):632–642

Obladen M (2012) Guttus, tiralatte and téterelle: a history of breast pumps. J Perinat Med 40 (6):669–675

Page-Wilson G, Smith PC, Welt CK (2007) Short-term prolactin administration causes expressible galactorrhea but does not affect bone turnover: pilot data for a new lactation agent. Int Breastfeed J 2:10–18

Peaker M, Wilde CJ (1996) Feedback control of milk secretion from milk. J Mammary Gland Biol Neoplasia 1(3):307–315

Perrin MT, Fogleman AD, Newburg DS, Allen JC (2016) A longitudinal study of human milk composition in the second year postpartum: implications for human milk banking. Matern Child Nutr 13(1):1–12

Prime DK, Geddes DT, Spatz DL, Robert M, Trengove NJ, Hartmann PE (2009) Using milk flow rate to investigate milk ejection in the left and right breasts during simultaneous breast expression in women. Int Breastfeed J 4(10):1–10

Perrin MT, Fogleman AD, Newburg DS, Allen JC (2017) A longitudinal study of human milk composition in the second year postpartum: implications for human milk banking. Matern Child Nutr 13(1):1–12

Petraglia F, De Leo V, Sardelli S, Pieroni ML, D'Antona N, Genazzani AR (1985) Domperidone in defective and insufficient lactation. Eur J Obstet Gynecol Reprod Biol 19(5):281–287

Picciano MF (2001) Nutrient composition of human milk. Pediatr Clin N Am 48(1):53–56

Powe CE, Knott CD, Conklin-Brittain N (2010a) Infant sex predicts breast milk energy content. Am J Hum Biol 22(1):50–54

Powe CE, Allen M, Puopolo KM, Merewood A, Worden S, Johnson LC, Corrine AF, Welt K (2010b) Recombinant human prolactin for the treatment of lactation insufficiency. Clin Endocrinol 73:645–653

Powe CE, Puopolo KM, Newburg DS, Bo L, Chen C, Allen M, Merewood A, Worden S, Welt CK (2011) Effects of recombinant human prolactin on breast milk composition. Pediatrics 127(2): e359–e366

Prentice AM, Roberts SB, Prentice A, Paul AA, Watkinson M, Watkinson AA, Whitehead RG (1983) Dietary supplementation of lactating Gambian women. I Effect on breast-milk volume and quality. Hum Nutr Clin Nutr 37(1):53–64

Prentice A, Addey CVP, Wilder CJ (1989) Evidence for local feedback control of human milk secretion. Biochem Soc Trans 17:122–128

Ramsay DT, Kent JC, Hartmann RA, Hartmann PE (2005) Anatomy of the lactating human breast redefined with ultrasound imaging. J Anat 206(6):525–534

Reddymasu SC, Soykan I, McCallum RW (2007) Domperidone: review of pharmacology and clinical applications in gastroenterology. Am J Gastroenterol 102(9):2036–2045

Riskin A, Almog M, Peri R, Halasz K, Srugo I, Kessel A (2012) Changes in immunomodulatory constituents of human milk in response to active infection in the nursing infant. Pediatr Res 71 (2):220–225

Rolland P, Piscitelli DG, Ferrari C, Petrocione A (1991) Single dose cabergoline versus bromocriptine in inhibition of puerperal lactation: randomised, double blind, multicenter study. BMJ 302:1367–1371

Rusby JE, Brachtel EF, Michaelson JS, Koerner FC, Smith BL (2007) Breast duct anatomy in the human nipple: three-dimensional patterns and clinical implications. Breast Cancer Res Treat 106:171–179

Russo J, Russo IH (2004a) The breast as a developing organ, Chapter 2. In: Molecular basis of breast cancer. Prevention and treatment. Springer, pp 1–48

Russo J, Russo I (2004b) Development of the human breast. Maturitas 49:2–15

Shawahna R, Qiblawi S, Ghanayem H (2018) Which benefits and harms of using fenugreek as a galactogogue need to be discussed during clinical consultations? A delphi study among breastfeeding women, gynecologists, pediatricians, family physicians, lactation consultants, and pharmacists. Evidence-based complementary and alternative medicine. Article ID 2418673, 13 p. https://doi.org/10.1155/2018/2418673

Smilowitz JT, O'Sullivan A, Barile D, German JB, Lönnerdal B, Slupsky CM (2013) The human milk metabolome reveals diverse oligosaccharide profiles. J Nutr 143(11):1709–1718

Smolina K, Mintzes B, Hanley GE, Oberlander TF, Morgan SG (2016) The association between domperidone and ventricular arrhythmia in the postpartum period. Pharmacoepidemiol Drug Saf 10:1210–1214

Stork D, Timin EN, Berjukow S, Huber C, Hohaus A, Auer M, Hering S (2007) State dependent dissociation of HERG channel inhibitors. Br J Pharmacol 151(8):1368–1376

Tolkunov Yu A, Markov AG (2005) The physiology of the alveoli of the breast. Saint Petersburg, Nauka, p. 219 (in Russian)

Ueda T, Yokoyama Y, Irahara M, Aono T (1994) Influence of psychological stress on suckling-induced pulsatile oxytocin release. Obstet Gynecol 84(2):259–262

Vorontsov IM, Fateeva EM, Hasenson LB (1993) Breastfeeding children. Ppmi, SPb, p 198 (in Russian)

Walker M (2005) Breast pump and other technologies. In: Riordan J (ed) Breastfeeding and human lactation, 3rd edn. Jones and Bartlett Publishers, pp 323–365

Wan EW, Davey K, Page-Sharp M, Hartmann PE, Simmer K, Ilett KF (2008) Dose-effect study of domperidone as a galactagogue in preterm mothers with insufficient milk supply, and its transfer into milk. Br J Clin Pharmacol 2:283–289

Watson CJ (2006a) Post-lactational mammary gland regression: molecular basis and implications for breast cancer. Expert Rev Mol Med 8(32):1–15

Watson CJ (2006b) Involution: apoptosis and tissue remodelling that convert the mammary gland from milk factory to a quiescent organ. Breast Cancer Res 8(2):203–207

Watson CJ, Kreuzaler PA (2011) Remodeling mechanisms of the mammary gland during involution. Int J Dev Biol 55(7–9):757–762

Wickes IG (1953a) A history of infant feeding. I. Primitive peoples, ancient works, Renaissance writers. Arch Dis Child 28:151–158

Wickes IG (1953b) A history of infant feeding. I.I. Seventeenth and eighteenth centuries. Arch Dis Child 28:232–240

Wickes IG (1953c) A history of infant feeding. III. Eighteenth and nineteenth century writers. Arch Dis Child 28:332–340

Wickes IG (1953d) A history of infant feeding. IV. Nineteenth century continued. Arch Dis Child 28:416–422

Wickes IG (1953e) A history of infant feeding. V. Nineteenth century concluded and twentieth century. Arch Dis Child 28:495–502

Wilde CJ, Calvert DT, Daly A, Peaker M (1987) The effect of goat milk fractions on synthesis of milk constituents by rabbit mammary explants and on milk yield in vivo. Evidence for autocrine control of milk secretion. Biochem J 242(1):285–288

Wilde CJ, Addey CV, Casey MJ, Blatchford DR, Peaker M (1988) Feed-back inhibition of milk secretion: the effect of a fraction of goat milk on milk yield and composition. Q J Exp Physiol 73 (3):391–397

Witkowska-Zimny M, Kaminska-El-Hassan E (2017) Cells of human breast milk. Cell Mol Biol Lett 22(11):1–11

Zanardo V, Nicolussi S, Carlo G, Marzari F, Faggian D, Favaro F, Plebani M (2001) Beta endorphin concentrations in human milk. J Pediatr Gastroenterol Nutr 33(2):160–164

Zinaman MJ, Hughes V, Queenan JT, Labbok MH, Albertson B (1992) Acute prolactin and oxytocin responses and milk yield to infant suckling and artificial methods of expression in lactating women. Pediatrics 89(3):437–440